THE NATIONAL LABS

THE NATIONAL LABS

Science in an American System, 1947–1974

Peter J. Westwick

HARVARD UNIVERSITY PRESS
Cambridge, Massachusetts
London, England · 2003

Printed in the United States of America

Library of Congress Cataloging-in-Publication Data

Westwick, Peter J.
The national labs : science in an American system, 1947–1974 / Peter J. Westwick.
p. cm.
Includes bibliographical references and index.
ISBN 0-674-00948-7 (cloth : alk.paper)
1. Laboratories—United States—History—20th century.
2. Science and state—United States—History—20th century.
I. Title.

Q180.6.U54 W47 2003
507′.2073—dc21 2002027317

For Medeighnia

ACKNOWLEDGMENTS

It is a pleasure to thank the many people who helped me write this book. I owe the deepest intellectual debt to J. L. Heilbron, whose wide knowledge and historical taste have inspired and guided me since I first started to study history of science. Daniel J. Kevles provided expertise on American science and politics and then found me a place to work. Todd LaPorte led me through the literature in political science and to the best restaurants in Los Alamos. I thank Alexi Assmus for important early encouragement, Zuoyue Wang for many conversations about Cold War science, and especially Cathryn Carson for valuable feedback and advice. Several historians with special knowledge of the national labs generously shared it: Barton Bernstein, Robert Crease, Sybil Francis, Barton Hacker, Gregg Herken, Lillian Hoddeson, Jack Holl, Adrienne Kolb, Roger Meade, Russell Olwell, Karen Rader, Michael Riordan, Rebecca Ullrich, Catherine Westfall, and, in the Department of Energy History Division, B. F. Cooling, Marie Hallion, Terence Fehner, and Skip Gosling. Albert Teich provided very useful manuscripts based on his early explorations of the labs. Robert Seidel in particular welcomed me to the subject with many discussions, suggestions, and drafts of papers. Maurice Goldhaber, Julius Hastings, Tom Row, Glenn Seaborg, Gen Shirane, Gerald Tape, Alvin Weinberg, and Herbert York recalled for me their experiences in the labs.

I have profited from conversations with many other historians, among whom I must single out Lawrence Badash, Roger Hahn, David Hollinger, David Kaiser, John Krige, Stuart W. Leslie, Peter Neushul, Elizabeth Paris, Joan Warnow-Blewett, and Spencer Weart. Gene Rochlin and Jessica Wang commented insightfully on portions of this book presented at conferences. My fellow graduate students at Berkeley provided intellectual stimulation and social distraction; in addition, Sonja Amadae, Ethan Pollock, and Susan Spath offered helpful perspectives from their own work. Diana Wear and the Office for History of Science and Technology at Berkeley and Michelle Reinschmidt and the Division of Humanities and Social Sciences at Caltech provided logistical and moral support. Michael Fisher at Harvard University Press shepherded the book to publication and made sure that it was not longer than it is, for which readers will no doubt thank him.

Many people made my archival research possible: Corene Wood and Robert Crease at Brookhaven; Carol Whitley at Associated Universities, Inc.; Roger Meade and the Freedom of Information Act staff at Los Alamos; Dave Hamrin, Juli Stewart, Yvonne Leffew, Shirley Adcock, and Ted Davis at Oak Ridge; John Stoner and Trina Baker at the lab in Berkeley; Beverly Bull at Livermore; Mark Masek at Argonne; Bill Roberts, David Farrell, and Lauren Lassleben at the Bancroft Library in Berkeley; Martin Tuohy and Donald Jackanicz at the National Archives branch in Chicago; Marjorie Ciarlante at the National Archives in College Park; and the staff of the Department of Energy History Division. I thank Alvin Weinberg and the Oak Ridge Children's Museum for access to Weinberg's papers. Portions of chapter 3 originally appeared in *The Bulletin of the Atomic Scientists* (Nov/Dec 2000), 43–49, and in *Minerva*, 35:4 (2000), 363–391. The National Science Foundation supported research for this project under grant number SBR-9619203; I also received support from the Mellon Foundation and the Department of History at Berkeley.

I must finally thank my family and friends for support and forbearance. I owe particular thanks to my mother, Dorothy K. Westwick, who instilled a love of learning and then helped me pursue it. My son Dane competed with the manuscript at its later stages for my attention, and usually won. I could not have written this book without Medeighnia, my wife.

CONTENTS

THE NATIONAL LABS

INTRODUCTION

In the spring of 1953 the weapons laboratory at Livermore tested its first fission weapon. The device fizzled: it left the shot tower standing, an icon of failure on the Nevada desert floor. Sympathetic scientists from the rival lab at Los Alamos began to document the event in photographs for posterity before Livermore crews could drag down the structure with a jeep. Pictures of the undamaged tower still adorn the walls at Livermore as a reminder of the failure and of its implications for the lab's competition with Los Alamos.[1] This rivalry for results and programs extended to the other large labs of the U.S. Atomic Energy Commission (AEC)—Argonne, Berkeley, Brookhaven, and Oak Ridge—that would earn designation as national laboratories.

Competition stemmed from the organization of the national laboratories as a *system* of several labs, as opposed to a single central facility. The systemic arrangement is a crucial characteristic: one cannot understand the history of any one of the laboratories without consideration of the other sites in the system. *Systemicity,* to coin a convenient shorthand, refers to the connections among the national laboratories and the ways in which interactions among the labs influenced their evolution. Systemicity required negotiation of research programs among the several sites, and changes in the program of one lab had a ripple effect on the rest of the system. The systemic structure had important consequences. It fostered specialization and diversification of lab programs and ensured the responsiveness of the labs to national priorities, both specific and general. The national labs might seem to contradict American democratic ideals: they embody government intervention in science through large-scale projects; secrecy; selective subsidies; and decisions made by scientists outside the political process. But the labs also represent American ideals of laissez-faire competition that resonated with special strength in the context of the Cold War.

WHY STUDY THE LABS?

A visitor to any of the national labs is struck by the sheer scale of the facilities. Examination of laboratory budgets and the rolls of prominent lab scientists conveys a similar awareness of the physical, financial, and intellectual resources that the labs commanded. These resources guaranteed the national laboratory system a central place in the landscape of postwar American science, a place they continue to occupy.

As an indicator of the effort of the labs, consider fiscal year 1958, when lab budgets were flourishing after Eisenhower's early austerity but before they took off in the wake of Sputnik. That year American colleges and universities spent a total of about $109 million on basic physical science research (about 20 percent of that from AEC contracts); the corresponding amount spent by the AEC in its multipurpose labs was $50 million. In terms of money, which translated fairly directly into manpower, the six labs performed almost half as much basic research in the physical sciences as all academic institutions combined. The national labs had a smaller though still significant role in the life sciences. The six large labs spent $13 million in fiscal year 1958 on biomedical research, close to a tenth of the $148 million spent in colleges and universities (about 5 percent of which came from AEC contracts).[2] In some fields, such as genetics, in which the AEC supported up to one-third of all research in the United States (two-thirds of which took place in the labs), the percentage was much higher.[3] The total amount spent on research and development in the national labs that year, $206.3 million, far outpaced the R&D commitment of the largest industrial corporations.[4] By the mid-1960s the United States had spent about $4 billion on research in the national lab system since the war and had invested comparable sums for physical facilities.

Economic investment paid intellectual and technological dividends. Lab scientists and the tools they developed—radioisotopes, research reactors, particle accelerators—changed our understanding of nature, from the structure of matter to the process of photosynthesis, from the creation of new chemical elements to the pathways of human metabolism. Their colleagues recognized their achievements with Nobel Prizes and other high honors. The resources of the labs gave them the power to shape the terrain of science, to build up certain disciplines and neglect others: the fields of high-energy physics, solid-state physics and materials science, and nuclear medicine and radiobiology derived crucial support from the lab system. The national labs had methodological as well as intellectual impact: they exemplified the large-scale, capital-intensive, multidisciplinary approach of what came to be known as "big science," and propagated the notion of the fruitfulness of interdisciplinary research.

The development of the national lab system forced fundamental readjustments in the institutional framework of science. The government sponsored the labs but contracted their operation to industrial firms, universities, and university consortia, and thus promoted a hybrid of public and private institutions. The policy could raise difficult questions for academic contractors, as at the University of California, a state institution, which operated two weapons laboratories, one in a distant state. The support of basic research and training at the labs encroached on the traditional domain of universities and forced continual negotiation of the roles of both institutions. The national labs blurred boundaries in the opposite direction of applied research by driving the growth of high-tech industries as both developers and consumers of technology. The labs often provided a sort of double subsidy: the federal government supported the development of new technologies in the labs, then had the labs buy the results after industrial firms took over their production. The labs provided a market for electron microscopes, radiation sensors, lasers, and linacs; in some cases, such as with electronic computers, the developmental push and purchasing pull from the labs catalyzed an industry.

The lab system also occupied a new niche within the federal government. It emerged from the ferment of postwar debate over the proper function and location of science in the government and became a main mode of federal support for research and development. Connections extended through the AEC to other executive agencies and to congressional committees, all of which had to adjust their framework and policies to accommodate the labs. The contingency of the evolution of the lab system reflected the difficulty in resolving two goals: to provide scientific autonomy while ensuring political accountability. The multiplicity of multiprogram labs complicated the process and repeatedly foiled attempts to impose rationalized organization.

The influence of the national laboratory system extended abroad. The labs variously invoked international competition or cooperation to advance their research programs. High-energy physicists claimed that their field transcended national boundaries and sought to collaborate with Soviet colleagues on a large accelerator project, but at the same time based appeals for the funding of new machines on considerations of Cold War competition. The effort in controlled fusion started out as a secret national program and evolved into an open international endeavor. Foreign scientists visited the national labs and disseminated the fruits of their experience in their home countries, and the labs sent their own scientists and plans abroad, including the concept of the national laboratory.

The importance of the national laboratories goes far beyond science and technology. The labs are the foremost manifestation of the new place that science occupied in American society after World War II and the development of atomic weapons. The products of their research programs helped define the diplomatic policy and military strategy of the United States. The artifacts of the labs contributed on a deeper level to the cultural context of Cold War America. The design and testing of nuclear weapons by the national labs stoked the fear of nuclear war and fallout. The promise and, later, the peril of nuclear power or the hint of cancer cures provoked similarly acute responses. The history of the national labs illuminates the pathways by which the results of lab programs reached the wider world and, in turn, those by which national concerns were transmitted to scientists in the labs. It thus connects the social context with the scientific; that is, national priorities with the emphases of the scientific programs in the laboratory system.

The existing historical literature does not do justice to the importance of the national labs and the scope of their activities. More important, it neglects a central characteristic of the national labs: their systemic structure and the resultant interactions. The official history of the AEC provides the perspective of policy makers in Washington, D.C., but neglects the ground-level view from the labs themselves.[5] Histories of particular research programs, such as particle accelerators, weapons design, or controlled thermonuclear fusion, may capture the competition among the labs, but by focusing on single programs they miss one of the main consequences of competition: diversification.[6] Another approach studies individual laboratories, which balances the view from Washington but also imposes an artificial isolation on each lab and thus fails to capture the crucial connections to the system.[7] There has been no systematic attempt to use archival material at the lab level and integrate it into a study of the system as a whole, despite promising exploratory surveys.[8]

But that is not all. A system-level view requires not only a wider scope but also a new approach. The historiographical lacuna on the lab system betrays a lack of curiosity about the systemic structure itself. Why did the labs exist as a system? What forces sustained systemicity? What effects did systemicity produce? The answers to these questions are to be found in the historical context. This book aims to provide them.

The history of the labs contributes more broadly to the historiography of American science in the Cold War and to postwar U.S. history in general. Historians have noted the enlistment of science for national security: the crash projects of World War II, on radar, the atomic bomb, and the proximity fuze, which realized the potential contributions of science to war and accustomed scientists to lavish budgets and large, bureaucratic organizations; postwar attempts to maintain high levels of federal support of science and debates over whether scientists or politicians should control it; adaptation of university departments and laboratories to new levels and modes of support; renewed scientific mobilization in the deepening Cold War for developing the hydrogen bomb and quantum electronics; the uneasy accommodation of scientists within a McCarthyite security complex; a redoubling of research support in reaction to Sputnik.[9] The distinctive features of the lab system reveal important characteristics of the American response to the Cold War; in particular, the desire to adhere to long-held values of capitalist democracy in principle even as Americans acceded to their violation in practice. Scientists also sought to adhere to the values of science, which did not always agree with the demands of national security.

The lab system provides a test case for a central problem in the historiography of postwar American science: the extent to which scientists aligned their research with the needs of national security.[10] The evidence supports both sides of the debate. The labs consistently responded to national priorities, not only in defense but also, for example, in the space race. The response was systemwide: scientists at all of the labs, including those in more basic programs at Berkeley and Brookhaven, sought to contribute to the national defense in the emergency of the early 1950s. Nor was this response confined to physical scientists: biomedical researchers enlisted in the effort to solve the problem of radioactive fallout and later took up space biology. Systemicity encouraged responsiveness. Political scientists have noted that competition within government bureaucracies allows policymakers at high levels to intervene in the allocation of resources among competing interests.[11] Competition among the national labs thus ensured that they would answer to national priorities; the structure of the laboratory system constrained the strategies of those who populated it.[12]

Lab scientists were constrained, not confined, by the systemic structure. They did not dangle like puppets at the end of federal purse strings, but managed to pull some strings themselves and convince their overseers that certain programs, such as high-energy accelerators, were in the national interest. They could justify basic research programs either as training grounds for future scientists or as a means to keep experienced scientists on tap for emergencies. Discretionary budgets gave lab directors seed money for interesting research programs that could start small but grow into major endeavors. Other programs of what one might call small science flew under the radar of budget examiners outside the AEC and hence had room to maneuver. Finally, the labs did not always have to compete: they could cooperate to reach a formal or tacit division of labor with other labs, which gave scientists freedom to roam within certain fields or approaches. These outlets encouraged the enterprise of lab scientists and directors; it is difficult to ascribe passivity to entrepreneurs such as Ernest Lawrence and Alvin Weinberg.

The emergence of the national labs expresses a central theme in American history. Industrialization in the late nineteenth and early twentieth centuries spurred the reconstitution of American society, from small-scale, informal, local groups to large-scale, formal, bureaucratic organizations. Large industrial corporations pursued strategies of diversification and functional specialization, balanced conflicting goals of institutional stability and technical innovation, and relied on a new class of mid-level managers for coordination of decentralized enterprises.[13] The national lab system brought these same characteristics to a large segment of American science. Scientific research, like other facets of society, did not easily adapt to complex organization. The difficulty met by lab scientists and the AEC in imposing a rationalized organization on the national lab system supports recent refinements in the organizational literature: historians have aimed to dispel an "aura of inevitability" and instead highlight the indeterminacy of the evolution of large organizations, by treating organizations not as rational, efficient actors but as collections of subjective, and fallible, individuals.[14]

The systemic structure of the labs, however, also illuminates interorganizational dynamics, and thus suggests a combination of the often internal focus of organizational literature with the external emphasis of the so-called systems approach.[15] The concept of large technical systems emphasizes the effect of social, political, and economic factors on the evolution of technological systems, and thus appears to apply to the lab system.[16] The systems approach, however, focuses on single technologies, whereas the labs incubated a variety of technologies, from computers to radiation sensors. Furthermore, although nuclear weapons and reactors would seem to exemplify large technical systems, they intersect but do not coincide with the national lab system. The laboratories formed an institutional, not a technological, system. Weapons and reactor research, like other lab programs, extended links to AEC production plants and single-purpose labs, university scientists, military branches, other government agencies, and large industrial corporations. These external relationships tended to increase with time. The national labs, in other words, did not form a perfectly insulated system. The permeability of the boundaries complicates sharp definition of either an institutional or a technological system.

Historians of the national labs, and of many areas of postwar American science, must confront the problem of classification. Much material on certain important subjects, especially nuclear weapons and reactors, remains classified. No classified material was consulted for this study. That did not unduly hamper the work. The Freedom of Information Act has opened up some files for research, and special initiatives have done so for other collections, such as biomedical records relating to human radiation experiments. And, like any historian who lacks texts, the historian of the laboratories can make plausible inferences across gaps in the record.

Despite the barriers of secrecy, the general outlines even of classified programs such as weapons and reactors, which would otherwise leave a gaping hole in the history of the labs, may be descried. In any case, the presence of weapons and reactor research in the national labs does not seem to require explanation. What needs explaining is the presence of all the other programs, such as high-energy physics, biomedicine, solid-state science, astrophysics, and meteorology. What were the labs doing in these fields? How did they get there? These are the questions that motivated this study.

Attention to systemicity and its consequences helps to answer these questions, though at the cost of neglecting other developments, including the evolution of individual laboratories and particular programs, except as they impinge on or illuminate systemicity. Similarly, this book concentrates on lab programs and how they were determined, rather than on the results of lab research, except as it influenced the subsequent evolution of programs. It thus necessarily neglects the very considerable achievements of scientists in the national labs. Fortunately, the studies of individual laboratories and programs mentioned earlier may be consulted for descriptions of scientific results.

WHAT WERE THE LABS?

The national labs were intended to serve two main purposes: provision of large equipment for basic research and of secure facilities for developing technologies for national security. The AEC agreed to the notion that modern scientific research "requires the work of men in teams of large size and the use of very expensive instruments." The scale and expense of research reactors and accelerators put them beyond the reach of individual universities; the national labs would furnish them to researchers from regional academic and industrial institutions. The national labs would thus provide the research results expected to underlie future technological advances, while also training future generations of scientists in the new art of nuclear science. At the same time, there were technologies in the labs awaiting development. The onset of the Cold War put an even higher premium on nuclear weapons and reactors and ensured that the labs would continue their lead role in nuclear technology; the labs provided "the answer to the problem of how to get the men and the machines together in places where atomic energy work can be done in security."[17]

The term "national lab" presents the same problems of definition as the names of scientific concepts and disciplines or other historical terms. Not all of the sites that count today as national labs were so considered, or even existent, in the period under study; nor was consideration as a national lab limited only to the labs that received the title. Lab scientists and the AEC, however, clearly assigned certain labs to the category of "national lab," which they took to mean the large, multiprogram laboratories. I follow their categorization.

For three decades only three labs bore the "national" title: Argonne in Illinois, Brookhaven in New York, and Oak Ridge in Tennessee. But the AEC also supported large labs at Berkeley in California and Los Alamos in New

Mexico; smaller sites such as the materials research program at Ames, Iowa; and two large reactor labs, the Knolls Atomic Power Laboratory, operated by General Electric in Schenectady, New York, and Bettis, operated by Westinghouse near Pittsburgh. After initial uncertainty, membership in the national lab system soon stabilized around the core group of Argonne, Berkeley, Brookhaven, Los Alamos, and Oak Ridge, with the addition of a second weapons lab at Livermore, California, in 1952.[18] The AEC and other branches of the federal government regarded these principal labs as a special, separate category, and the group was cohesive enough for their directors to form a "club" with regular, rotating meetings.

Three conditions helped to identify the national labs. First and most obvious was size: staffs of several hundreds or even thousands, of whom up to half might be scientific personnel, and annual operating budgets in the millions of dollars characterized each of these labs. Second was the pursuit of diverse programs in basic research. Knolls and Bettis, and other AEC sites such as Hanford and Savannah River, focused on production of nuclear weapons material, the weapons themselves, or reactors. Although it is difficult to draw a sharp line between research and development—some of the national labs strayed into development, while other labs such as Hanford and Sandia made forays into research—one may still make a general distinction, as the AEC and lab scientists did. The distinction did not prevent other labs from eventually crossing the line, as Hanford and Sandia would later do to join the category of national labs.

The pursuit of basic research alone, however, did not define the national labs. The AEC supported basic research at dozens of sites throughout the country, at universities and in industry, through the research contract, including special facilities such as accelerators and research reactors paid for by the AEC. What set the national labs apart was their pursuit of multiple programs, which allowed and encouraged them to diversify over time. All of the principal labs, except for Ames, maintained or developed broad research programs in physics, chemistry, engineering, biology, and medicine. The AEC came to consider multiprogram status as the defining characteristic of the national labs.

These first two conditions justify the one exception to our synchronic definition of the national labs. The AEC sometimes included the laboratory at Ames in studies of the multiprogram labs, and Ames's director, Frank Spedding, was in the "Lab Directors' Club." But Ames was less than one-fourth the size of the other large labs, and it confined almost all of its effort to a single field, metallurgy and materials science, from which it failed to diversify. Ames seems to have earned its membership in the club by virtue of Spedding's long association with atomic energy and the other lab directors, dating back to the work at Ames on the Manhattan Project.

If the AEC referred to the principal labs as its "multiprogram laboratories," a third characteristic—the provision of facilities for visiting researchers—played an increasing role in their definition. The multiprogram labs fulfilled this purpose to varying degrees. The classified weapons programs at Los Alamos and Livermore limited their ability to provide facilities for uncleared visitors; in the words of its director, Los Alamos was "not legally a National Laboratory." Ernest Lawrence's Radiation Laboratory at Berkeley (UCRL), which predated the war, played a curious hybrid role as both AEC lab and campus research unit. Berkeley scientists admitted that "UCRL does not consider itself a national Laboratory, but to a first approximation will accept all comers. . . . UCRL is not a National Laboratory in the sense that BNL [Brookhaven] is."[19] Yet none of the "national" labs adequately fulfilled this function either: classified reactor work interfered with access to Argonne and Oak Ridge, and even Brookhaven, where in-house staff greatly outnumbered visitors, fell far short of the ideal. The AEC and lab scientists could continue to consider Berkeley, Livermore, and Los Alamos in the same category as the nominal national labs, and they would eventually earn the title themselves in the 1970s and later. By that time, provision of facilities had joined multiprogram status as a defining characteristic, and special-purpose facilities such as Fermilab entered the club of "national" labs.

THE APPROACH

The book is divided into four parts of two or three chapters each. Part I lays out the institutional framework of the system: the constitution of each lab and its situation within the organization and policies of the AEC and the federal government. The framework transmitted forces from the external environment, described in Part II. These external forces encompassed the allocation of resources, institutional interactions, and cultural influences, and evolved through three main periods. Part III describes the two chief responses of the labs to the environment: specialization and diversification. The last part provides an epilogue and conclusions.

The Framework

Systemicity acted through the institutional framework of the national laboratory system. Each lab had distinctive features that imparted individuality within the system. At the same time, the programs and budgets of the labs were coupled, so that changes in the program of one lab would affect the programs at the other sites. One might think of it like the solar system. Like planets, the labs follow individual trajectories under the influence of a central body, the federal government; but each planet also acts on the others. Planets in adjacent orbits, or labs with similar programs, are coupled more tightly; at certain times planets, or lab programs, may align. Even the appearance of alignment could affect the attitudes of those casting horoscopes—that is, divining the future of lab programs—for a particular planet. There are also smaller satellites: moons, comets, asteroids, like the labs at Ames, Sandia, and Knolls. And like the sun, the federal government is very large, with a complicated structure and internal forces and much hot air. But there is a crucial difference: the inhabitants of our labs are human beings who could themselves influence the path of their planets. Much of this history turns on the complicated problems in celestial mechanics that resulted.

Chapter 1 examines the initial constitution of the national lab system. While historians have neglected the implications of systemicity, they also have not sought to examine its origins. They tend to accept the common explanation that the labs derived from the Manhattan Project; the exigencies of war determined their location and organization and their administrators perpetuated them after the war. It is a necessary but not sufficient argument. Only Argonne, Los Alamos, and Oak Ridge were wartime creations; the Berkeley Rad Lab predated the war, Brookhaven appeared after it, and Livermore several years after that. Amid the general postwar flux there was no assurance that facilities assembled for a wartime crash project would be sustained in peacetime, nor that scientists would stay to staff them. Even if the labs survived, they need not necessarily follow the wartime patterns of a centralized, compartmentalized military operation based in government-owned, contractor-operated facilities. And the concept of the national laboratory itself required formulation, a process that continued even as the system emerged in the postwar period.

Our laboratory solar system, in other words, did not spring into place fully formed; each lab coalesced under the influence of particular historical forces that ensured its individuality and gave it a measure of independence. That independence allowed each lab to act in its own best interests, which could include striking out on its own in a new direction, challenging another lab for a particular program, or leaving certain regions for others to explore. Chapter 2 investigates the various sources of individuality of the labs.

One source stemmed from the geography of the system. Hundreds and thousands of miles separated the labs from one another and from Washington. Considerations of safety and security, as well as power and water requirements, dictated the isolation of the labs in rural locations; only Berkeley, which predated the war, could be said to enjoy a metropolitan environment. Once in place, the labs proved difficult to move. Continued geographic isolation precluded close collaboration between scientists at separate sites and entailed the duplication of expensive facilities and support services; it also limited the availability of the labs to outside users and led them to build up permanent research staff, which likewise acted to decouple the labs.

Geography reflected available transportation. Air travel was unreliable enough during the war for project leaders to refrain from flying.[20] Without fast, safe travel, the United States was too big to have a single lab convenient to all potential users. Hence at the end of the war scientists and Manhattan Project managers developed the concept of regional labs, to provide big, expensive equipment to scientists within a reasonable rail journey from the site. The founders of Brookhaven did not try to locate it close to an airport, a fact later lamented, and the criterion for Argonne's location stipulated "an hour's automobile or train distance" from midwestern universities; by contrast, the selection of the site for Fermilab, after the advent of commercial jet travel in the late 1950s, required "proximity to a major airport having frequent service to major U.S. cities."[21]

The AEC enhanced the individuality of the labs through its policy of contracting for their operation with industrial firms, individual universities, or university consortia. Each of the labs had a different contractor (with the exception of Berkeley, Los Alamos, and Livermore), which often imparted its character to the lab environment, from the industrial cast of Oak Ridge to the self-consciously academic atmosphere of Brookhaven. Aside from some initial flux after the war, each lab would retain the same contractor throughout the period of this study. The diversity of lab contractors entailed duplication of administrative staff and hampered coordination of lab programs. It also precluded easy interchange of scientists among the sites, owing to the different personnel policies of the contractors. Although the labs shared the same labor pool and often competed for the same scarce scientists, each lab built up its own in-house staff and perpetuated its distinct character.

Despite their individual characteristics, the labs did not lack links. Chapter 3 describes the various couplings that bound the labs into an interdependent system. Lab scientists turned one potential barrier between the labs into a junction: the effect of security restrictions and classification on scientific research, another consequence of the mobilization of science. Secrecy contradicts the ideals of both science and democracy; but in the climate of the Cold War, scientists and politicians compromised these ideals in the interests of national security. Secrecy and classification threatened to isolate the labs by cutting off the flow of information among them; without knowledge of what was going on at other sites, competition and cooperation were impossible. To overcome the barriers of secrecy, the AEC and the labs created a classified facsimile of the open, self-regulating scientific community replete with special conferences, publications, and committees. The classified community provided formal recognition of the need to ensure communication among the interdependent labs, and the exclusive environment within the gated community bound the labs even more tightly.

Their common sponsor also coupled the labs. All the labs worked for the AEC, which brought them together in fields for which the AEC had statutory responsibility. The presence of several labs, each of which maintained multiple programs, forced them to compete for resources within the confines of the AEC mission. Intellectual curiosity and the scientific reward system, with its emphasis on priority of discovery, fostered competition, especially when aligned with personal ambition and political acumen: instead of conceding fields to their colleagues, scientists might instead challenge them, as Brookhaven refused to accept Berkeley's preeminence in accelerators. Common problems, however, could encourage the labs to cooperate as well as compete to advance the goals of science and the AEC. A lab that lacked resources for a particular program could turn to other sites for help. Thus, for example, scientists at Brookhaven, Los Alamos, and Oak Ridge looked to Berkeley for guidance in accelerator physics; in the other direction, and at the other end of the programmatic spectrum, Los Alamos appealed for assistance to Argonne, Berkeley, and Brookhaven to build the hydrogen bomb. The scientific spirit aided cooperation, as did the shortage of scientists, which forced labs to look outward for capabilities they could not find in their own staffs.

The AEC did not easily accommodate the several multiprogram labs within its organization. The tangled line of authority from the labs to the AEC became known as the Laboratory Problem, a Gordian knot that consistently escaped solution. The labs also enjoyed, or endured, common oversight in the executive and legislative branches of the federal government. The budget process provided the strongest coupling force within the lab system. Although each lab compiled its proposals and budgets separately, when lab budgets reached Washington they were blended into the AEC budget, and individual labs disappeared into AEC programs. AEC program managers thus served as the arbiters of systemicity, weighing the programs of each site against the rest of the system before sending their budgets up the line.

The Laboratory Problem and the dispersive budget process persuaded the lab directors to organize in a common front against the AEC. The lab directors had common personal backgrounds in the Manhattan Project and often the same disciplinary training in the top schools of American physics in the 1930s. Their cohort would later acquire an identity as the Los Alamos generation, which marked the coming of age of American physics. Most of the directors were young, since the contractors tried and failed to hire more established directors: Norris Bradbury was only supposed to lead Los Alamos until a permanent director could be found (he ended up as a twenty-five-year stopgap); Alvin Weinberg assumed the helm of Oak Ridge after a succession of prominent physicists turned down the job; and Argonne turned to Norman Hilberry to succeed Walter Zinn after rejection by several more distinguished names. Only Ernest Lawrence at Berkeley, and his successor Edwin McMillan, had Nobel Prizes, and Lawrence's established reputation bolstered Berkeley's position in the AEC and in Congress.[22] Unsubtle digs at the qualifications of the directors by the eminent members of the AEC's General Advisory Committee would not have soothed any of their insecurities.[23] Lacking individual strength, at least until they proved their ability as leaders, the lab directors banded together in what became known as the Lab Directors' Club, which increased their leverage with the AEC and their connections with one another.

The Environment

Our solar metaphor captures the couplings and decouplings among the labs, but it suggests that light years of distance isolated the system from external forces. So let us switch metaphors. The labs displayed the individuality and interdependence of organisms. They also shared an external environment, which acted upon them and which they in turn influenced. As the environment changed over time, the labs had to adapt, shedding less useful appendages and growing new ones. In the competition for resources the labs could specialize to occupy a particular niche within the environment. At the same time, overspecialized organisms stand less chance of adapting to rapid environmental changes; lab scientists recognized this and sought diversified programs. Instead of natural selection of genetic mutations, however, we find a sort of Lamarckian evolution through acquired characteristics and artificial selection of both programs and labs. Although our ecological metaphor runs the risk of anthropomorphizing the labs instead of recognizing each as a collection of individuals and interests, it appropriately highlights the interconnectedness of the labs with one another and with the context, and the implications of these connections for the evolution of the system. Part II of this book describes the changing environment of the national labs and their interactions with other institutional organisms.

This history covers in detail the period from 1947 to 1962, roughly the "long decade" of the 1950s that can be dated from the onset of the Cold War to the Limited Test Ban Treaty.[24] For the national labs this period is bounded at one end by the transition from the Manhattan Project to the AEC in 1947. Although the roots of the labs extend to the Manhattan Project and earlier, their history as national labs begins with the shift from a wartime crash project under the military, with the single purpose of developing an atomic bomb, to a peacetime civilian program empowered to pursue myriad applications of atomic energy and long-term research into its properties. At the other end of the period lie the weapons test ban treaty, the emergence of the power reactor industry, and the completion of the second generation of postwar high-energy accelerators, the last to be located at the multipurpose labs. These developments reflected threats to the original missions of the labs, which lab scientists faced by finding new fields to cultivate. The diversification of the labs complicated the inherent organizational problems and raised questions about lab missions. These matters came to a head around 1960 and induced close examination by the AEC and Congress of the purpose and organization of the lab system.

On 17 March 1954 the front page of the *New York Times* carried the headline "Atom Smasher Sets Record; Japan Gets Radioactive Fish."[25] The news marked milestones at opposite ends of the spectrum of lab programs. The achievement by the Berkeley Bevatron of a beam of 5 billion electron volts (BeV) capped the development of the first generation of postwar accelerators. The hot fish came from the Japanese fishing boat *Lucky Dragon,* which unluckily wandered into the waters near Bikini atoll during the first test of a compact thermonuclear weapon. The Castle-Bravo shot marked the end of the crash program for the hydrogen bomb and "a complete revolution" in nuclear weapons; henceforth weapons development became a matter of refining existing designs.[26] Castle-Bravo also started the fallout controversy that flared through the late 1950s and led to a moratorium on aboveground weapons tests and then, in 1963, to the formal test ban.

Several other events suggest the significance of 1954. The Cold War entered a new phase after the end of the Korean conflict and the death of Stalin. The Soviets cut into the American lead in the nuclear arms race and tested their own hydrogen bomb only months after the United States, foreshadowing a military standoff and a shift to sociopolitical confrontation. In 1954 McCarthyite security fixations reached their zenith, or nadir, in the Oppenheimer case. That same year Congress passed a revised Atomic Energy Act that attempted to open atomic energy to industrial development. The revised act eased access to atomic energy research by loosening controls on information and materials; more generally, it signified the further politicization of atomic energy, which would increase external pressures on the labs. On a still wider scale, in 1954 President Eisenhower began to implement his Atoms for Peace scheme, which brought the labs an increasing number of international contacts. We may thus take 1954 as a turning point in the history of the lab system for reasons both social and scientific.

A third period in the history of the national laboratory system stretches from the early 1960s to 1974, the year of the energy crisis. In this period programmatic challenges to the central missions in weapons, reactors, and accelerators were compounded by federal budgetary pressures arising from the space race, the war in Vietnam, and President Johnson's Great Society. Science in general, and nuclear science in particular, came under public criticism for its contributions to military technology and its neglect of social problems. Lab scientists again found opportunity in the very context that constrained them and diversified further—especially into environmental research and non–nuclear energy sources. The continued diversification of the labs helped produce the transformation in 1974 of the AEC into the Energy Research and Development Administration, the predecessor of today's Department of Energy.

The response of the lab system to the fiscal and social challenges of the 1960s and early 1970s confirms and extends conclusions based on their earlier history. Unfortunately, a thorough analysis of the third period cannot be pursued at the same level of detail as for the earlier periods. As one moves closer in time to the present, more archival material remains classified, and important collections for the labs from the late 1950s forward are off-limits to uncleared researchers. Enough material exists, however, to furnish an epilogue (Chapter 8) sketching the history of the labs for this period and indicating exemplary developments.

The periodization is not exact: the starting point is marked by a gradual transition from the Manhattan Project to the organization of the AEC, which lasted from the end of the war through 1948; the turning point of 1954 encompasses diverse developments over the course of a year or so; and the end of the long decade is blended into the first few years of the 1960s. The energy crisis in late 1973, which precipitated the end of the AEC, had its roots in the 1960s, as did the response of the labs that prepared the transition to a new federal agency. Nor is the periodization complete: there are important continuities through each of the historical markers in policies, programs, and people.

These standard caveats of the historian do not diminish the insights that periodization may provide. Our periodization highlights changes in the external environment over time and the evolution of the lab system through successive equilibria, each with certain characteristics. Chapter 4 describes the adaptation of the labs to the winter of the Cold War, from 1947 to 1954, culminating in the national emergency of the early 1950s. In this period lab scientists and managers struggled to formulate the structure and missions of the lab system in the face of the dominant force of national security. Both of the main purposes of the labs—development of nuclear technology for the military and industry, and provision of large facilities for outside researchers—raised the basic question of competition versus centralization. Was competition the surest route to progress, at the cost of duplicated programs, or did centralization focus resources and energy to produce results more quickly? Programs in weapons, reactors, and accelerators all forced consideration of this question. Amid the chill of the Cold War, national security pressures provided a centrifugal force, overcoming arguments for centralization and instead supporting diffusion of programs throughout the system; hence the formation of Livermore, the revival of reactor programs at Oak Ridge, and the addition of accelerators at Brookhaven. National security also helped justify a third mission of the labs, the pursuit of small-scale nuclear science, whether as the foundation for future applications, a means to keep top scientists on tap for national security research, or a way to train new generations of scientists.

Chapter 5 describes the false spring of the Cold War, from 1954 to 1962. The apparent easing of tension after the Korean War and the death of Stalin, and proposals such as Atoms for Peace, encouraged schemes for international cooperation. The internationalist ideals of science suggested a model for cooperation, and the national labs provided an institutional mechanism. But while the national emergency ebbed, the Cold War persisted, albeit extended beyond military confrontation to society at large, including science. The national labs exemplified the difficult combination of internationalist ideals and nationalist competition. At the same time, the easing of security restrictions increased external interactions by the national labs at home. They thus had to negotiate boundaries on three sides. Universities protested the encroachment of the labs on their prerogatives in basic research and education, even as they began to acquire the sort of big equipment the labs were supposed to provide. Industry meanwhile questioned the pursuit of applied development in the labs and sought earlier transfer of technologies. Industry provided more unified opposition; universities organized, if at all, on a regional basis, while industrial firms formed nationwide groups such as the Atomic Industrial Forum. Finally, as the labs diversified, the already indistinct border between them and the work of other government agencies blurred further. The boundary disputes on all three sides—academia, industry, and the federal government—underscore the difficulty of defining the borders of the system.

Consequences

Competition for resources forces organisms to specialize and diversify within an environment, and the labs were no exception. Chapter 6 describes the adaptive strategies pursued by lab scientists. Because federal budget managers sought to keep a lid on lab budgets, fiscal pressure encouraged specialization. Accelerators, research reactors, fusion, and nuclear rocket propulsion all illustrate how lab scientists differentiated their designs and programs to avoid duplication. Specialization extended from specific programs to general philosophies, from the conservatism of Los Alamos and Argonne to the go-for-broke attitude of Livermore and Oak Ridge. At the same time, lab scientists diversified into new fields, driven by both intellectual curiosity and the need to reduce their exposure to variations in the AEC budget. They first sought alternatives within the AEC's purview, but the decline of their original missions drove the labs into new ones for survival. Hence, for example, Los Alamos scientists eyed high-energy physics, solid-state science, molecular biology, and the space program as fertile fields to offset declining prospects in weapons design.

Specialization and diversification both enhanced systemicity, from opposite directions. The centrifugal diffusion of programs appears to have decreased systemic interactions by breaking down the strong coupling between, for example, Argonne and Oak Ridge for reactors and Brookhaven and Berkeley for accelerators. But if each lab was less strongly related to a single counterpart over time, the cumulative effect of additional links tied them more strongly to the system as a whole. Oak Ridge developed a weaker link to Argonne in the power reactor program but stronger ties to Los Alamos, Brookhaven, and Livermore; meanwhile it became coupled with Los Alamos and Livermore through the fusion program and with Berkeley, Argonne, Brookhaven, and Los Alamos through accelerators. From the initial breakdown of the system roughly into two-lab subsystems for weapons, reactors, and research, the labs ended up in a single integrated system. Likewise, albeit from the opposite direction, specialization seems to have cut links between labs by reducing competition. But as Adam Smith pointed out long ago, specialization entails interdependence. Specialization increased the ties among the labs by forcing them to cooperate: a particular lab had to rely on the others for a function it did not possess.

Specialization and diversification redefined the meaning of "national" labs. "National" at first meant regional: the labs were intended to provide facilities to researchers from nearby institutions, not throughout the whole country. Each lab could thus justify a wide complement of facilities, at the cost of duplicating those of the other sites. Specialization entailed unique facilities at each site and hence service of a single, national constituency. Diversification meanwhile broadened the missions of the labs beyond atomic energy, so that lab scientists and their overseers felt free to address national priorities in any field. Thus the labs responded to the space race in the late 1950s, environmental concerns in the 1960s, and the energy crisis of the early 1970s. In both their constituency and their missions, the labs grew into the "national" designation.

Chapter 7 examines two areas that exemplify the addition of new fields to the missions of the labs, roughly following our periodization: biomedicine for 1947 to 1954, and solid-state and materials science for 1954 to 1962. Lab scientists justified these paste-on missions for their relevance to programmatic research—biomedicine for the effects of radiation from both production and use of nuclear weapons and reactors, solid-state science for radiation-tolerant materials for weapons and reactors—but both also seemed the sort of small science best left to university researchers. Their smallness, however, allowed them to escape detailed budget scrutiny and thus gradually emerge as major programs. Biomedicine derived additional political support from its portrayal as a peaceful, humanitarian use of atomic energy, and solid-state science typified the interdisciplinary research that scientists were claiming as a distinctive feature of the labs.

Conclusions

The consequences of systemicity help explain the distribution of research programs, but they do not explain the existence of the national labs or their characteristic structure. The AEC did not have to support a system of multiprogram labs. Scientists after the war deserted Oak Ridge and Los Alamos in droves, and those who remained considered relocating the labs; Brookhaven was a postwar creation, and Berkeley joined the system as a special case. The AEC tried to remove the linchpin of the Oak Ridge program in order to centralize reactor development at Argonne, and the premise of centralization likewise led the AEC and its advisers to resist the creation of Livermore. But the AEC also adopted the attitude expressed by Lewis Strauss: "When we set out to discover new continents over uncharted oceans, the chances of arrival at a distant beachhead are greatly enhanced if the voyage is undertaken by more than one ship's company."[27] The constant tension between centripetal and centrifugal forces in the system would ensure the lability of the labs amid changing contexts.

Despite the seemingly haphazard interplay of interests in the history of the system, strong forces drove its evolution. The federal government proved willing to spend millions of dollars every year on each of the labs, demonstrating the value placed on scientific research by the government and the American public in the aftermath of World War II; the United States, enjoying unprecedented prosperity after the war, could afford to back up its values. The flow of federal funds stemmed from the connection of science to national security. The scientific community, like the rest of the nation, quickly remobilized to meet the challenge of the Cold War, but the military implications of nuclear science and technology made the lab system an especially high national priority.

The Cold War contributed to the sustenance of the system on a deeper, cultural level. The national laboratory system from its inception invoked the benefits of competition to justify the maintenance of multiple laboratories. Competition, lab scientists and managers proclaimed, would stimulate scientific and technological progress and ensure the efficient use of resources. The application of laissez-faire ideals to the marketplace of ideas came as the United States compromised free market principles in the economic sphere; the lessons of the Great Depression and its cure, World War II, entrenched Keynesian theory as government policy after the war. But the tactical abandonment of Adam Smith came at an inconvenient time; it would not do for the United States to admit to interventionism in the midst of the socioeconomic death struggle between capitalist democracy and communism.

Hence the importance of the rhetoric of competition in the Cold War: competition was what distinguished the United States from the Soviet Union and guaranteed eventual victory. The AEC, "an island of socialism" in the United States that managed entire communities and formulated five-year plans for reactor development, yet whose nuclear weapons corralled the communist hordes, was perhaps especially sensitive to its situation. It may not be coincidence that appeals to "the traditional American spirit of competition," as Edward Teller put it, abounded in the nuclear weapons program in particular. The national labs provided a natural outlet for the rhetoric of competition and an example of its benefits in practice: an American high-energy physicist could remark that the Soviets were making "serious mistakes" in their accelerator program by concentrating their efforts and avoiding competition. When the United States followed the same path itself a couple of years later in the steps that led to a single-purpose accelerator facility at Fermilab, the director of Argonne warned against the creation of a "National Monopoly Corporation for High Energy Physics."[28]

The labs represented another cherished American tradition, a corollary of competition: decentralization. Decentralization took two forms. The first, institutional, followed in part from the dispersal of facilities under the Manhattan Project. The second, administrative, followed the political philosophy of the first chairman of the AEC, David Lilienthal, who embraced decentralization as a way to return power to local governments and individual citizens. He implemented it through the administration of AEC facilities like the national labs by managers from local field offices of the AEC. Although institutional decentralization allowed each lab to remain independent, administrative decentralization and delegation to the field gave the labs some distance from Washington and hence enhanced the connections among the labs relative to their connection to central authority; the labs could rely on informal negotiations among themselves to avoid program decisions handed down from above.

The national lab system thus expressed American political ideology in a scientific institution. As for scientific ideology, according to the convention current in the postwar period, science was best left to individuals pursuing basic research in isolation from social or political pressures.[29] But the labs themselves contradicted this laissez-faire scientific individualism by integrating scientists in large organizations under the federal government. Just as the structure of the lab system appealed to political ideals in the Cold War amid interventionism in practice, so did it preserve scientific ideals of pluralism and autonomy at the institutional level even as it undermined their original conception.

Appeals to competition and decentralization sometimes stemmed less from deeply held American values and more from the need to make a virtue of necessity. Although some cases, such as the creation of Livermore, reveal genuine appeals to competition, in others the main goal was to satisfy particular interest groups. For example, centralization of the reactor program foundered not on its subversion of competition but on the opposition of lab scientists, particularly at Oak Ridge, and of the military services—the navy, which required more effort on naval reactor propulsion than Argonne alone could offer, and the air force, which sought its own reactor propulsion program for aircraft. Where competitive rhetoric did play a role, it was not the only player: the creation of Livermore was also prompted by Cold War hawks in the scientific community and in Congress and by the desire of the air force to preserve its lead strategic role.

Invocations of competition to justify the appeasement of special interests betray the fact that there was not perfect competition. Although programs could die out, institutions survived; all of the original labs, and several new ones, persist to this day, even as their initial purposes have dwindled or disappeared. The AEC chose to sustain the labs, much as federal regulatory agencies and social welfare programs mitigated the effects of unbridled economic competition on individuals and the environment. In other words, interventionism extended to the national lab system. In both the invocation of competition and its implementation in practice, the labs reflected the political culture of Cold War America.

I

THE FRAMEWORK

1

ORIGINS

The national laboratories trace their origins to the Manhattan Project in World War II. Argonne, Los Alamos, and Oak Ridge derived from the project, which also brought the Radiation Laboratory at Berkeley under its program and spawned a new lab at Brookhaven immediately after the war. But the Manhattan Project itself stemmed from work at Berkeley and a laboratory at Chicago, of which Argonne was an annex, and plans to establish Los Alamos and Oak Ridge predated the appearance of the Manhattan Engineer District (MED). Although the MED played a crucial role in its early history of the national labs, some of the labs preceded it, and the lab system itself evolved throughout the war and the interim period. But if the wartime labs differed from the eventual national labs, they already demonstrated the interplay of centripetal and centrifugal forces. The project was overcoming geographic centralization and diffusing outward to new sites even as the MED assumed centralized administration of the project.

MANHATTAN PROJECT

Long before the United States entered World War II, university scientists had begun probing the implications of nuclear fission after news of its discovery reached the United States in 1939. The story is familiar. These efforts soon came to concentrate on the campuses of Columbia, Chicago, Princeton, and Berkeley under the aegis of the Office of Scientific Research and Development (OSRD), an executive agency formed in June 1941 at the initiative and under the direction of Vannevar Bush.[1] Bush and OSRD soon seized on results from Berkeley and Britain that indicated the feasibility of a nuclear bomb and focused the fission program on the pursuit of such a weapon. Pearl Harbor further pushed the pace of research and led scientists to examine the difficulties of coordinating the far-flung programs, the delays resulting from duplication, and the advantages of centralizing research on fission at one site.

In late December 1941, Henry Smyth proposed the creation of a central laboratory for nuclear research; as chair of the Princeton physics department, Smyth thought that universities would sooner grant leave to faculty to work at a central lab than at rival universities. Only a few days later Smyth would endorse a plan advanced by his Princeton colleague Milton White, prepared with the advice of Karl Compton and Lee DuBridge, whose association with MIT's Radiation Laboratory for radar research had familiarized them with large central labs. White urged the creation of a single laboratory for nuclear energy research, to be run by the government "in an industrial region."[2] A few weeks afterward the representatives of the various nuclear research groups met and debated the advantages of centralization. Arthur Compton, Karl's brother, who supervised physical research on fission for OSRD from his base at the University of Chicago, considered the merits of the various sites and toyed with the idea of concentrating research at Berkeley. Finally and not surprisingly, he decided to let others make the move and come to Chicago.[3] He called the centralized program the Metallurgical Laboratory, or Met Lab for short, a code name intended to obscure its true purpose.

Through 1942 the Met Lab accumulated staff, coordinated experiments that continued elsewhere, and began experiments on criticality in a uranium pile. The Met Lab program included research on separating plutonium from irradiated fuel elements from the pile; some work on electromagnetic separation of uranium isotopes continued at Berkeley. As project leaders began to look beyond research and development to production, they faced factors that would overcome the impulse to centralize. To make an atomic bomb would require the production and separation of uranium and plutonium isotopes in sprawling plants that might cost more than $100 million to build and run. OSRD realized it was not up to the task and transferred responsibility for the production plants to the Army Corps of Engineers in June 1942; in September the army appropriated complete control of the atomic bomb project, including research, and put Leslie R. Groves, an army engineer who had directed the building of the Pentagon, in charge of it. The corps set up an office in New York to oversee the construction and called it the Manhattan Engineer District.[4]

One of Groves's first acts was to acquire a site for the proposed production plants. Scientists at the Met Lab had earlier recognized the hazards of building a nuclear reactor in the center of a large city, and space and security requirements had also dictated a more remote location. Arthur Compton, on a springtime horseback ride, had selected a spot in the Argonne Forest, twenty-five miles southwest of Chicago.[5] The army acquired the land in August, but labor troubles at the Argonne site led Met Lab scientists to build the first nuclear pile surreptitiously in a squash court under the university's football stadium.[6] The Met Lab still planned to use the Argonne site for a larger pilot plant for the eventual full-scale production complex, but this, too, soon outgrew the site. Groves acquired land for the production plants in the Clinch River valley in east Tennessee, twenty-five miles northwest of Knoxville. The site had access to electric power, available through the Tennessee Valley Authority, and water to cool the piles, provided by the Clinch, and its isolation provided greater safety and security than an alternative area in Michigan proposed by Compton.[7]

In September an OSRD committee recommended that the Met Lab use the Tennessee tract for the pilot plants, to which Compton agreed; the Met Lab's first reactor would still move to the Argonne site. Groves, meanwhile, brought in Du Pont to take over the design and construction of the pile and separation plants in Tennessee. By December even the Tennessee site seemed too small, and Groves and Du Pont resolved to seek another location for the full-scale uranium-graphite piles and plutonium separation plants, which they found in Hanford, Washington. Du Pont would still build the semiworks reactor and plant in Tennessee with the help of reluctant Met Lab scientists, for whom there remained only the smaller reactor at Argonne.[8]

So no sooner had the project coalesced in Chicago than it again dispersed to new locations for safety, security, space, and the supply of power and water needed by the experimental reactors; the process of dispersal began by the spring of 1942, before the MED came on the scene. Through 1943 physics, chemistry, and health physics labs arose near the town of Oak Ridge in Tennessee, at what came to be called the Clinton Laboratories. The Clinton pile went critical in November 1943; meanwhile, Chicago scientists rebuilt their original reactor in February 1943 at what they already referred to as the Argonne Laboratory, under the direction of Enrico Fermi. The Argonne Laboratory remained a division of the Met Lab until May 1944, when it attained organizational independence (although its staff remained on the Met Lab payroll).[9] The institutional independence of Clinton and Argonne did not imply programmatic independence: the Clinton pile furnished plutonium and other radioactive materials to the Met Lab and other sites in the Manhattan Project; the abundant neutrons from both the Argonne and Clinton piles tested materials for the Hanford reactors; and personnel from the Met Lab staffed both labs.

Reactors were not the only force driving the dispersal of atomic fission research. One of the sites to which Clinton provided plutonium and Chicago sent staff was a new laboratory on a remote mesa in New Mexico. Unlike the research on slow-neutron chain reactions for piles, fast-neutron research had remained scattered among several labs in addition to Chicago. J. Robert Oppenheimer, a physicist in charge of fast-neutron research, had criticized the lack of coordination in the program, where experimenters in the various labs measured cross-sections and often arrived at different results. He urged in September 1942 that fast-neutron programs be centralized in a new lab, separate from Chicago, to concentrate work on the development of the bomb itself. Oppenheimer's predecessor as head of fast-neutron research, Gregory Breit, had arrived at the same conclusion for the opposite reason: Breit complained upon resigning in May of the lack of security at Chicago and the need for tighter compartmentalization, which he felt could be achieved by isolating the work on a weapon outside Chicago. Breit's complaints had reached James Conant, scientific adviser to Groves; when Oppenheimer's complaints also reached Groves, he approved the creation of a new lab.[10]

Security and safety again dictated the location: explosives experiments required isolation, and security from external attack and espionage, perhaps more important, demanded inaccessibility. The general had initially considered locating the new lab at Oak Ridge, but Oppenheimer did not want such a close association with the production complex.[11] The search turned to the Southwest and settled on a site occupied by the Los Alamos Ranch School; the students were summarily evicted, and Groves selected Oppenheimer to direct the lab.[12] The importance of remoteness in the location of the wartime labs would affect recruitment and programs in the future.

The lab at Los Alamos, arising from the same impetus toward centralization that had established the Met Lab in Chicago, thus contributed instead to the multiplication and dispersal of laboratories in the Manhattan Project. Both Groves and Oppenheimer assumed at the outset that in order to ensure security, Los Alamos would be a military laboratory, and that the army would commission key scientific staff as majors and Oppenheimer as a lieutenant colonel. The prospective colonel set out to staff his lab with little consideration for the other sites in the project: "We should start now on a policy of absolutely unscrupulous recruiting of anyone we can lay hands on."[13] But scientists balked at the idea of working within a military chain of command, which might compromise their autonomy. Groves and Conant tried to allay such fears by allowing for operation by a civilian contractor, the University of California, during the initial phase of the work, just as the other labs worked under contractors. For the final phase, sometime after the end of 1943, they expected the lab to revert to the military and its scientific staff to become officers. When the time came, however, the MED did not invoke this option and Los Alamos remained under civilian operation, a non-decision with important consequences for the history of the national labs.[14]

POSTWAR PLANS

By the time the MED arrived on the scene, the Metallurgical Laboratory had begun to go to seed. The MED nourished the new labs that sprouted at Argonne, Clinton, and Los Alamos, and by mid-1943 each had taken root. The tide had turned in the war, and even as they pushed the bomb project to completion, scientists and administrators alike contemplated what might become of the labs at war's end. Only weeks after Pearl Harbor, Milton White, in his proposal for a central government laboratory, had conceived that it should "carry on after the war to develop both power and destructive aspects" of nuclear energy. By early 1943 Groves was certain that the government would support nuclear research beyond the end of the war. Compton assumed that this would entail a central laboratory in addition to university research, with the lab located near a large city like, say, Chicago.[15] Speculation quickened as the military outlook brightened and representatives of other sites made their own plans. Eugene Wigner envisaged the expansion of Clinton Laboratories from its staff of about 800 to a postwar total of 3,500.[16] One Berkeley booster proclaimed, "Since the whole project started here[!] and since we have assembled a highly competent and complete staff for this type of research, it would seem that our University campus is the logical place for such research. . . . The Government . . . must continue to put up substantial sums to fully realize on the vast sums already spent."[17]

Those who administered the sums perceived military gains differently from sanguine scientists and saw instead an opportunity to begin demobilization.[18] Groves and his deputy, Colonel Kenneth Nichols, limited long-range research programs at the Met Lab in June 1944 and a month later warned Compton of the possibility of staff cuts of 25 to 75 percent. Compton had sought to promise Fermi the directorship of Argonne after the war in order to entice him from Columbia permanently; Columbia's physicists protested to Bush at OSRD, who castigated Compton and the University of Chicago for proposing such a plan and threatened to move Argonne "away from Chicago's covetous fingers." Compton decided that it was "politically unwise at this time to spend government funds for fundamental research."[19]

The episode highlighted concern at the Met Lab and Argonne over the future of the two institutions. They had suffered from the staffing of Clinton and Los Alamos at their expense, and Groves did not allow them to hire replacements.[20] Those scientists who remained found themselves with less and less to do. They took up, among other things, the question of what to do with the institutions and organization of the Manhattan Project after the war. In July 1944, Zay Jeffries, a Met Lab scientist, proposed the preparation of a prospectus on postwar nucleonics, or nuclear research and development. Compton appointed a committee to assist Jeffries, and the resultant report appeared in November 1944. It advocated continued work after the war on nuclear weapons and the creation and support of laboratories, by the government, at universities.[21]

Faced with the agitation of Met Lab staff, Groves wanted "to convince the scientists that we were not forgetting the postwar policy" and established a committee to consider the issue. The committee, under Richard Tolman of Caltech, issued its report in December 1944, likewise recommending a comprehensive postwar program.[22] But neither the Jeffries nor the Tolman report appeared to change the policy of the MED. Compton continued to plan for the long term; Groves continued to support work relevant only to winning the war, and Nichols reminded Compton of "the present uncertainty concerning our plan for Argonne Laboratory." By mid-1945, however, the issue could not be evaded. In February 1945 Bush urged the secretary of war, Henry Stimson, to create an advisory committee for high-level planning. Stimson obliged in May and formed the Interim Committee, which included a Scientific Panel composed of Compton, Lawrence, Oppenheimer, and Fermi. The panel set up its own subcommittees on various topics, including one on organization under Walter Bartky of the Met Lab.[23]

Bartky and his committee canvassed scientists at the Met Lab, Argonne, Clinton, and Berkeley in June 1945 for views on the postwar organization of nuclear research. The proposals indicate the tension between centralization and dispersion, competition and cooperation, and basic and applied research that would characterize the national laboratories. Most of the responses envisaged regional laboratories sponsored by some form of national nucleonics authority, with centralized labs considered only for the case of military research; in other words, some variant of the status quo, from which the responses emanated. The idea of regional labs had been floating around the Met Lab for at least a year: Smyth, the earlier champion of centralization, had suggested at a meeting in April 1944 the establishment of regional laboratories after the war, with Argonne to serve the Midwest.[24]

The idea now found a forum: the national authority should "maintain and operate at least four regional laboratories" to provide reactors and other equipment too expensive for universities; it "should avoid the pitfall of setting up expensive centralized installations with staffs of hired employe[e]s."[25] The labs would be run by nonprofit corporations, with boards of directors drawn from the science faculties of nearby universities.[26] But if one respondent could claim that "we now have in effect several cooperating regional laboratories at Berkeley, Ames, Chicago, Argonne, and Clinton," Walter Zinn, acting director of Argonne, invoked competition: "One of the main reasons for establishing a number of laboratories would be to stimulate competition and thus eliminate the likelihood of all the laboratories wasting the funds granted."[27] Clinton scientists suggested a combination of small and large "Central Research Laboratories," with two or three smaller ones near academic centers to provide expensive tools to visiting scientists from local faculties and one or two larger labs (such as Clinton) for big developmental projects such as reactors.[28] Others proposed a variation on this theme: a single lab for weapons and several regional labs to provide expensive equipment for researchers.[29]

Most of the responses to Bartky's committee contained some version of the idea of regional laboratories, funded by the government, to provide expensive facilities and house large-scale projects. The only apparent exceptions came from biomedical scientists, who objected to the support of regional labs and instead sought the return of all research to the universities.[30] These dissenters perhaps did not perceive the need for big, expensive equipment for biomedical research, unlike their brethren in the physical sciences. They would come to change their tune. In the meantime, the opinions of the majority—that is, of the physical scientists—carried the day. Bartky's preliminary report agreed that much research should be left to universities; but it also recommended the creation of five or six "regional laboratories," perhaps to be run by a corporation of nearby universities. The regional labs would provide facilities too expensive for a single university; undertake large-scale developmental projects for the government; disseminate information to academia, government, and industry; and furnish special materials (such as radioisotopes) to local universities, from which they might "borrow" researchers to supplement their permanent staff.[31]

The Scientific Panel did not act immediately on Bartky's recommendations, but it urged that the MED be allowed to spend up to $20 million a year to support long-term programs and that the government in the future support basic and applied nuclear programs to the tune of about a billion dollars a year. The Interim Committee agreed only to the recommendation regarding the MED; it left future policy to the national nuclear authority that it would itself help to plan.[32]

POSTWAR REALITIES

Demobilization

The general demobilization in the United States at the end of World War II affected the laboratories of the Manhattan Project. The departure of atomic scientists from the labs was not due just to crises of conscience after the dropping of the bomb, worry over military control of nuclear research, or the desire to return to academic research. Federal plans for postwar reconversion had long entailed the termination of wartime contracts, presumably including research contracts such as those of the MED.[33] Demobilization of some of the labs was already well under way by V-J Day, enforced by the administrators of the Manhattan Project. The Met Lab and Argonne had been dwindling for over a year despite persistent efforts by their scientists to perpetuate them. In Berkeley, Ernest Lawrence had focused on winning the war; not until May 1945 did he accept that extensive federal support would be desirable to sustain the postwar Rad Lab and begin to plan accordingly. Even so, the demobilization of the Berkeley Rad Lab began three months before Hiroshima: staff by 1 September numbered 513, down from the wartime high of over 1,200; the budget that summer of about $4 million a year was half that of the year before, and Lawrence expected it to drop to $1 million a year after the war.[34]

The MED did not neglect the task at hand and Los Alamos and Clinton pushed through to the end. Staff levels at Los Alamos continued to climb through July 1945, but most of the late additions were military personnel (by the end of the war only half the staff were civilians).[35] Oppenheimer acted immediately after Japan's surrender to reduce the staff, though without abandoning the project.[36] The Clinton Laboratories had fulfilled their purpose and aided the design and construction of the Hanford reactors. The University of Chicago withdrew from the contract for Clinton, and the Monsanto Chemical Company took it over on 1 July 1945, the first industrial organization to operate one of the MED's large research labs.[37]

The lack of long-term certainty over the future of the labs encouraged the flight of their scientific staff. Clinton and Los Alamos suffered from their distance from academic and metropolitan centers. Of the three thousand employees at Los Alamos at the end of July, only a third remained by the end of the year. "Everybody wanted out," including Oppenheimer, who returned to academe and, with Groves, designated Norris Bradbury interim director. Bradbury did not help matters by, to his later chagrin, holding recruiting sessions for industry and universities at Los Alamos that fall.[38] Security restrictions—censorship of personal mail, limits on visits by friends and relatives—"present[ed] one of the most frequently listed causes leading to departure from the project."[39] Living conditions on the mesa, in temporary wartime housing, contributed to the exodus, especially after water supply pipes froze that winter and produced a water shortage.[40] By the next spring, one in three of the remaining Los Alamos residents thought the town would not be a satisfactory permanent site, and some suggested relocating the lab to southern California.[41]

Conditions at Clinton were little better. The presence of Du Pont personnel during the war had given the lab a different character, "more that of a Du Pont pilot plant than a University of Chicago research center."[42] During the war Met Lab scientists had referred to it as "Down Under," and Du Pont staff called it "the Gopher Training School"; the lab's new management, Monsanto, mocked Oak Ridge in official telegrams as "Dogpatch."[43] The lab formed a "Location Committee" to consider whether it might be better to move the lab to a more favorable location, such as a metropolitan area in the northeastern United States. The committee polled the senior scientific staff, who almost unanimously recommended moving, particularly the chemists and physicists: "A strong majority does not contemplate staying with the laboratory at its present location for more than two years."[44]

Groves and the MED had long refused to promise postwar support of the labs, but legislation to replace the MED with some sort of atomic energy commission stalled in Congress amid a lengthy debate over civilian versus military control of atomic energy. In the absence of legislation, Groves now perceived the need to preserve the MED's investment and ensure future national security and to reassure its scientists. In October 1945 the MED established a Research Division to separate research from production, which the pressures of war had combined, and to strengthen coordination of programs.[45] Oppenheimer, after a visit with the general, informed his senior scientists less than a week after V-J Day of pending staff reductions but also assured them that the project would continue.[46] Groves agreed to pay to improve living conditions at Los Alamos, "despite the fact that this will commit to some extent at least any future control body. . . . We should not count on atomic bomb development being stopped in the for[e]seeable future."[47] The lab began replacing departed scientists, and Bradbury ended the policy of paying travel costs for terminating employees. His "shaking of the tree" had the desired effect: it forced staff either to commit to stay by September 1946 or to pay their own way off the mesa.[48] The Crossroads weapons tests at Bikini in 1946 gave Los Alamos a concrete objective, although it diverted scarce lab staff, who sought to palm off some Crossroads tasks, such as data analysis, on Oak Ridge.[49]

Some Berkeley scientists would also participate in Crossroads, "a nice vacation as you have very little to do."[50] But the focus at Berkeley, picking up where Lawrence had left off at the outbreak of war, was accelerators. Lawrence's plans for after the war, revised in September 1945, envisaged the completion of the 184-inch cyclotron begun before the war with Rockefeller Foundation funding; the design and construction of two new accelerators, a synchrotron and linear accelerator, both conceived by Berkeley scientists on wartime assignments; continuation of biomedical and chemistry programs from the war and before; and a nuclear reactor. Lawrence confidently assumed that the government would pay for his plans, in the long run if not immediately.[51] Groves agreed only to authorize pilot plants for the new accelerators and continue the subcontract for biomedicine and chemistry: for the big cyclotron he asked "why the government should bear all or a part of the cost of the work"; "considerable additional study must be given to the method and extent of government subsidy." Groves added that final decisions had to await creation of a new agency.[52] His hesitation was short-lived. By December, while Congress cogitated, Groves would approve a revised program for the Rad Lab, including $170,000 for the completion of the 184-inch cyclotron, although the reactor remained only a possibility.[53] After convincing Nichols of the need to accelerate further the work of the Rad Lab the next spring, Lawrence wired back to Berkeley: "There [is] no limit on what we can do but we should be discreet about it."[54]

Lawrence had offered to continue to make the 60-inch cyclotron available to groups at Chicago and Clinton, and, as Los Alamos hoped to do with Crossroads data analysis, he sought to relinquish production of radioisotopes to Clinton. The MED did support radioisotope production at Clinton after the war, along with continued development of chemical extraction processes and a new reactor for higher neutron fluxes.[55] The sharing of programs and equipment with other sites indicates the continued interconnectedness of the labs through the organizational uncertainty of the postwar period.

Constitution of the System

The MED's postwar support of Berkeley, Clinton, and Los Alamos preserved the status quo, without implying fundamental institutional innovations. Scientists in Chicago had propagated the concept of laboratories as regional facilities providing equipment too expensive for universities. Argonne became the test for the idea on 2 December 1945, at the recommendation of a committee appointed by Nichols of scientific representatives of midwestern universities. The University of Chicago agreed on 11 February 1946 to administer an Argonne Regional Laboratory, which would take over the program, facilities, and staff of the Met Lab.[56] In the meantime, Groves had appointed an Advisory Committee on Research and Development, composed of several prominent scientists, to suggest policy and programs and specific steps to implement them. The committee solicited suggestions from MED scientists prior to its meeting in March 1946.[57]

Farrington Daniels, who had assumed direction of the Met Lab, submitted the plan for Argonne, which expanded its ambitions and put the "national" in the names of the national labs. Daniels's plan for the "Argonne Nucleonics Laboratory" called for "a national Nucleonics Laboratory near

Chicago. This laboratory would have a twofold purpose of carrying on nuclear research for the Government, and serving to develop and aid in the nuclear research program of universities throughout the country, particularly those located in the middle west." Daniels's proposal for a "central national laboratory" near Chicago perhaps reflected, and sought to repeat, the origins of the Met Lab in the centralization of slow-neutron research during the war. The regional concept persisted, however, in his emphasis on the Midwest; despite the availability of Argonne to universities "throughout the country," he proposed a board of directors for the lab selected from representatives of midwestern universities.[58] Argonne's director, Walter Zinn, had supported several regional labs in his submission to Bartky's panel. The ambiguity in the Argonne plan, between one central, "national" lab and several regional ones, reflected the continued interplay between centrifugal and centripetal forces that characterized the labs from their inception.

The Argonne proposal faced competition from a new quarter. A coalition of northeastern universities and their scientists had assembled to advance their own interests. The coalition stemmed from the frustration of physicists returning from wartime assignments over the settlements received by universities associated with the Manhattan Project, often at the expense of northeastern institutions. I. I. Rabi and Norman Ramsey at Columbia, "slightly jealous due to the fact that . . . the University of Chicago . . . ended up with a very nice reactor [and] the University of California was ending up with a nice high-energy accelerator," decided to pursue a reactor for themselves and gathered a group of New York–area universities for that purpose.[59] In January 1946 they wrote to Groves proposing that a "regional research laboratory in the nuclear sciences" be established near New York City.[60] Meanwhile, scientists at MIT and Harvard had mobilized another effort for a northeastern regional lab with a reactor, this one for the Boston area.[61]

Groves sent Nichols to inform representatives of the New York group that the MED would support only one northeastern lab and suggest that the two groups pool their proposals; he added that they had better act quickly before the MED's authority gave way to a new agency that might delay a new lab.[62] The New York and Boston groups heeded his advice and in March formed the Initiatory University Group (IUG) of nine universities—Columbia, Cornell, Harvard, Johns Hopkins, MIT, Pennsylvania, Princeton, Rochester, and Yale—under the leadership of George Pegram, the dean at Columbia. The IUG would incorporate in July as Associated Universities, Inc., the contractor for the northeastern lab. The group submitted its proposal to Groves's advisory committee on 3 March 1946, calling for "the establishment in this area of a nuclear laboratory including a major chain reacting pile" to "complement those already in existence."[63] As at Argonne, Brookhaven's founders confused regional service with national aspirations: their proposal for a "Northeastern Regional Laboratory" became, in the final version, a "National Nuclear Science Laboratory."[64]

When the Advisory Committee for Research and Development met on 8–9 March 1946, it took the national name and applied it to the regional concept. National laboratories, it suggested, should provide equipment too expensive for universities or private labs; the national labs would pursue programs of unclassified basic research guided by a board of directors selected from participating universities and other institutions. It approved the establishment of such national labs at Argonne and in the Northeast.[65] The committee's concept of a national lab did not include classified developmental research, as Argonne's proposal had done; hence Los Alamos and Clinton did not qualify and instead entered the committee's report as places for classified work on military or industrial applications. The committee suggested that Lawrence's Radiation Laboratory at Berkeley might deserve the status of a special type of national laboratory.[66] Nichols, Groves's deputy, extended the committee's report to encompass these other labs and developed a rough division of labor among them: Argonne would focus on reactors and a regional lab in California on accelerators, the northeastern lab would develop both reactors and accelerators, Los Alamos would pursue weapons, and Clinton would handle industrial programs.[67]

The scientists on the advisory committee embraced the concept of "national," that is, regional, laboratories and "agreed in general terms that establishment of national laboratories in other regions of the United States should be encouraged."[68] Groves feared the possible multiplication of facilities. He told an AUI representative that "it would be best not to present [Brookhaven's] case as a claim for a 'regional' laboratory which might lead to various other claims for 'regional' laboratories."[69] His fears were well founded, as other groups came out of the woodwork with suggestions for new sites. Chief among these was a group from southern California, led by Lee DuBridge of Caltech, who late in 1946 submitted a proposal for a regional lab with a reactor, to be run by Caltech, UCLA, and USC. Although the MED responded favorably, by that time it was giving way to the AEC, which, as Nichols had predicted, was not eager to make new commitments.[70] Plans for "the third national laboratory on the West Coast" made it into the congressional budget, but DuBridge's proposal died on the vine.[71] Other groups also sought to tap the MED's keg of dollars: the National Bureau of Standards suggested that the MED should provide it with a reactor, and even those with a stake in one of the labs sought support for themselves, as Columbia did for "an active expanding program in nuclear science," including two accelerators.[72]

New sites offered competition to extant ones. The New York group's first communication to Groves, advanced in the midst of demobilization and talk of recession, assumed a zero-sum mentality and suggested that the MED create a northeastern regional lab instead of expanding Clinton Labs.[73] The cosmopolitan Rabi took his case to Clinton's Location Committee: Clinton's remoteness would deter potential staff, students, and visiting scientists, he argued, and if the production plants closed, "Oak Ridge may deteriorate into a small village." Rabi urged that if the MED were to support only one reactor, it ought to build it near New York. The Clinton Research Council agreed to consider Rabi's view and voted down a resolution to build the next pile at Clinton "regardless of other possibilities."[74] Charles Thomas, Monsanto's manager of Clinton, responded that a new northeastern lab would lack housing owing to the postwar construction shortage and that "a regional laboratory must be maintained for the southern universities." But the bulk of Thomas's argument appealed to inertia: Clinton had the staff and facilities in place, while Los Alamos and the Met Lab were in disarray, and until an Atomic Energy Commission could make new commitments, Clinton provided the best opportunity for expansion. But Thomas also claimed that "Clinton Laboratories is not in a competitive position with other regional laboratories. The country should have a great many piles. It is possible that a laboratory near a large city might do entirely different kinds of work from ours."[75]

Some northeastern scientists were less magnanimous: "The situation is somewhat competitive in that some other sections of the country, by force of past circumstances, have a head start, and we may find that the available resources to support such work are all devoted to these regions if we are too slow in getting our specific proposals before the Manhattan District."[76] Others, including Jerrold Zacharias of MIT, preferred plurality and argued against moving Clinton.[77] Inertia and plurality prevailed. Nichols indicated that Clinton would not move for two or three years, if ever.[78] The MED, for the moment at least, was not running a zero-sum game.

The advisory committee applied the national name to regional labs at the same time that it relegated programs of military or industrial—that is, national—importance to the nominally non-national labs at Los Alamos and Clinton. The United States already had a "national" laboratory in place that might have assumed the function. The federal government had established the National Bureau of Standards in 1901 to generate and maintain accurate scientific standards in support of industry, especially in electricity and optics. The bureau provided for basic research in support of its metrological mission; its scientists perceived it as, according to the head of its electrical division in 1905, "the American National Physical Laboratory, using the word physical in a liberal sense, as its work includes both chemistry and engineering." Its overseers saw otherwise; in 1903 the bureau and its lab had transferred from the U.S. Treasury to the new federal Department of Commerce and Labor, at the urging of a congressman: "The newly created National Bureau of Standards . . . necessarily goes into a department primarily devoted to manufacturing and commercial interests." The first secretary of commerce deleted the "national" from the title of the Bureau of Standards, and the lab over time became preoccupied with programmatic work of immediate industrial importance.[79] The bureau in 1934 regained the "national" moniker, but the decline of basic research had left it unfit to develop atomic energy, although it played a small role during the war and, as we know, attempted to obtain a reactor after it.[80]

The advent of atomic energy and its importance to national security convinced both scientists and politicians that the government should continue to provide labs for its development. Scientists also persuaded politicians that the government should support fundamental research as a source of potential applications. The aspect of the national labs elaborated by the advisory committee, the provision of facilities for basic research by visiting scientists, had a statutory precedent in the federal government: congressional resolutions in 1892 and 1901 made facilities in various government agencies available "to scientific investigators and to duly qualified individuals, students, and graduates of institutions of learning."[81]

The Bureau of Standards under these statutes had invited research associates to work in its labs, with the additional justification that the bureau would thus train new technical experts; but given the emphasis of the bureau's program, these researchers came mostly from industry to work on particular practical problems.[82] The idea had appeared in another context as an element of the Rockefeller Foundation's program in the 1930s, which included Lawrence's plans for a 184-inch cyclotron at Berkeley. The cyclotron, insisted Warren Weaver, the director of natural sciences at the foundation, would be a national, or international, facility, "built for all science," akin to the 200-inch telescope that the foundation was building on Mount Palomar, which Weaver called its "national laboratory."[83] The MED's advisory committee intended the new national labs to serve this end and isolated national security programs at Los Alamos and Clinton. Nichols, however, merged the federal and the philanthropic conceptions into labs with dual identities: as homes for the development by permanent staff of applications, often classified, for national security; and for the provision of expensive facilities for basic research by visiting academic scientists. The national labs thus represented a new settlement in the long struggle in American science between the pursuit of knowledge and its uses in society, between the pure and the practical. The ability of the labs to balance these goals would determine their effectiveness and survival for the next fifty years.

2

INDIVIDUALITY

On 1 January 1947, the Atomic Energy Commission took over the vast enterprise of the Manhattan Engineer District. It inherited large research laboratories at Argonne, Brookhaven, Berkeley, Los Alamos, and Oak Ridge, the first two designated as national labs. The AEC would consider whether to sustain all of the sites; in the meantime, it continued to support geographic decentralization. In addition to geographic separation, three factors helped to preserve the individuality and independence of each site within the system. The AEC chose to rely on contractors to carry out its operations. Each of the labs had a different contractor, except for Berkeley and Los Alamos, which imparted distinct character, degree of involvement, and administrative policies to each lab. The second and third factors stem from these policies. The labs adopted different forms of organization, whether based on academic disciplinary departments, large technical projects, or forceful individuals, or on centralized or decentralized support staff. Each contractor also imposed its own personnel policies; different salaries, benefit packages, and tenure possibilities at each site effectively restricted exchange of personnel within the system. These factors decoupled the labs and limited the advantages of systemicity, which the AEC recognized and tried to overcome.

CONTRACTORS

Why Contractors?

The AEC's policy of contract operation followed that of the MED, which had inherited the policy from its predecessor agencies: the National Defense Research Committee (NDRC) and OSRD had adopted the contract to take advantage of existing facilities, organizations, and research groups, based on Vannevar Bush's experience with the National Advisory Committee for Aeronautics and his earlier interactions with industry at MIT.[1] General Groves had continued the policy in order to avoid having to assemble a large staff and to maintain the pace of the project, although the MED did not seek contract operation in all circumstances, as the initial plans for military operation of Los Alamos indicate.[2] Among the contracts carried over to the MED were those for the Met Lab at Chicago and the Rad Lab at Berkeley. When Arthur Compton had centralized work in the Met Lab in February 1942, the University of Chicago agreed to administer the lab as its part in the war effort. Said its vice president, E. T. Filbey, "We will turn the University inside out if necessary to help win this war. Victory is much more important than the survival of the University." Filbey need not have worried about the survival of the university; on the contrary, wartime research contracts helped revive budgets after the Great Depression. The university handled only the fiscal and business aspects of the lab and left scientific and technical decisions to Compton.[3] The addition of the Argonne laboratory within the Met Lab brought it under the aegis of the university. The Manhattan Project tested the patriotism of the university when it then asked Chicago to operate the pilot plant that Du Pont was building in Tennessee. Compton, Filbey, and William B. Harrell, the university's business manager, wondered why Chicago should take responsibility for a hazardous operation, with little connection to its academic purpose, in a distant state; but patriotic duty overcame their scruples.[4]

In Berkeley, Ernest Lawrence had obtained in December 1941 one of the first OSRD contracts, for work on electromagnetic separation of isotopes. Contracts for chemical and biomedical research and for Oppenheimer's fast-neutron work continued under OSRD even as electromagnetic separation grew to industrial scale and the giant racetracks it required were built under other MED construction contracts in Tennessee. The University of California handled only the business aspects of the contracts and let Lawrence and his scientists handle the technical decisions, perhaps to a greater degree than Chicago did with its contracts, since Lawrence and his lab had long enjoyed a relatively independent position within the university.[5] The trust and experience that the university accumulated under the MED with these contracts and Oppenheimer's connections with Berkeley may explain why Groves asked the University of California to manage the new lab at Los Alamos in December 1942.[6]

By virtue of their early involvement in atomic energy research, the universities of Chicago and California found themselves operating large hazardous technical projects hundreds of miles from their home states. The University of California, a state institution, faced particular problems in an interstate enterprise. Lab staff in New Mexico were UC employees and hence were required to contribute to California's state retirement system. The university and the army haggled in negotiating the first contract about the amount of university control over purchasing, disbursement, and hiring.[7] Security restrictions limited the administrative oversight of the university: the original contract was classified "Secret," and uncleared university officials were left in the dark, including the attorney for the university whose job was to review contracts.[8] Robert Underhill, the university's business manager who negotiated the Los Alamos contract, had to ask what state the lab was in so that the university could get liability insurance for personnel. In late 1943, several months after he negotiated the contract, Underhill received a visit from Lawrence, who "came in and he shut the door and he saw it was locked and practically pulled the curtains down and everything else. . . . He said, 'You know what they're doing down in Los Alamos?'" Underhill did not. He had a vague idea that it was some sort of physics project.[9] University officers and Regents were not allowed to visit Los Alamos, even though they were ostensibly managing it. Hence Oppenheimer's judgment on the university after the fact: "The contractor during the war years was an extremely helpful and able contractor, but was really distinguished primarily by his absence."[10]

At the end of the war the universities of California and Chicago, like the scientists in the labs, debated their desire to continue. Chicago relinquished the Clinton contract in July 1945 to Monsanto, which had produced weapons components for the Manhattan Project at its lab in Dayton, Ohio. Groves tended to go with the familiar, and Monsanto's strength in chemical technology, a focus of the Clinton program, as well as the industrial character of Clinton made Monsanto an easy choice. For its part, Monsanto hoped to use the Clinton contract as an entry into the new field of nucleonics.[11] As for Argonne, Compton had assumed from the outset that the University of Chicago might give up the contract after the war. In the postwar organization of Argonne as a national laboratory, the university's chancellor, Robert M. Hutchins, and business manager, William B. Harrell, tentatively agreed to administer the lab in the interim, although the university and the representatives of participating midwestern institutions assumed that the long-term arrangement would not involve the university.[12]

Chicago had a fallback position for nuclear physics. In the wake of the Fermi fiasco, Compton had proposed that the university establish its own laboratory for nuclear physics. The university approved an Institute for Nuclear Studies in the summer of 1945, to be funded by philanthropies, along with similar institutes for the study of metals and radiobiology. The new institutes foiled the original purpose of the Met Lab and Argonne: to make use of the talent of the University of Chicago in the MED. The AEC later noted that Chicago nuclear scientists affiliated with the institutes—Samuel Allison, Fermi, Herbert Anderson, Warren Johnson, and others—"devote only insignificant amounts and in several cases none of their time to the Argonne program," and wondered whether "the investment in the Institute isn't buying talent away from Argonne?"[13]

The University of California could have arranged a similar outcome. Neither the equipment in Lawrence's Radiation Laboratory nor the land it sat on belonged to the government, unlike the pattern obtaining in the government-owned, contractor-operated laboratories of the Manhattan Project. The Rad Lab could have reverted to its prewar modes of funding and its role, as Lawrence planned in early 1944, as "a Division of the Department of Physics . . . [with] a small permanent staff." Instead Lawrence changed his mind, acquired money for his big plans from Groves, and thus secured the Rad Lab its special place in the postwar lab system.[14] The university, however, could not accommodate Los Alamos so easily. The initial Los Alamos contract called for it to lapse six months after the end of hostilities with the Axis nations. "It was clearly understood," said Underhill, "that we didn't intend to stay in Los Alamos forever."[15] By March 1944 the university planned to start "tapering off" its association with the weapons lab.[16]

Scientists at Los Alamos did not encourage the university to stay. They had complained about its personnel policies and business operations, conducted long-distance and hence often inefficiently from an office in Los Angeles.[17] Oppenheimer sometimes took matters into his own hands, to the alarm of Underhill and university administrators struggling to maintain some measure of control over a distant secret enterprise. "Very little attention has been paid by certain persons at the site to the requirements of the Regents," Underhill complained.[18] The absence of the contractor was physical, not fiscal, and the university tried to exercise some oversight over the business operation of the lab. The situation degenerated into "a position of feud," "lack of sympathy," "conflicts," and "essential distrust" between Oppenheimer and the university.[19] Underhill recognized that "they were never charmed with me down in Los Alamos"; the sense of Los Alamos staff that the university was "not a very easy-going parent" led to the movement at the lab around the end of the war to transfer the contract to Caltech.[20]

Underhill did not enjoy the prospect of the "credit" for the bomb accruing to a rival university. The university's Regents perceived benefits in the research potential and public service function of its operation of Los Alamos, provided that the university could retain the contract on its terms. Underhill insisted that it had the right to approve all personnel contracts and that such contracts could not extend past the expiration of the university's contract. Bradbury complained that Underhill was hindering recruitment.[21] When Nichols proposed in early 1946 that the university relinquish administrative control over Los Alamos employees, Underhill replied that the university could not protect itself against claims of third parties if it were not responsible for staff. The university had the whip hand: Underhill suggested "that the Regents determine immediately to serve notice on General Nichols that they will not renew the contract in any form after June 30, 1946, thus forcing the Government to manage this project under the Army—a method which is distasteful to the employees."[22]

With the passage of legislation creating the AEC, the University of California agreed to give the new agency "a chance to breathe and decide what they want to do with the project," despite president Robert Sproul's statement to the Regents that "if we get rid of bomb-making, plutonium, and New Mexico, I would be very happy."[23] The AEC took a deep breath, resuscitated Los Alamos as a permanent laboratory, and asked the university to keep the contract. Placed again in a strong bargaining position, backed up by Lawrence's Radiation Laboratory, the university negotiated on its terms. These included the right to terminate the contract with 120 days' notice to the AEC, owing to "the rather peculiar situation at Los Alamos where the University of California is operating a laboratory without being qualified to do business under the laws of the State of New Mexico."[24] For its part, the AEC obtained an "easier academic atmosphere" at Los Alamos, as one scientist put it, adding, "many of the key men there would probably not be available in a Government-operated laboratory."[25]

As the contractors wavered in their desire to continue with the work after the war, so did the AEC question the legacy of contract operation. The organic Atomic Energy Act of 1946 allowed the AEC "to conduct research and developmental activities through its own facilities," and it soon considered the possibility of doing so.[26] The AEC asked the University of California to continue as the contractor for Los Alamos in late 1947 only after first discussing "what contractor could take over the work, or how it could be operated directly by the Commission."[27] Earlier that year the GAC had proposed a central lab for programmatic research and recommended "that it be operated directly by the Commission itself—but not under Civil Service. This would avoid problems of incompetent or uninterested contractors, would assure intimate contact with the Commission, and adequate and liberal arrangements for personnel."[28] Such attitudes were not confined to Washington; around the same time, scientists working for Monsanto at Oak Ridge petitioned the AEC to take over operation of their lab.[29]

The AEC maintained the status quo but continued to consider alternatives to contract operation. Congressional questions gave incentive to further consideration: congressmen, especially those with responsibility for the federal budget, feared the devolution of responsibility from federal agencies and the potential for abuse of contracts. In the spring of 1949 the House Appropriations Committee reviewed the AEC's contract policy; that summer members of the Joint Committee on Atomic Energy accused the commission of "incredible mismanagement" and launched their own investigation.[30] David Lilienthal, who had managed direct federal operations under the Tennessee Valley Authority, soothed congressional concerns over the contract system: "It is not a sacred cow . . . and the whole contract system is, I think, one of the big questions about the future."[31]

That September AEC staff drafted a paper considering direct AEC operation of its facilities, including the national laboratories. It listed several advantages: "Direct AEC operation would mean that the laboratories would be national in the finest sense—available to all qualified scientists. No criticism could be made that operations of such a laboratory favored the few over the many. Direct operation would also relieve the universities of an administrative burden which generally they would prefer not to have. Employment of universities for management services, usually expected of Government or business, is almost unprecedented."[32] Yet the inexperience of the federal government in the administration of basic research programs, the desired participation of academic students and faculty and private industry in atomic energy research, and the difficulty of separating some labs, such as Berkeley, from their contractors argued against direct operations. The main drawback, however, would be the difficulty of recruiting staff: "It seems generally accepted that scientific personnel consider direct employment by the Government highly undesirable." AEC staff and commissioners bought into this theory despite the willingness of Oak Ridge scientists and the AEC's General Advisory Committee (GAC) to consider direct operations and the fact that "staffing does seem to have been accomplished by other agencies," not to mention the AEC itself, which managed to avoid civil service regulations.[33]

The recruitment issue reemerged a year later when commissioner T. Keith Glennan met with AEC staff to reconsider the problem. Motivated by the notion that "the AEC might be importuning the universities into a field of activity that was a bit off the beaten path of their prime function as institutions of higher learning," they cited other benefits from direct operation that might offset the personnel problem, including easier exchange of staff and closer coordination of programs, and elimination of duplication of administrative staff.[34] That is, common direction of the labs by the AEC might overcome the decoupling effects of contracting.

Consequences of Contractors

The variety of contractors gave each lab a special character, from Brookhaven's self-consciously academic atmosphere to the industrial environment of Oak Ridge. Thus the several labs maintained their individuality under the aegis of a single agency. The various contractors also had different levels of involvement in the labs. The University of Chicago set business policy for Argonne, but "scientific policy [was] the almost exclusive province of the [lab] director and the AEC."[35] In Berkeley, the same insistence on oversight that alienated Oppenheimer from the university also irked the accountants at the Rad Lab.[36] Unlike Chicago, the University of California was a public institution accountable to the state legislature, which subjected it to external pressure for fiscal control. In addition to oversight from business managers, the Regents of the university kept tabs on the labs through the Special Committee on AEC Projects, consisting of three cleared Regents. The committee reviewed the contracts and general policies but left decisions concerning the scientific program to Lawrence and Bradbury and their staffs.[37]

The operation of Brookhaven represented a new arrangement. Each of the nine members of Associated Universities, Inc. had two representatives on its Board of Trustees, usually one administrator and one scientist who split off onto an Executive Committee and a Scientific Advisory Committee. These committees took an active role in considering lab policies, budgets, and programs; as one of the trustees noted, at both Berkeley and Argonne "the governing bodies of the two contracting institutions are very much farther removed from contract operations" than at Brookhaven. Other trustees criticized the activism of AUI, especially its involvement in the scientific program: "I wonder," said Hans Bethe, "whether the Executive Committee is not taking the position of the Associated Universities somewhat too seriously."[38]

Academic administrators played important roles in the laboratory system. Underhill negotiated the Los Alamos contract and rode herd on the lab; the business officer at Chicago, William A. Harrell, served a similar, if gentler, function for Argonne; George B. Pegram, dean of Columbia, helped establish Brookhaven and AUI before handing over the reins to Lloyd Berkner.[39] The sources of irritation for all these men included what they took as the meddling of the AEC in their affairs and the reconciliation of the academic function of their nominal charges with the administration of the labs. AUI haggled with first the MED and then the AEC over the initial three-year contract for Brookhaven; as negotiations dragged on through 1947, AUI held out for control of salaries, lab records, patents, and classified research, and in general protested "too much management and control in the contract . . . too much expensive red tape."[40]

The University of California faced special problems. The Rad Lab represented a hybrid between an organized research unit on the Berkeley campus and an AEC laboratory and hence blurred the line between the AEC and the university, between research and teaching. The Rad Lab was a division of the physics department on campus, although Lawrence recognized by 1947 that the arrangement was "largely nominal."[41] The university eventually placed Lawrence and his lab under the university president and then under the newly established chancellor at Berkeley, at the same level as the deans of the various colleges and professional schools.[42] University faculty split time with the Rad Lab, but since the AEC paid part or all of their salaries, the university's budgetary apparatus lacked its usual control. The university on occasion had to remind part-timers of their duties as faculty, namely, teaching.[43]

Graduate students working in the Rad Lab and paid by the AEC also served as teaching assistants for a lab course on campus, as they were the only qualified students available; the chair of the university physics department pointed out that the AEC in effect was paying for teaching assistants. The lab course "started with a great deal of borrowed apparatus (mostly belonging to the Atomic Energy Commission—a technically illegal procedure)."[44] Los Alamos lacked the close association with a college campus and thus avoided similar problems. The Rad Lab and Los Alamos were "entirely independent," with separate contracts, although Bradbury assured the University of California Regents that there was "a great amount of scientific collaboration."[45] Bradbury served directly under the president of the university and so above Lawrence, who reported to the Berkeley chancellor.

A prime problem for the university in particular, and the labs in general, was patent policy. The Atomic Energy Act of 1946 required the government to retain patent rights for inventions in the fields of nuclear weapons and nuclear power. The AEC went further and insisted on retaining the rights to all inventions by contractor employees; in 1949 it made these inventions available to American industries through non-exclusive, royalty-free licenses, and set up the Patent Compensation Board to determine suitable rewards for the inventors. By late 1951 it had accumulated nearly three hundred patents but had compensated only one inventor.[46]

In the early 1930s universities had recognized the increasing interrelations of science and industry and, spurred by the depression, began to adopt policies to patent faculty inventions. The University of California used the nonprofit Research Corporation as a holding company for faculty patents; one of the first patents the company acquired from Berkeley was from Ernest Lawrence, who in return secured financial support for his Rad Lab.[47] The university solicited inventions from its faculty, filed for patents, and bore the expense; the inventor received a cut of any royalties, based on an announced scale, and the university plowed the rest into its research budget. The university thus protested the AEC's patent policy. Other federal agencies, notably the military, left patents to the contractor and required only that the government keep the right to use any inventions without royalties. The University of California for years sought a similar policy from the AEC, to no avail.[48]

Overhead and indirect costs represented, and represent, another sticky wicket for lab administrators and their overseers to negotiate. The wartime research contracts had included provisions for overhead expenses under the principle of no profit, no loss. Overhead covered the costs of administration, increased utility and maintenance bills, and depreciation of equipment, such as the 60-inch cyclotron at Berkeley, which ran twenty-four hours a day for the Manhattan Project "if they couldn't find 26 hours."[49] During the war the OSRD advanced money to the universities and counted on postwar accounting to recover any excess owed the government; the $18 million it recaptured in this way measured the magnitude of the problem.[50] After the war the AEC negotiated overhead for each laboratory contract, generally as a fixed percentage of labor costs in the lab budget. The contractors came to appreciate the additional funds, which helped support administrative units of the university. Chicago considered resuming the Clinton contract in part because its president, Hutchins, "wanted the overhead." At Berkeley, the Regents regarded overhead on government contracts as "continuing income."[51] The situation did not escape the notice of AEC staff: one of the arguments for direct operation of the labs was the "reduction in the overhead now paid to the contractors."[52] The AEC sought more precision by calculating a general cost per scientist, which in turn required definition of a "scientist."[53]

Another term requiring definition was "overhead" itself, and its close cousin "indirect costs." The gradual resolution of these terms reflects the education of scientists and university administrators in the new modes of government support of research.[54] During the war OSRD had calculated overhead rates at 30 to 50 percent of labor costs for university contracts. Except for Berkeley, however, the labs were government owned and did not require compensation for depreciation of facilities nor for personnel borrowed for contract work. The labs did have to maintain nonscientific support services, such as shops, security, fire and safety, maintenance, purchasing, accounting, and administration, which they budgeted as "indirect costs." The labs achieved a gradual reduction in the overhead rate by shifting overhead to indirect costs; thus, for example, Berkeley could reduce its overhead charge to 25 percent of wages and salaries in 1948 and to 12 percent for 1949. Around the same time Oak Ridge maintained separate allocations for indirect costs, which covered the shops and engineering departments, and overhead, which covered lab administration and other support services and amounted to about 15 percent of direct costs.[55] The labs had to work out ways to distribute indirect costs, which could amount to about half the total lab budget; for instance, whether they should keep indirect costs a separate budget item or try to distribute them to research groups.[56] The arbitrary allocation of indirect costs would thus provide a form of discretionary budget for lab directors to support marginal or pet projects.

While indirect costs remained within the lab, overhead went to the contractor to cover the costs of administering the lab. The AEC maintained "cost-plus-overhead contracts" with AUI, Chicago, and California, under which the AEC advanced funds for overhead to the contractors, who voluntarily refunded whatever they did not spend at the end of the year. The proportion of that amount returned by the contractor varied widely: AUI and California ended up with about 1 to 2 percent of direct costs as overhead, while Chicago ended up with 4 or 5 percent.[57]

The cooperation of the contractors became a central issue when the initial postwar contracts came up for renegotiation around 1950, and the AEC's controller sought to negotiate a harder, lower figure for overhead.[58] The AEC demonstrated its resolve by taking a hard line even with AUI, which charged far less than the other contractors and took a more active role, but whose first reactor project had become mired in delays and overruns. Although AEC staff thought that Brookhaven charged too much for overhead, AUI sought an additional discretionary fund on top of the current overhead to handle unexpected administrative expenses. AUI and Brookhaven did not have a university to run and hence did not depend on overhead, but the Brookhaven contract was also AUI's only resource. After dickering over the details for a year, the AEC agreed in January 1951 to grant a management fee to AUI, which would return unused funds to the AEC; in return, the AEC got a reorganization of AUI.[59]

The University of Chicago was not so successful. It charged far more for overhead at Argonne than the other university contractors while seeming to give only limited administrative attention to the lab. The AEC gave it first shot at the contract beyond 1952 on the condition that it address "the deficiencies in Laboratory administration" and its indirect expenses. The university responded with a proposal that was, according to AEC staff, "clearly out of line with the management services provided by the University": 5 percent of direct costs, with 3 percent tied to specific administrative units plus 2 percent for items that the university claimed "cannot be isolated or measured." The AEC, noting that personnel at Argonne, not the university, performed most administrative work, threatened to pull the Argonne contract from Chicago.[60]

The committee of AEC staff members considering the Argonne contract thought an industrial contractor might be best for the lab. The omission of Oak Ridge from the conclaves of academic administrators highlighted its difference as the lone lab under an industrial contractor (although other special-purpose AEC labs, such as the Knolls Atomic Power Laboratory, operated under industrial corporations). The lab at Oak Ridge, still known as Clinton Labs, had almost returned to the academic fold at the outset. The environment at Clinton had not changed under Monsanto: one AEC member viewed it as "a factory town" and added, "Research doesn't belong in that atmosphere." In early June 1947 the GAC advised that the AEC not locate a new high-flux reactor there.[61] Monsanto's contract expired at the end of that month; the company had already insisted that it did not want the Clinton contract without a new reactor; the AEC called its bluff, and Monsanto left the table. In this vacuum Clinton scientists made their proposal that the AEC operate the lab directly. The AEC stuck to its policy of contracting and in September 1947 announced that the University of Chicago would resume the contract for Clinton.[62]

The imminent return of an academic contractor might have earned the lab the "national" title: while the AEC negotiated with Chicago, it also designated the subject "Clinton National Laboratory." Scientists at the newest national lab hailed the announcement and predicted "more intimate contacts between the national laboratories at Argonne and Oak Ridge."[63] There remained the details of the contract, which raised many of the issues that bedeviled academic contractors: personnel policy; overhead, indirect costs, and contractor's fees; and patent rights. Chicago also struggled to persuade a succession of prominent scientists to direct the lab.

Lilienthal, beginning to suspect that the university still had no great desire to run Clinton, worried about the exclusion of industry from atomic energy research. In November he sounded out Union Carbide about the Clinton contract. The company, which already operated the production plants in Oak Ridge, had its own reasons to consider consolidating the Oak Ridge contracts: employees at one of the production plants noticed that their wages and benefits were lower than for comparable staff at Clinton and went on strike in December. With the Clinton contract, company officials reasoned, they could impose uniform personnel policies across the Oak Ridge site and pacify their employees. The AEC agreed to give Union Carbide the contract. It informed the stunned staff at both Clinton and Chicago of the decision at what became known as "Black Christmas" 1947.[64] The new contractor and its affiliation with the production facilities in Oak Ridge reinforced the industrial cast of the lab; perhaps in recognition of the affiliation, the lab in early 1948 was renamed Oak Ridge National Laboratory.[65] The "national" in the name remained from the episode.

The newest national lab thrived enough to change the AEC's attitude toward industrial contractors, so that AEC staff in 1951 could contemplate handing Argonne over to an industrial contractor. Around the same time the president of AUI perceived "a group in the Commission which favors industrial management of all national laboratories."[66] The same AEC staff report that studied direct operation by the AEC noted that "universities are not known particularly for their excellence in the field of business management," and "one of the most satisfactory management relationships exists at Oak Ridge where the contractor is an industrial concern."[67] Union Carbide had operated the production plants under a cost-plus-fixed-fee contract, which the AEC agreed to implement for the lab contract. The fee amounted to 4 percent, comparable to the overhead rate for which Chicago came under criticism. But Union Carbide imposed its own detailed budgeting requirements on Oak Ridge from the start; its active management and perceived success in fulfilling the AEC's programmatic goals earned it the fee in the AEC's eyes and made it a model for other contractors.[68]

Evolution of Devolution

After the AEC toyed with the possibility of operating the labs itself, haggled with its contractors over the terms of their contracts, and mulled over the industrial option, it reached a modus vivendi that would last through the 1950s. It kept the same stable of contractors, renewing their contracts at four- or five-year intervals. In 1952 the University of California assumed operation of the new weapons lab at Livermore. The operation of Brookhaven by an academic consortium provided a blueprint for other multi-institutional collaborations in the United States and, abroad, in the accelerator laboratory CERN (European Council for Nuclear Research), built in Switzerland by several European governments.[69] The University of Chicago retained the Argonne contract despite the opposition of a consortium of midwestern universities (the Midwestern Universities Research Association, or MURA), also on the AUI model, that criticized Chicago's management and complained of their lack of access to Argonne's facilities.[70]

Despite their occasional criticism of the contract system, congressmen, especially conservatives, could appreciate the role it provided for private enterprise. Contracting allowed the AEC to rely on the experience and administrative economies provided by a "chosen few" institutions and get by with a small staff.[71] Nevertheless, the AEC felt compelled after the investigations of 1949 to review its policy for Congress, emphasizing its accounting and reporting requirements, and to exert stricter controls over its contractors.[72] The policy required a delicate balance between delegation and accountability. Congress continued to monitor the balance and saw it swinging to the other extreme: in 1953 the House Appropriations Committee chided the AEC for excessive "interference" in the work of competent contractors. An internal AEC review of contract administration defended the trend and the staff necessary to continue it, concluding that the AEC "must exercise control and supervision to insure economy in the use of Government funds and conformance with Government procedures and standards of doing business."[73]

The overhead problem resurfaced from time to time. The AEC eventually applied to the other labs the solution of a negotiated fixed fee in lieu of overhead.[74] The AEC continued to object to high overhead fees, spurred by external pressure from Congress and the Bureau of the Budget. In 1958 the University of California decided not to seek a raise in its fee lest it attract attention to the AEC.[75] At the same time, however, the university's use of overhead raised eyebrows and questions at the AEC. In 1958 the California state legislature cut the appropriation for the University of California and directed it to use half of the AEC overhead fee to support the general campus budget.[76] The AEC did not know that the provocative legislation would only codify current practice; the university for years had salted away overhead from lab contracts in a campus reserve fund.[77]

Internal and congressional reviews did not alter the basis of the AEC's contract policy. After serious consideration of direct operation, which the organic Atomic Energy Act allowed, the AEC did not operate any of its national labs with government personnel, unlike the military services and, later, NASA.[78] It worked with a small coterie of contractors, and when it considered changing contractors for a lab, or finding a contractor for a new lab, it first approached one of the chosen few, as with the Clinton contract and the University of Chicago in 1947, or Livermore and the University of California in 1952. Instead of opening contracts to bid, as it did for some construction projects, the AEC selected particular contractors for the labs and negotiated terms of contracts.[79] The large research labs kept their contractors through the 1970s.[80]

The contracting system allowed delegation of responsibility to experienced institutions but also invited AEC supervision and congressional scrutiny and introduced problems such as the reimbursement of overhead and indirect costs. It offered university contractors access to elaborate facilities, overhead, and, in Berkeley's case, shared staff. It provided more intangible rewards in the achievements of lab scientists that redounded to the credit of their employers: the University of California gained from the Berkeley lab "much prestige and favorable publicity at what may be termed 'bargain rates'"; the university Regents complained when they did not receive enough public credit after the Castle-Bravo nuclear test.[81] But contracting also forced universities to reconcile their pedagogical purpose with off-campus research, some of it classified and far off-campus indeed. Meanwhile, industrial corporations gained entry into the new field of nucleonics, where they could scout business opportunities and develop trained staff who might transfer to other operations within the company.

Most important, the continuation of the contracting system perpetuated the difficulties identified early by the AEC: duplication of administrative staff, obstacles to exchange of personnel, and the individuality of each lab that hampered coordination of lab programs—all of which decoupled the labs from the other members of the system and reduced the possible benefits of systemicity.

LABORATORY ORGANIZATION

The individuality of each lab extended also to its mode of organization. The presence of accountants on the staff of Berkeley's Radiation Laboratory invites examination of the structure of the laboratories themselves. The size and complexity of each lab created an organization more familiar to the corporate boardroom than the research laboratory. In this, too, the labs owed much to the wartime experience. The one lab that predated the war, Berkeley, had brought organizational hierarchies into the laboratory; the scale and secrecy of the war work formalized the bureaucracy and demonstrated its advantages.[82] At the Met Lab in Chicago, Arthur Compton relied on advisory committees and an overall research council, and the flexible, informal arrangements of the early war years gave way to a military chain of command under the army.[83] A similar system of scientific committees under a lab-wide governing board composed of division leaders assisted Oppenheimer at Los Alamos, with the addition of a Liaison Committee to overcome the effects of compartmentalization.[84]

After the war the labs, like the rest of the country, retooled to meet new goals, including the initiation or resumption of basic research suspended in the push to produce a military weapon. None of the labs, however, neglected the needs of the AEC, and their administrators struggled to accommodate basic and programmatic research, scientists and support staff, big and small science, and multiple disciplines and interdisciplinary programs within a single organization.

Each of the labs had a strong director appointed by the contractor. All of them were physicists. At Berkeley, Lawrence still reigned over his Rad Lab with entrepreneurial enthusiasm. Bradbury would stretch his interim appointment as Los Alamos director into a twenty-five-year tenure. Walter Zinn had assumed direction of Argonne during the war after Fermi departed for Los Alamos. Zinn had completed his Ph.D. in physics at Columbia in 1934 and had collaborated with Leo Szilard and Fermi on early chain reaction experiments, accumulating experience with reactor engineering as Lawrence had with accelerators. Philip Morse, a professor of theoretical physics at MIT, guided Brookhaven through its first years before relinquishing the reins to Leland Haworth, a physicist who had headed Brookhaven's accelerator and reactor projects.

Oak Ridge was again an exception. In the interim period it divided administrative and research direction between James Lum and Eugene Wigner. When Wigner returned to Princeton in the fall of 1947, Union Carbide appointed C. Nelson Rucker as director after several top scientists rejected the job, and assigned Alvin Weinberg as associate research director. Weinberg had written his dissertation at Chicago on mathematical biophysics, but by dint of his subsequent work on piles at the Met Lab had acquired the patina of a reactor physicist. In practice Weinberg's enthusiasm and activism held sway and he directed the research program; in 1955 Oak Ridge formalized his duties as lab director.[85] After the initial postwar flux, lab directors tended to remain once installed; only Argonne had much turnover in its directors, and this only after Zinn left the lab in 1956.

For about a decade after the war the same group of physicists led the national labs. Most of the contractors left the direction of scientific research to the labs and their directors. The multipurpose nature of the labs required coordination of their various programs and distribution of budgets. This unenviable responsibility fell to the lab director, who would then coordinate his own programs with those at other labs through the various AEC program managers; the control of limited discretionary funds by the lab director gave them some leeway in the determination of relative levels of effort.

Below the director each lab divided its work into several departments or divisions. The labs differed in their organization of these internal groups and struggled to accommodate both research and development. Brookhaven chose an academic model and set up "departments" along traditional disciplinary lines: chemistry, physics, biology, and medicine. Less self-consciously, Oak Ridge followed a similar disciplinary arrangement, and both Brookhaven and Oak Ridge formed separate, additional groups to handle large projects and developmental problems. Lawrence's Rad Lab, despite its association with the Berkeley campus, shunned traditional disciplinary departments and congregated instead around individuals and their machines, such as Luis Alvarez's linear accelerator and Edwin McMillan's synchrotron. At Los Alamos, Bradbury inherited a more functional arrangement, based on the lab's primary mission on weapons and the disciplinary backgrounds of its scientist; divisions were designated by initials, from A (administration) to Z (ordnance engineering). This structure masked a flux of groups within the divisions to accommodate shifts in the lab program and research interests of its staff.

The presence of multiple programs, the characteristic that distinguished the large national labs from other AEC labs, presented special problems and opportunities in their organization. Some of the programs at the labs shaded into interdisciplinary research, a transient phenomenon difficult to define, since scientists in cross-disciplinary collaborations could retain their disciplinary identities. When they did not, and instead submerged their identities and merged their interests, new disciplines could result, which the labs had to accommodate. In addition to functional divisions organized around reactors, accelerators, or weapons, lab-wide committees and committees for individual projects helped effect the integration of multiple programs.

Two groups played special roles. The focus in the Manhattan Project on production of specific reactors and weapons had placed a premium on engineers, which the postwar mission continued. Although Lawrence initially hoped to scale back his engineering staff to a half dozen or so after the war, by 1948 Berkeley boasted eighty engineers, or about a quarter of the total scientific staff. Brookhaven took Berkeley as a model and struggled to duplicate its strong engineering groups. The labs took various approaches to organizing their engineers: Berkeley set up separate developmental groups for mechanical and electrical engineering for accelerators; Los Alamos had its weapons engineering division; Brookhaven and Oak Ridge had divisions for their reactor engineers but forced accelerator designers to cultivate their own engineering support. Although engineering could fall into peripheral service slots on organization charts, its integration with the scientific program was central to the fulfillment of lab missions.[86]

The provision of big machines and technology, however, seemed to provide no place for theorists. Berkeley, Los Alamos, and, until 1950, Argonne maintained separate groups of theoretical physicists. These groups functioned as calculators for the rest of the lab: at Berkeley, for example, theorists under Robert Serber calculated the energies and properties of particles produced in accelerators; the theoretical group at Los Alamos crunched complex hydrodynamic equations for weapon design. The computers developed by some of these theorists, notably those at Los Alamos, embodied this calculational function.[87] Oak Ridge made the service role explicit in spinning off the mathematics and computing group from the Physics Division in 1948 into a Mathematics Panel that fell in the organization charts under administration or, later, with other support functions such as health physics and instrumentation.[88]

If it was difficult to distinguish research from development, some of the latter mingled with production functions. Los Alamos maintained the Z Division, for ordnance engineering, at Sandia Base near Albuquerque. By 1948, Z Division had grown to rival its parent lab in size. Its duties in producing nuclear weapons from Los Alamos designs, or, as Bradbury put it, "packag[ing] the bang," as well as its location, separated it from the rest of the Los Alamos program. The University of California was uncomfortable running the isolated production operation and sought to escape the contract. The AEC's Division of Military Application also recommended that an industrial contractor would be more appropriate, and in 1949 the AEC selected AT&T to run Sandia Laboratory.[89] Sandia assisted Los Alamos with weapon hardware and would later open up a branch in Livermore for the same function. As a single-purpose, production-oriented facility, Sandia belonged to a different class of labs within the AEC, although in the late 1950s it would expand into research.

Instead of spinning off facilities associated with production, Oak Ridge acquired them. The production plants at Oak Ridge included the Y-12 area, built during the war for electromagnetic separation of uranium isotopes. Union Carbide maintained research divisions there after the war for isotope research and production, electromagnetic separation, and chemical research. As Oak Ridge National Laboratory grew through the late 1940s, it had eyed Y-12 as lebensraum; the AEC's expansion of production in 1950 made Y-12 space too precious for the lab to acquire, but the research divisions there merged that year with the lab. They would remain at the Y-12 site but report to the lab instead of the Y-12 managers.[90] The association of Y-12 with production would later lead lab staff to suggest an official reorganization and explicit separation, or quarantine, of heavy reactor engineering, isotope production, and materials chemistry at Y-12 and the concentration of basic research at the lab on the X-10 site.[91]

Each lab employed thousands and spent several millions of dollars a year. The administration of these enterprises itself required substantial effort; hence lab organization charts sprouted branches for accountants, purchasers, patent lawyers, and financial planners. Even with a liberal definition of scientist that embraced everyone from lab technicians to senior scientists, or everyone with a bachelor's degree or higher, scientists and engineers made up only one-third or one-fourth of the total staff at the labs. At Clinton, where research staff numbered 575 out of a total of 2,185 in 1947, administrators noted that Los Alamos and Argonne as well as industrial labs such as GE, RCA, and Bell Labs had a better tooth-to-tail ratio. But the average ratio throughout the AEC was even lower, less than one-fifth, and it would hover around that level for years.[92] Scientists noted the preponderance of administrators and complained about their dominance on laboratory committees: at Brookhaven and Oak Ridge, administrators outnumbered scientific department heads two-to-one on top-level policy committees.[93] Administrators at Berkeley worked under the "rather tight controls" of the university and hence held the scientific group leaders, otherwise "fairly autonomous," on a "fairly tight rein" financially.[94] The division of the directorship at Clinton produced "constant wrangling between research and administrative personnel."[95] Despite the efforts of lab directors to increase the relative numbers of research staff, the ratios remained more or less constant as the labs grew in size.

The support apparatus was not limited to administrators. Fire departments, health monitors, and security officers assumed added importance in the hazardous and secret work on atomic energy. The scale and complexity of the labs extended also to their physical plant. Accelerators and reactors involved intricate mechanical and electrical systems and huge water and power supplies. To provide them, and more mundane equipment, the labs relied on engineers, who counted as scientific staff, and a motley assortment of shop personnel including draftsmen, machine operators, mechanics, and glassblowers. At Berkeley such staff outnumbered the numerous engineers. Oak Ridge at its formation counted two hundred craftsmen in its shops (including a blacksmith), or about 10 percent of total staff.[96]

The independence of the labs implied duplication of support services, both engineering and administrative. Since support functions made up the bulk of each lab's effort, duplication did not come cheaply. It also did not occur precisely. Brookhaven admired the centralized engineering organization of Berkeley's Rad Lab but could not recreate it. Systemicity then worked to their advantage: Brookhaven could draw on support from the Berkeley groups in the design of several of its early accelerators.[97]

The individuality of the labs thus extended to their organization, from the disciplinary departments and decentralized engineers of Brookhaven to the informal individualism and centralized support of Berkeley. Their organizations need not align with one another nor remain static; Oak Ridge and Argonne proved willing to create and disband groups with alacrity, while at Los Alamos an apparent stability masked movement at lower levels. Nor did the organization of individual labs necessarily match that of the AEC and its program divisions, which left some lab groups with no obvious sponsor at the AEC level.

MANPOWER

The AEC's reliance on contractors, mostly academic ones, and the disciplinary organization of individual labs were designed in part to attract scientists. The character imparted by the contractor, as with the industrial management of Oak Ridge or the universities that ran Brookhaven, affected the marketability of the lab to scientists used to working in a collegial atmosphere. Personnel policies, research potential, and factors such as housing and lifestyle helped to attract academics. Competition among the labs for scientific staff in addition to programs enhanced the importance of marketability. The flexibility given top scientists within the organization of some of the labs and the determination of the AEC to advance its program by supporting such scientists increased the need to attract high-quality staff. The labs would have to do so within the confines of their system and the context of the postwar United States.

Seller's Market

The systemic structure of the national labs forced them to compete with one another for scientific staff. From their inception the labs relied on raids of the other sites in the system: the Met Lab recruited staff from university programs, including Berkeley, only to watch its scientists depart for Clinton and Los Alamos, the latter thanks to Oppenheimer's "absolutely unscrupulous recruiting" which earned him admonishments against overly depleting the Met Lab and MIT's Radiation Laboratory.[98]

The competition for staff accelerated after the war with the addition of new sites in the system. Brookhaven, which started from scratch, looked first to the other labs for expertise in the new field of nucleonics. The inter-university planning group sought to recruit Wigner and then Zinn to direct Brookhaven, earning them a scolding from the MED's area engineer and a plea from Nichols to "refrain from competitive bidding for key personnel now located at other laboratories."[99] His words fell on deaf ears. Brookhaven in mid-1946 lured Lyle Borst, supervisor of Clinton's reactor research, to help design its reactor; Borst promptly started asking his former Clinton colleagues to join him.[100]

The competition compelled the area engineer to convene a discussion between Morse and Wigner, whose disaffected scientists at Clinton were a tempting target. Wigner agreed that "no attempt should be made to bar discussion between [Brookhaven] and any member of the Clinton Laboratories, who might be interested in changing his place of work . . . after all this was a free country." For his part, Morse agreed to let Wigner know which Clinton staff he was after so that Wigner might make a counteroffer or make a case for the importance of the individual to the Clinton program. Morse thanked Wigner for "the spirit of cooperation with which you [Wigner] are endangering your laboratory's research program in order to help us get started." The Brookhaven campaign extended to Argonne, Berkeley, and Los Alamos, and produced similar agreements.[101] The network of leaders of wartime projects had built up a reservoir of bonhomie that eased their relations as lab directors, imposed some scruples, and prevented cutthroat competition for staff.

The labs did not form a perfectly insulated system: competition for staff reached outside its borders to universities and industrial labs. Los Alamos and Clinton lost most of their departing staff to academia. The University of Chicago, Groves observed, "was probably the worst offender, but others were also engaged in wholesale recruiting campaigns."[102] Turnabout is fair play. Brookhaven pursued academic scientists with equal ease, and also raided industry—including some companies working on AEC contracts, prompting a rebuke from Carroll Wilson, the general manager of the AEC.[103] Wilson's own staff in Washington, however, was also trolling for lab scientists. The acting director of research in 1948 sought "the best places to look for the few chemists, physicists and metallurgists we require" and identified "several obvious advantages in getting them straight out of our own laboratories." Lab scientists would know the people, the programs, and the organizations from the start. "On the other side is the disadvantage that the good men we want are not men who will not be missed by the laboratories they would be leaving." He hoped that the laboratory directors would "agree with the view that competent staffing of the Research Division is important to you, so important that you can afford to contribute one of your valuable men to it occasionally."[104]

A shortage of qualified scientists compounded competition. The war had interrupted the training of young scientists just as it made them a more valuable commodity. Groves's Advisory Committee on Research and Development warned that "the most serious limitation to the conduct of the work in the nuclear energy field was the shortage of personnel."[105] In this seller's market certain scientists found themselves in particular demand. Theoretical physicists played musical chairs, but with more chairs than players, and the labs joined in the game. After Berkeley hired Robert Serber from Los Alamos in the fall of 1945, Los Alamos lured Stanislaw Ulam to supplement a succession of consulting theorists. The shortage of theoretical physicists (and scientists in general) would be a central issue in the decision to support a new laboratory at Livermore.

Biomedical researchers, particularly those familiar with the effects of radiation, also saw their stock rise after the war. The AEC warned Congress that "there were, and are, desperately few competent scientific and technical workers and teachers in the field of the effects of radiation on living things."[106] Wigner's protests protected Clinton from the predations of Brookhaven when it tried to hire away Karl Morgan, head of the Health Physics Division. Morse looked further down the ranks and appealed to their common cause: "It would obviously be foolish to denude your laboratory of health physicists in order to take care of our problems, which are immediate. Nevertheless . . . it might well be the policy of the Atomic Energy Commission to spread these important health physics people over their projects as evenly as possible."[107] Brookhaven also looked to Berkeley and its thriving group of medical physicists, as did Oak Ridge.[108]

The pursuit of large, multidisciplinary programs also put a premium on scientists who could work within large groups and on others who could lead them. For the latter, Raemer Schreiber, a group and then division leader and eventually an associate director at Los Alamos, noted the departure of "spiritual leaders, the Bethes, the Weisskopfs, the Oppenheimers," after the war, and the growing dependence on "team leaders" like himself and Darol Froman. The team leader "doesn't necessarily have the brand new ideas, but he kind of knows how to go about getting something done about it. . . . The type of work which was done here and in other wartime laboratories . . . gave more recognition to the sort of team approach."[109]

The team approach permeated diverse aspects of the work. In 1956 Samuel Goudsmit, head of the physics department at Brookhaven, informed his staff of the attributes they would need to work in teams: "In this new type of work experimental skill must be supplemented by personality traits which enhance and encourage the much needed cooperative loyalty. . . . I shall reserve the right to refuse experimental work in high energy to any member of my staff whom I deem unfit for group collaboration." Norman Hilberry, who succeeded Zinn as director of Argonne, applied a metaphor from his own background in reactors: "Just as one achieves large power outputs by resorting to dilution within the [reactor] core, so in a laboratory one supplements the endeavors of the brilliant with the contributions of highly qualified yet less creative scientists and engineers." Lawrence, whose lab at Berkeley provided an early exemplar of cooperative research, recognized that lab scientists and administrators would have to adjust their standards for recruitment and promotion to accommodate team players.[110] In this the national labs reflected the ascension of "the Organization Man," at a time when industrial labs found "No Room for Virtuosos" and a Monsanto recruiting film proclaimed, "No geniuses here; just a bunch of average Americans working together."[111]

Inducements

The labs relied on a variety of factors to induce scientists to work in their new arrangements. Clinton provided a tempting target for recruiters in part because of its location. The desire of its scientists to escape Dogpatch was matched by attitudes in Washington, where members of the new GAC predicted "Clinton will not live even if it is built up."[112] Scientists accustomed to collegial campus life found the Tennessee backwoods dull. Clinton scientists perceived that the community had not "developed much of a culture of its own"; "it would be desirable if Oak Ridge could somehow attain dignity and stature in its general intellectual life."[113] Rabi, raised and residing in New York, taunted Clinton scientists: "You fellows out here in the wilderness, why don't you move the laboratory to civilization?"[114] But Brookhaven was not metropolitan either. The lab's planners had intended to find a site easily accessible from northeastern universities but remote enough to ensure safety. After a search and much haggling, they finally settled on Camp Upton, a former army base halfway out Long Island, equally inaccessible to universities and certainly remote. Morse described his first visit to the site as a "flattening experience," and the planning committee accepted it only as an "equalization of disappointment."[115] Some northeastern academics classed Brookhaven as one of the "rural atomic energy laboratories," and AUI trustees complained, "Top flight personnel are difficult to obtain in outlying communities."[116]

Los Alamos was if anything even more remote than its eastern counterparts, with the added disadvantages of distasteful (to some) weapons work and strict security. A Manhattan Project administrator cited the "the frontier type life and isolation" as a barrier to recruitment.[117] The natural beauty of the site and the romantic legacy of the wartime effort may have compensated, and many lab scientists would come to embrace the southwestern lifestyle, down to the big belt buckles and bolo ties. There were other inducements: the sense that Los Alamos "was by this time one of the best-equipped laboratories in the world," with more on the way and the time to use it. "It was a rather heady feeling," said Schreiber, "to go out here and be in charge and write orders for $100,000 worth of stuff. I would go back there [to Purdue] and be maybe at best an associate professor fighting for research money."[118] If many were eager to return to academia, others remembered teaching loads and grant proposals and thought twice.

Universities, however, represented the traditional home for most scientists, especially amid the uncertainty of the immediate postwar period. In a survey of some six hundred American scientists in 1948, about half preferred academic employment, less than a third industrial labs, and only 11 percent government labs.[119] The sentiment did not bode well for the national labs, although it provided an argument for academic contractors for them (and the support for industry was significant). The "easier academic atmosphere" provided by the University of California helped attract scientists to Los Alamos, and Brookhaven affiliates likewise expected the lab to maintain a "University atmosphere."[120]

The labs could not exactly recreate the university environment, but they gave it the old college try. One of the main attractions of academia was the prospect of tenure. The operation of the labs through short-term contracts, of at most five years, seemed to preclude the possibility of tenure for lab staff: a contractor who granted tenure and then lost the contract would be left holding the bag. In addition, one of the main justifications for the national labs was to provide facilities for university scientists, which argued against the maintenance of a large in-house staff. Brookhaven, which inherited no developmental work and hence was the most committed to the provision of research facilities for outside users, initially decided to hire senior staff only on a leave-of-absence basis and to keep permanent staff to a minimum and at the junior level.[121] In early 1948 the lab reversed field and considered granting tenure to senior staff, with junior staff granted short-term appointments. Tenure would apply only within the limits of the contract and subject to security clearance. One Brookhaven scientist complained that the plan set up a "caste system" between research and technical service staff and contradicted the intention to serve visitors. Haworth nevertheless adopted the tenure policy, to provide security for senior staff and continuity in the research program.[122] Shortly thereafter Oak Ridge entertained a proposal that Ph.D. scientists with five years of service be considered for tenure. The lab's Personnel Committee, educated by a copy of Brookhaven's policy, considered the proposal and recommended it with minor changes. The plan met with "considerable disagreement," however, in the Research Council and failed to pass.[123]

Neither Argonne nor Los Alamos chose to follow Brookhaven's lead. In the case of Los Alamos, the University of California Regents "simply said, 'You cannot give tenure on a government contract.'"[124] They made one exception for Bradbury, to fend off recruiters from other schools, but only over the objections of the university's Budget Committee that "academic titles and consequent membership in the Academic Senate should . . . be reserved for those appointees who are genuinely concerned with academic instruction. . . . The Los Alamos laboratories are not a part of the academic University."[125] The same objections could have applied to the Radiation

Laboratory, whose hybrid status would lead the same committee to complain about its independence within the university.[126] The Rad Lab precedent provided the solution for the case of Bradbury. The university had appointed Robert Thornton and Robert Serber as "Professors of Physics, Radiation Laboratory"; although each would earn his title by teaching courses on campus, the initial appointments were intended only to retain their valuable services as research physicists in the Rad Lab. The university applied the same solution to Bradbury, who became "Professor of Physics, Los Alamos Laboratory."[127] Zinn had a similar appointment at the University of Chicago; Bradbury and Zinn would remain the only members of their respective laboratories to hold academic appointments at the contracting university.[128]

If Argonne, Oak Ridge, and Los Alamos did not adopt an official policy of tenure, neither did they make any concerted effort to clear deadwood from their staff rolls, which continued to expand throughout the 1950s. Argonne administrators would note that "while it is true that there are no tenure positions in the Laboratory, it is equally true that no one is fired after a few years' service." As a consequence, less competent junior staff occupied positions that the lab would otherwise have offered to more promising recruits.[129] Tenure, whether de facto or official, had unintended consequences. Haworth noted in the mid-1950s that Brookhaven's scientific staff, about one-quarter of whom enjoyed tenure, exhibited a demographic bulge in the thirty-five to forty-five age group, thanks to the initial rush of recruitment at the formation of the lab. As this group aged and approached retirement, the lab might lose some vigor and face another recruiting crisis. Argonne and Oak Ridge displayed similar demography. Despite concerns that tenure might confine the research program to the interests and competencies of senior scientists, Brookhaven continued its policy, and Argonne would again consider adopting one in the early 1960s as a way to retain outstanding senior staff.[130]

The labs continued to maintain or consider tenure because of a persistent seller's market for qualified scientists. They also turned to more concrete inducements in their competition with academia, industry, and one another. In 1946 the MED had appointed a committee to study its wage and salary policies. The so-called Loomis-Tate committee proposed a standard salary scale for scientists, but found "no reason to urge that the present laborato[r]ies with established and satisfactory scales of their own should change."[131] The committee reconvened the next year to update its report for the AEC. Wilson, the general manager of the AEC, asked it to consider, among other things, the variation of salaries among the labs: he noted the "very urgent and very delicate problem" of the recruitment of scientists by other AEC labs, which resulted in "disruption of the research program but also in a spiraling of the salaries paid." He asked whether the AEC should impose a standard salary scale for the whole system or even set up a central clearinghouse for scientific personnel from which the labs would draw staff.[132]

The committee recommended instead laissez-faire: "In the long run, it will be actually healthy if a sort of survival-of-the-fittest process takes place whereby the scientists shift, by individual decisions, to laboratories where conditions . . . are most attractive and favorable to work." It recognized the possibility of bidding wars for the best scientists but hoped that "informal communications" among the lab directors could avoid them.[133] The labs did not always exercise restraint. The individuality of the labs and their contractors encouraged competition and counteracted attempts like Wilson's to use the systemic structure to advantage. For instance, AUI and the University of California insisted on setting wage and salary scales and retaining approval of all salaries above a certain level, and Los Alamos and Oak Ridge had internal committees to review personnel decisions to more junior grades.[134]

Wilson's worry over interlab recruitment and potential feedback in salary scales also applied to universities and industry. A Los Alamos scientist noted the fear that high salary scales at the labs might disrupt the scales at academic contractors such as California or Chicago. Los Alamos's leaders did, on occasion, discuss "the apparent difficulty of reconciling Laboratory and University salaries," especially when university departments tried to recruit Los Alamos staff. But, the same scientist noted, "our main competition is with industry."[135] Los Alamos based its salaries on a nationwide survey of industrial labs, not universities, that it conducted every year or so. It often included the other labs in the survey and its distribution. The first one in 1947 found the labs lagging behind the average salary paid by industry to scientists.[136]

Los Alamos found itself regularly ratcheting its salary scales to keep up with industry. The Los Alamos personnel office circulated to industrial research organizations the results of the survey, in exchange for their participation, thus ensuring feedback and further spiraling of scientific salaries.[137] Competition with the other labs also quickly drove up salaries, especially for fields in particular demand, and established scientists commanded offers far beyond base rates: in 1948 Brookhaven offered $10,000 a year to Hardin Jones and Oak Ridge $7,500 to Cornelius Tobias to try to lure them from the medical physics group at Berkeley.[138]

The offers to Jones and Tobias provide some comparison with academic salaries, since they both had appointments on campus at Berkeley. At the time both were earning $4,800 a year, although the university responded with raises of its own to keep them. Bradbury noted that salaries at Los Alamos (and by implication in industry) generally outstripped academic scales: academic faculty could augment their income through book publication or consulting, and Los Alamos staff worked through the summer. The latter condition implied a multiplier of 12/9 in theory, but in practice the lab had to offer 3/2 of academic scales to recruit faculty, owing to the "greater leisure of academic work[!], the freedom to publish, the absence of programmatic research, and the opportunity to establish a permanent base in a University community."[139]

Similarly, Morse's initial policy at Brookhaven multiplied the salaries for visiting academic scientists by 6/5 or 4/3, plus another 10 to 20 percent to cover the short-term change of residence; salaries for permanent lab staff were even higher to compensate for the initial lack of tenure. Even with these adjustments, and after the adoption of tenure, Brookhaven found itself several years later losing out to industry, especially in the case of engineers.[140] Berkeley and Livermore likewise had trouble attracting engineers in the mid-1950s, especially with the expanding aerospace and electronics industries, along with a few nucleonics firms such as General Atomics, in California at that time.[141] The lab directors in 1956 worried that lab salaries were lagging about 20 percent behind industry, especially aerospace; Berkeley and Livermore instituted across-the-board raises of 22 percent, Los Alamos doled out a 12 percent raise, and Brookhaven also gave up to 12 percent raises in order to keep pace.[142]

The continued shortage of personnel, concomitant spiraling salaries, and the costs of sustaining staff on a semipermanent basis revived Wilson's proposal of a central clearinghouse for scientific personnel. The idea recirculated in the early 1950s in proposals by the GAC and its counterpart, the Advisory Committee for Biology and Medicine, to promote exchange of personnel among the labs.[143] Such exchange would capitalize on systemicity by allowing each lab to draw on the resources of the other labs in the system, and hence reduce the duplication of expert and service staff at every site. As a further benefit, the GAC noted, "the interplay caused by the interchange of promising young men and of leaders in the various fields keeps the water stirred so that an able administrator can assure a certain turnover and has a chance of avoiding the consequences of inbreeding and stagnation." The director of the Division of Research passed the recommendation on to the labs.[144]

But the diversity of contractors and the resultant individuality of each lab limited the possibility of personnel interchange. To swap a scientist would require the coordination of different salary scales and policies, including tenure, but also vacations and retirement and health plans. Hence, when the United States scurried to resume nuclear weapons tests after the Soviets broke the moratorium in September 1961, the GAC sought to transfer scientists from Oak Ridge, Argonne, and other AEC labs to expand the weapons development programs at Los Alamos and Livermore, but had to consider ways around the barriers to exchange.[145] Although one of the justifications for the labs was to provide a pool of scientists on tap for national emergencies, the various personnel policies limited the potential to draft scientists from other sites, and the subsequent sustenance of large in-house staffs and competition between sites watered down the quality of talent at each individual lab.

Truly National Labs

The assembly by the national labs of large in-house staff, a fair fraction of whom enjoyed official or de facto tenure, exacerbated the shortage of qualified personnel about which the labs complained. The retention of top scientists at the labs instead of universities kept them from training new generations of scientists to fill the gap. The original concept of a national lab was to provide facilities for visiting researchers from universities, not for in-house staff. The extent to which the labs accommodated visitors defined their "national" status. Oak Ridge staff in 1948 claimed to be "making every effort to cooperate with the member universities in establishing the Laboratory as a true National Laboratory." The claim to fidelity to the national concept, appearing here for the first time, would become a popular refrain for all of the labs in the years to come.[146]

The GAC the next year ran down the list of labs and graded their adherence to the concept. The presence of classified, programmatic work at Argonne and Oak Ridge gave them only "minor" roles as national labs:

"Reactor work at Argonne is interfering with its role as a national laboratory"; "Oak Ridge is a complex of production, programmatic research and development, and national laboratory." The University of California Radiation Laboratory and Los Alamos Scientific Laboratory had not yet acquired official "national" designation. The Rad Lab's affiliation with the Berkeley campus, and the almost exclusive use of its machines by local staff, rendered the definition problematic: "It is unclear to what extent Berkeley is a national laboratory either in matters of policy or in the minds of scientists. As a matter of principle it would appear that the expenditure of Federal funds in the creation of nearly unique national scientific facilities would imply that these facilities should be accessible to American scientists."[147] Los Alamos would take early steps toward nationalization and establish a university affiliation plan and admit properly cleared graduate students for research at the lab.[148] Brookhaven presumptuously appropriated the national notion and dubbed it "the Brookhaven idea." The lab initially hoped to have half, but only half, of its scientific staff composed of visiting scientists and students.[149]

Brookhaven, the lab most committed to the national ideal, could not fulfill its good intentions. The quick adoption of tenure undermined its rhetoric, the construction of research facilities fell far behind schedule, and the lab did not meet its quota. By the summer of 1952, however, with a research reactor up and running and the Cosmotron proton accelerator coming on-line, the lab could count over 150 scientists spending the summer at the lab (91 of them undergraduate or graduate students), or about half of the permanent scientific staff. Yet during the academic year, the number of visitors fell to about half of the summer totals.[150] In other words, twice as many permanent staff as visitors used Brookhaven's facilities in summer, and over four times as many the rest of the year.

The other labs were even less successful. Argonne had only twenty-one visiting researchers in 1948, or about 5 percent of its scientists. Zinn, asked in 1955 how much visitors contributed at Argonne, responded, "Not much." Classified weapons programs at Los Alamos and Livermore and reactor programs at Argonne and Oak Ridge limited the opportunities for visitors to those labs.[151] The systemic structure also contributed to their failure to fulfill their service role. As they did for permanent staff, the labs competed also for visitors, particularly for those, such as theoretical physicists, in short supply.[152] The duplication of facilities such as the large accelerators at Berkeley and Brookhaven reduced the demand for each one.

More mundane matters proved just as crucial to the national lab ideal. Housing, in particular, proved a persistent problem: the labs had to accommodate visitors physically as well as scientifically. In 1946 Morse emphasized that for Brookhaven, "*the housing situation is our No. 1 problem.*" One of Brookhaven's "major purposes is to secure a sizeable number of scientific personnel on loans and leaves of absence. They could not be expected to purchase homes in this area. Rentals are still virtually nonexistent." The AEC, however, "was very loath to spend money for housing at Brookhaven."[153] Or anywhere else. Clinton staff warned of the "serious," even "critical" housing shortage in Oak Ridge. Lilienthal complained to Congress that Los Alamos staff were living in "shacks. This was a kind of scientists' slum."[154] The labs were trying to implement the national ideal and house their visitors at the same time the nation as a whole was undergoing a serious housing shortage, as returning veterans, aided by federal mortgage subsidies and flush with wartime savings, flooded the housing market. The shortage for the labs, however, outlasted that of the nation as a whole. The reluctance of the AEC to provide housing probably stemmed from the consistent grilling it received in Congress over the communities it operated in Oak Ridge, Hanford, and Los Alamos, which smacked of socialism to Cold Warriors.

Through the early 1950s the labs struggled to house their visitors. Oak Ridge administrators wondered whether the lab should just drop its summer research program for lack of lodging. For Haworth at Brookhaven, still hampered by a housing shortage, "The whole future of Brookhaven as a cooperative Laboratory may hinge on this point." The lack of housing determined the pace of programs at Los Alamos, where "authorization of new hires was frequently delayed pending a housing vacancy." The GAC traced Argonne's lack of university researchers to its need for housing and noted that the transfer of employees among the labs would also require provision of transient housing.[155] The AEC would eventually allow the construction of housing at Argonne and Brookhaven in the late 1950s, but even this would prove insufficient.[156]

3

INTERDEPENDENCE

The individuality of the labs ensured that each would protect its own interests within the national laboratory system. But several forces coupled the labs together, not only in programmatic competition but also in a united system within the same framework. They were linked through the AEC's organization and its mission; the AEC made the linkage explicit through its budget process, which combined lab budgets within AEC programs. The process over time encouraged AEC staff to consider the labs less as integrated institutions and more as a collection of projects; in response, the lab directors organized to present a common front against the AEC. Throughout, lab scientists worked within an exclusive classified community that bound them together and provided the communication necessary to sustain systemicity.

SECURITY

The work of the national labs on programs for national security, in particular nuclear weapons and reactors, brought an elaborate apparatus to protect the physical security of lab facilities and materials, classify the information they produced, and ensure the loyalty of the scientists who worked in them. The labs evolved under the tension between the promotion of national security in the context of the Cold War and the preservation of the values of democracy and science. The preclusion of an open, self-regulating scientific community had particular consequences for the national labs because of their organizational and geographical separation: barriers to the flow of information could lead to expensive duplication and the pursuit of problems down dead ends. AEC administrators and lab scientists struggled to accommodate the requirements of security within the systemic arrangement of the labs.

The Origins of Secrecy

The national labs did not originate secrecy, but, like big science, it found its fullest expression in them. Secrecy in science predated the labs in three forms. Personal secrecy—scientists keeping unpublished results to themselves—stemmed from the reward system of the scientific community, which emphasized priority of discovery.[1] Industrial secrecy derived from the rewards of proprietary priority, which, for example, prevented physicists in the nascent radio industry of the early twentieth century from publishing all their results and induced AT&T to compartmentalize information within its laboratories. Similar proprietary concerns have stifled peer review and fostered duplication in current genomics research.[2] Finally, secrecy for national security had an equally long history: for instance, in the work of Lavoisier and other French chemists during the Revolution on new forms of explosives, which they pursued in their secret weapons laboratory.[3] As the last example demonstrates, secrecy is not necessarily forced on scientists by national governments: the usual, excusable ignorance of military and political leaders of the details of the latest scientific and technical developments, combined with the technological conservatism of the military, has often required scientists to assume the initiative in developing new technologies for national security. After the discovery of fission, nuclear physicists in America, many of them émigrés, imposed self-censorship on further work in their field lest the Nazis get any ideas.[4]

This did not prevent them from protesting when General Groves and the Manhattan Project institutionalized censorship to satisfy President Roosevelt's insistence on secrecy. In addition to forbidding the publication of sensitive research results, Groves compartmentalized information within the project, thus preventing any disloyal scientist from learning about all aspects of it and conveying the knowledge to a rival nation. Compartmentalization kept scientists from knowing what was going on at other sites or even in other programs within the same lab, and hence in a few cases delayed the progress of the work. Considerations of these consequences spurred the establishment of Los Alamos.[5]

The end of the war did not immediately imply the end of restrictions, but with demobilization the Manhattan Project began to explore relaxing them.[6] On 12 August 1945 a report on the project was issued, written by Henry Smyth and hence known as the Smyth report, which publicized some classified information but also served to preempt further release.[7] But Groves retained the policy that all research was born classified and then had to be reviewed by senior scientists for release. The procedure ensured that only a trickle would see the light of day; the reviewers had declassified only five hundred documents, a fraction of the total produced by the Manhattan Project, by the end of 1946.[8]

Philip Morse and Ernest Lawrence instead urged that the AEC declare certain areas unclassified, so that research in those areas would not have to await review before publication. Morse recommended that the lab directors aim low: "If we try to get all of the fields opened that everybody would like to have declared open now, we will have a long battle on our hands. . . . [It is] easier, safer and better strategically to get a few areas which are obviously unclassifiable declared so at present."[9] The AEC continued to declassify by the contents of each document, not by broad topics, which ensured a backlog of reports awaiting review. Most of those that the reviewers did read remained secret: 82 percent of the reports produced in the labs between November 1947 and November 1948 were deemed unpublishable. Los Alamos apparently did not attempt to report on much of its work. Only Brookhaven could publish most of its research. Even in biomedical research, seemingly distant from the technology of weapons and reactors, the vast majority of research reports were classified (table 1).

The security apparatus presented one of the main obstacles to both accommodating outside visitors and retaining permanent staff.[10] The labs also had to adapt to the AEC's policy of personnel clearances. All employees with access to classified data or facilities needed a "Q" clearance. This extended to the administrators of contractors; for instance, certain officers and Regents of the University of California.[11] Prospective employees had to fill out a questionnaire and undergo a full FBI investigation, which AEC security staff evaluated. The backlog of cases brought delays of two months in the clearance of new employees—quick by today's standards, but costly to recruitment in the tight postwar market for scientists.[12]

Cold War Climate

The rising tide of the Cold War swamped the public protests of scientists against the onerous conditions of classified research.[13] The Soviet Union's rejection of the Baruch plan for arms control before the United Nations and its intransigence in eastern Europe, reflected domestically in congressional criticism in the summer of 1947 of the AEC's security procedures, led the AEC to tighten security. Zinn could perceive the changes on a visit to Washington in November 1947; he warned Argonne's executive committee that clearances would only become more difficult to obtain.[14] In 1948, despite the AEC commissioners' public rhetoric in favor of intellectual freedom, the commission withdrew the clearances of twelve Brookhaven employees and suspended two Oak Ridge researchers and investigated six others. The charges against the last read, in one case, "You are reported to be intolerant of security regulations."[15]

In this climate lab administrators chose to tolerate restrictions. The

Table 1. Classification of research reports produced in the labs between November 1947 and November 1948

Research/labs	Classified	Unclassified	Total
Physical sciences			
ANL	100	8	108
BNL	4	37	41
UCRL	147	12	159
ORNL	200	43	243
LASL	43	3	46
Total	494	103	597
Health and biology			
ANL	21	3	24
BNL	2	3	5
UCRL	17	2	19
ORNL	22	12	34
LASL	4	0	4
Total	66	20	86
Total all research			
ANL	121	11	132
BNL	6	40	46
UCRL	164	14	178
ORNL	222	55	277
LASL	47	3	50
Total	560	123	683

Source: AEC, *Fifth Semiannual Report to Congress* (Jan 1949), app. 6.

trustees of AUI "recognize[d] the extreme difficulty under which the Commission is laboring in the matter of security clearances and that the Laboratory . . . is not the place in which to make an issue of personal liberties out of security clearance procedures." These procedures, which included hearings before a review board for individuals with questionable backgrounds (although the hearings did not include the right to confront accusers), were "a genuine and, on the whole, successful attempt to protect the rights of the individual."[16] Lawrence abandoned a proposal to declassify parts of his laboratory. At Berkeley, Donald Cooksey proclaimed, "we are convinced of the desirability from the viewpoints of the Laboratory and the Commission of securing top clearance for all personnel who are likely to have continuing access to the work of the laboratory, whether or not it may involve restricted data."[17]

In 1949 the GAC suggested to Kenneth Pitzer, the director of the AEC's Research Division, that the AEC consider declassifying Brookhaven and Berkeley, the labs with the least programmatic load. Lawrence thought it "clearly undesirable in the national interest [and, he did not add, his laboratory's interest] to contemplate freeing the Laboratory completely of restricted work."[18] Brookhaven's program awaited the startup of its research reactor, which by AEC policy would remain classified. Pitzer determined, and the GAC agreed, that "the Commission should be able to use both Brookhaven and Berkeley for classified work." Although most of its research was declassifiable, Brookhaven deferred to the desires of the AEC and agreed to seek Q clearances for all of its staff; those denied clearance could stay only if their contribution was "of prime importance in wholly unclassified areas."[19]

Restricted access to the facilities at all of the labs thus contributed to their failure to fulfill their purpose as regional labs for visiting researchers. For example, the reactor at Brookhaven, the lab's original raison d'être, was available only to those visitors with a clearance, whose results would have to be declassified before publication. There appeared to be no relief in sight. The Soviet Union exploded its first atomic bomb—called "First Lightning" by the Soviets, "Joe 1" by the Americans—in August 1949. Truman announced the decision to pursue the hydrogen bomb in January 1950. The next month Klaus Fuchs confessed to spying for Moscow, the FBI arrested the Rosenbergs soon after, and another atomic scientist, Bruno Pontecorvo, defected to the Soviet Union later that year. The outbreak of the Korean conflict that summer completed the context of a national emergency.

The pressure of domestic anticommunism reached the labs by various paths. The apparent loss of a small bottle of uranium oxide from Argonne in 1949 inspired a congressional investigation. The lab eventually produced a bottle, maybe even the missing one, but the affair resulted in even tighter controls on fissionable material, as well as on publications and visitors. Since the incident dragged the other labs into the investigation, it also isolated Argonne from the rest of the system: wary Los Alamos scientists started sending Argonne only older, redacted reports, and Zinn responded by cutting collaboration with Los Alamos.[20]

The same year the University of California, acting in part to preempt action by the state legislature, instituted a loyalty oath for all employees, which extended to those at the Rad Lab and Los Alamos. The action sparked a firestorm of faculty protest, an exodus of professors from the campus, and the formation of a faculty committee that would criticize the university's association with classified research. Although the exodus included some physicists who split their time between campus and the Rad Lab (theorists Robert Serber and Geoffrey Chew and experimentalist Wolfgang Panofsky), scientists at the Rad Lab and Los Alamos had already undergone investigation and clearance, and the oath does not seem to have inspired an outcry within the labs. Lawrence and several of his lieutenants in Berkeley were staunch anticommunists; Chew complained of an "unsympathetic atmosphere" at the lab for those who protested the oath.[21]

The national emergency of the early 1950s fostered McCarthyism and preoccupations with secrecy, culminating in the revocation of Oppenheimer's clearance in 1954. But it also contributed to an emphasis on results, with less regard for the niceties of process. Security did not come free. It was "a significant part of the workload" and consumed "a substantial amount of administrative expense" at the AEC.[22] Security expenses at the labs—for guards, secure facilities, and classification offices—had inspired the GAC's calls for the declassification of Brookhaven and Berkeley in 1949.[23] In 1953, at the height of McCarthyism, the House Appropriations Committee expressed "the belief that there is a very large waste of public funds" in the AEC security program.[24]

There were less tangible costs. After Joe 1, as Serber of the Rad Lab pointed out, the AEC could assume "that the facts in question must by now have been discovered" by the Soviets and that keeping them secret only handicapped American scientists. The AEC's director of classification called for less secrecy in order to win the race with the Soviet Union, which extended now to reactors.[25] The backlog of cases awaiting clearance persisted, with associated delays in recruitment and employment, and clampdowns on security hurt morale at the labs, as in the wake of the canceled clearances at Brookhaven and Oak Ridge in 1948.[26] Security standards were inconsistent. Los Alamos might classify certain work "Secret," while Oak Ridge considered "relatively innocuous parts of the same subjects" to be "Top Secret"; work left unclassified at Berkeley might be classified at Los Alamos.[27] The military established its own standards, which did not necessarily coincide with the AEC's and hence hampered the exchange of information on weapons: for example, only 10 percent of top staff officers at the headquarters of the Strategic Air Command had an AEC Q clearance.[28] The AEC also feared those effects of security exaggerated by the systemic structure of the labs; in particular, the duplication and inefficiency that resulted from ignorance of relevant research at other sites. It was not alone: a congressional investigation in 1953 perceived that "security restrictions play some part in fostering duplication and overlapping because it is extremely difficult for one group of scientists to find out exactly what other groups are doing."[29]

The criticism by Congress of overzealous AEC security practices marked a shift in attitude, soon apparent in the backlash against the Oppenheimer decision and the censure of Senator Joseph McCarthy. It received official expression in the legislative and executive branches. The revised Atomic Energy Act of 1954 eased industry access to classified data, especially for reactors. Late in 1953 Eisenhower presented his Atoms for Peace plan, which encouraged some reduction of security to foster international exchange.[30] The first Geneva conference on the peaceful uses of atomic energy in 1955, to which all of the labs sent representatives, fostered a new atmosphere of openness.[31] As criticism of federal security programs mounted in Congress, commentators sensed a fresh breeze; the head of Brookhaven's physics department pronounced himself "optimistic [that] a steady improvement appears now probable."[32]

The labs responded. Lawrence again proposed opening up parts of the Rad Lab for unclassified research. He noted that the establishment of Livermore as an annex of the Rad Lab had allowed the segregation there of much classified work, and also cited the effects of security on recruitment. His arguments and the new climate swayed the AEC, which approved the adoption of unrestricted areas at the Rad Lab.[33] In response to Atoms for Peace, the AEC again raised the possibility of declassifying all of Brookhaven's program so as to provide a place for uncleared foreign visitors to work. The Division of Research and the GAC did not wish to lose classified work in progress and also feared that such a step would isolate Brookhaven from the other labs. Nevertheless, the next year the AEC agreed to declassify Brookhaven's research reactor, a centerpiece of the program henceforth open to uncleared visitors.[34] Also by that time less than 5 percent of the AEC's biomedical program was classified, a far cry from the figures for 1948.[35]

The liberation of the labs in the mid-1950s, however, still left many lab programs restricted. The AEC and its scientists did not idly wait for public and political winds to shift. The national emergency of the early 1950s had encouraged them to find ways to reduce the costs of secrecy while satisfying the needs of national security. Lab scientists had already reached an accommodation with the security apparatus that stayed hidden from view, even after the labs let light in on some programs after the emergency ebbed.

The Security of Sherwood

An example will illustrate the evolution of attitudes toward, and effects of, security. The AEC supported research on controlled thermonuclear fusion for electrical power production from 1951; in 1953, with the strong personal interest of AEC chair Lewis Strauss, it expanded its effort and dubbed it Project Sherwood. The program was highly classified from the outset. Although the director of the Division of Research had in 1952 encouraged its declassification, in order to encourage recruitment, the AEC kept it secret.[36] The GAC considered its classification the next year but reached no consensus. Some, like Rabi, favored classification, since any practical machine that resulted could also produce neutrons and tritium, useful for fusion weapons. Others responded that it should be declassified until such a machine emerged: "To classify it at present would be like classifying space ships." Wigner noted that any program classified after first being left declassified would attract attention with the cessation of publication.[37] We see again that scientists, at least those in positions of political influence, did not abhor secrecy.

Associated with the issue of classification was that of compartmentalization. The AEC held classified conferences in June 1952 and April 1953 to foster exchange of ideas, but in July 1953, weeks after the appointment of Strauss as AEC chair, it began to insist on "strictest compartmentalization," and the classified conferences stopped.[38] Hence Robert R. Wilson developed some "important ideas" on fusion in isolation, and only after learning of them did the AEC bring him into the fold.[39] Consider also Nicholas Christofilos, the Greek engineer who had independently developed the principle of strong focusing for accelerators and had also come up with a scheme for fusion and filed a patent for it in Greece. He had meanwhile come to the United States, filed a secret patent application, and begun working at Brookhaven. Since he lacked a Q clearance, however, he could not pursue his research, and as he was an émigré, his application for clearance would take awhile.[40]

In the freer climate after 1954, the prospect of declassifying Sherwood resurfaced. The classified Sherwood conferences resumed in October 1954 and thereafter convened three times a year. In 1955 the GAC and a steering panel set up by the AEC on Sherwood debated the merits of declassification; again, some scientists sought to retain secrecy, in particular for any device that resulted, while others, including Edward Teller, argued for complete declassification. The GAC perceived that compartmentalization was impeding progress. The AEC noted that some of the labs were having trouble recruiting scientists for the program; even its existence was classified, and they apparently were reluctant to sign on to a program they had heard nothing about.[41] Strauss continued to insist on secrecy, but the next year his fellow commissioners outvoted him to open up part of the program. The impetus for complete declassification came from the second Geneva conference on peaceful uses of atomic energy, held in 1958. Thenceforth Sherwood could proceed with full interchange among the labs, universities, industry, and scientists abroad.[42]

A Classified Community

The organization of the national laboratories in a decentralized, dispersed system amplified the effects of classification. Compartmentalization for security reinforced geographical and organizational separation. One might expect two contradictory effects on scientific research. On the one hand, certain research programs that otherwise would not have survived peer review might persist: for example, the experiments that subjected human patients to injection with fissionable materials, in which some Berkeley medical physicists took part. On the other hand, the lack of communication might make survival of programs more precarious by stifling the interchange of new ideas. The isolation of Wilson and Christofilos in Sherwood exemplifies this possibility.

The staff of the AEC Division of Research recognized both possibilities in 1953 as they considered the pros and cons of a centralized laboratory for Sherwood. One argument for a centralized lab: "classified information can be held more securely and with less obstruction to its dissemination to those needing it." Decentralization, however, "provides the best assurance that the various schemes will be reviewed, criticized, and revised by scientists with different abilities and points of view." We will return to the other arguments offered in this debate; the significance of the ones considered here is their relative insignificance. The AEC staff report listed them in order of importance, and the implications of security ranked last.[43]

AEC staff attached little importance to the consideration of security because they had already overcome many of its problems through the creation of a "classified community." The term is apt but ill defined. Examples of the conferences, publications, and interlocking advisory committees employed at both the AEC and lab levels illustrate how the formation of a classified community overcame the effects of security restrictions and thus ensured the existence of competition and cooperation within the national laboratory system.

The wartime policy of compartmentalization put a momentary end to the conferences that scientists had used to coordinate fission research. A conference in Berkeley on the theory of fission and fusion weapons in the summer of 1942 was the last to gather scientists from a number of sites during the war. In the transition period of 1946, Groves agreed to the resumption of conferences. In June scientists and lab directors from the various sites gathered in Chicago to swap information about their research and outline new programs. That same month Groves acceded to a request from Bradbury for a conference at Los Alamos in August. Fifty-seven scientists, all with clearance, from Argonne, Oak Ridge, and Berkeley, along with a few consultants, joined Los Alamos staff to discuss classified aspects of nuclear physics.[44]

The success of the Los Alamos meeting inspired the lab directors to suggest, at another meeting in Oak Ridge that October, that the new AEC sponsor "information meetings" to promote interchange among the sites.[45] The AEC agreed and sponsored classified conferences, generally in the spring and fall of each year, whose locations rotated among the labs. The first conferences included sessions on physics, chemistry, biology, and reactors, and later came to include metallurgy, health physics, radioisotopes, and solid-state physics. There were occasional, additional conferences on particular topics: an information meeting on biomedicine at Oak Ridge in March 1948 drew close to three hundred scientists for the first day, whose program was classified; attendance dropped to two hundred for the second and third days with unclassified programs.[46] The AEC intended the information meetings to "parallel public professional society meetings"; they allowed for the discussion of classified results and gave scientists peer review and recognition of their work. As Bradbury put it, "These meetings are large enough . . . so that one feels very much a part of a large technical, scientific group, where one can make a contribution which can be talked about and whose impact on other laboratories considered."[47]

In addition to personal communication at conferences, scientists rely on publications to inform colleagues of their work and establish priority. Classification prevented open publication of research results in scientific journals, and compartmentalization had limited the circulation of classified reports during the war, especially those from Los Alamos and Berkeley.[48] As with conferences, the AEC tried to simulate the open literature. The AEC issued multiple copies of classified reports and sent them directly to its laboratories. For lab scientists who might not have received a relevant report, the AEC twice a month issued "Abstracts of Classified Research and Development Reports," which covered all classified work of the AEC except for certain highly secret programs. In 1949, 1,630 classified documents appeared in the "Abstracts." The AEC also published a form of classified textbook in the National Nuclear Energy Series, of whose 110 volumes about 60 were classified. Finally, for those reports that were declassified or, after the establishment of unclassified subject areas, born unclassified, the AEC published the *Nuclear Science Abstracts* of unclassified results.[49]

The combination of classified conferences and publications fostered a freer flow of information. Since almost all lab scientists had clearance and hence could plug into this network, there was little chance of isolation. The quick inclusion of Wilson and Christofilos in the Sherwood community exemplifies its efficiency. The AEC let the information meetings die out in the early 1950s. They had served their purpose, overcoming the compartmentalization of the Manhattan Project. In the meantime, the labs were developing another mechanism that promoted the flow of information through the system.

Historians have noted the elaboration of a new means to coordinate the technologically, financially, and administratively interdependent industries at the core of the postwar American economy: the interlocking directorate. Major corporations chose members of their boards of directors from outside the corporation, often from their counterparts in other sectors of the economy, and the same powerful people could sit on several such boards. Important information thus flowed through the top, strategic levels of American industry.[50] The national labs developed an analogous arrangement, though not consciously modeled on the industrial example and with several important differences: whereas representatives of, say, Ford would not sit on the board of General Motors, the labs encouraged representatives of the other sites to sit on their committees; the laboratory version included scientists from lower levels and permitted discussion of details of research programs in addition to strategy and policy; and the lab committees were only advisory.

The form of interlocking directorate developed by the national labs first emerged at Brookhaven. The Board of Trustees of AUI set up a Scientific Advisory Committee composed of scientists from each of the nine member universities plus two scientists from other institutions. The committee soon found that it could not cover the broad Brookhaven program by itself, and in late 1947 it split into three subcommittees for the physical, engineering, and life sciences, each with four members. The committees were intended to advise AUI on technical aspects of the lab's program, but also to inform university scientists about the program and induce them to visit. All of the members had security clearance.[51]

Two years later AUI expanded the arrangement and established a five-man "visiting committee" for each department, to visit once a year and report to AUI. (The visiting committees thus could not address interdepartmental relations—that is, interdisciplinary research—or the overall balance of the lab program.) AUI also explicitly encouraged the inclusion of representatives of other AEC labs.[52] Hence Weinberg from Oak Ridge served on the reactor science and engineering committee, joined later by David Hall from Los Alamos; Glenn Seaborg would visit with the chemistry committee; and his Berkeley colleague Emilio Segrè signed on for physics.

Union Carbide had meanwhile implemented a similar plan for Oak Ridge, inspired by the example of Brookhaven. Oak Ridge first formed a visiting committee for its Chemistry Division in 1954, which included Richard W. Dodson of Brookhaven and Seaborg among its five members.

The chemistry committee proved so successful that the lab created committees for its other divisions later that year, which again included representatives of the other labs. Oak Ridge extended informal invitations to scientists from the other labs to attend the committee meetings; the director of the AEC's Division of Research promoted the practice to the other labs as a means to integration.[53] The same year, 1957, Argonne followed suit and appointed review committees for each lab division, with similar membership: for example, the first committee for particle accelerators included G. K. Green from Brookhaven and Ed Lofgren from Berkeley.[54]

By 1957 Brookhaven, Oak Ridge, and Argonne had formed review committees with representatives from the other national labs. Only the University of California resisted the idea for Berkeley, Los Alamos, and Livermore, which might reflect its general lack of involvement in their programs. The visiting committees had little advisory influence: they usually acted more as partisan colleagues than critics, gushing about the work of the department and arguing for more support of it, their disciplinary allegiance stronger than their institutional ties.[55] They did, however, provide peer review and help the diffusion of information about the programs at each lab, including classified research. Brookhaven's committee members all held clearances until 1954, when relaxed restrictions allowed visitors to some departments to get by without one; nevertheless, even afterwards most members continued to hold clearances. In 1958 Oak Ridge administrators debated whether to require clearances for its visiting committees and decided "that in general it should be required, with exceptions granted when absolutely necessary."[56] The committees thus served as a classified conduit to colleagues at other labs.

The classified community extended beyond the boundaries of the national labs. At the level of the labs, the visiting committees included members from other AEC contractors, industrial corporations, and universities, and hence provided advocates for the labs in the wider scientific community. At the level of the AEC, the network of publication embraced other government agencies: in 1951 the AEC standardized procedures for cataloging, abstracting, and distributing research reports with the army, navy, air force, and NACA.[57] The interlocking committees of the labs had counterparts in Washington. The AEC's own General Advisory Committee and Advisory Committee on Biology and Medicine often included several scientists with ties to the labs: for instance, the first GAC included Rabi from AUI and Seaborg, with Oppenheimer, the former Los Alamos director, as chair. Many of these same prominent scientists found themselves on advisory committees for other government agencies. To cite two examples, Oppenheimer joined important advisory panels for the departments of Defense and State while he still chaired the GAC; John von Neumann, who served on the GAC from 1952 to 1954 before joining the AEC, chaired a committee advising the air force on missiles in 1953; Herbert York of Livermore and Darol Froman of Los Alamos joined von Neumann's committee the following year.

An example will illustrate how this extensive classified community could affect research programs. The possibility of the use of nuclear energy for rocket propulsion had first circulated soon after the war and then sat simmering on a back burner for several years. In 1954 some scientists at Livermore discussed the idea with air force staff; in the meantime, Los Alamos scientists, inspired by Froman's membership on von Neumann's committee, discussed the idea with colleagues at the Defense Department's Applied Physics Laboratory at Johns Hopkins. As a result of the growing interest, in October 1954 the air force convened a committee on the subject, chaired by Mark Mills of Livermore and including von Neumann. The Mills committee's recommendations led to programs on nuclear rocket propulsion at Livermore and Los Alamos which would engage a large fraction of the effort of both labs.[58]

The nuclear rocket program owed its origins to informal conversations and connections between air force staff and scientists at AEC and Department of Defense labs and to a formal recommendation from an air force committee chaired by a scientist from Livermore. The program thereafter had to develop the means to exchange information between the AEC labs working on nuclear reactors and defense contractors working on non–nuclear rocket components. At first even the existence of the program at Los Alamos was classified: as Raemer Schreiber, the head of the program, pointed out, "No blueprint is needed to see the problem of [Los Alamos staff] asking detailed questions about non-nuclear systems in the face of this ruling." The different classification policies of the two agencies complicated exchange. Nevertheless, Schreiber found he could obtain necessary information with "good prior coordination through official channels," and lab scientists in the program do not appear to have suffered from lack of communication.[59]

By the mid-1950s, as public and political attention to security reached its peak, the labs had developed ways to satisfy security requirements while maintaining communication among themselves and with other members of the national security apparatus. Public attention to McCarthyism and the Oppenheimer case masks the accommodation reached by scientists with the national security state. The simulation of the external scientific community within the confines of classification made possible their modus vivendi. Secrecy did not originate in the national lab system, but it reached new extents in the labs and had important consequences for them. Security barriers enhanced their geographical and organizational separation and increased the possibility of duplication. The development of a classified community circumvented these barriers and allowed the flow of information, and thus made possible the competition and cooperation that characterized the national laboratory system.

ORGANIZATION OF THE AEC

As demobilization at the end of World War II offered an opportunity to reconsider classification and compartmentalization, so did it bring a shift from a centralized military operation, largely sheltered from political oversight and geared to produce a military weapon in a crash program, to a decentralized, politically accountable, civilian organization in charge of research, development, and production in diverse aspects of atomic energy. When the AEC took over the vast enterprise of the Manhattan Engineer District in January 1947, it inherited not only several large research labs but also the administrative machinery that ran them. The AEC soon imposed its own organizational structure on the labs, imported new staff to oversee them, and reconsidered the policies of the MED, including the old problem of centralization of research.

The organic Atomic Energy Act of 1946 vested administration of the atomic energy establishment in a five-member Atomic Energy Commission, all of whom would be civilians appointed by the president and approved by the Senate. The president would also designate a chairman of the AEC; for the first one Truman chose David Lilienthal, the former head of the Tennessee Valley Authority (TVA). The appointment of the AEC marked a transition from a military organization with policy authority concentrated in the person of Groves to a civilian commission with power diffused among its five members.[60] Only one scientist, physicist Robert F. Bacher, served on the original commission; for scientific and technical advice the commission relied on its General Advisory Committee of nine scientists and engineers, many of whom were veterans of the Manhattan Project. Given the familiarity of the GAC with the atomic energy project and the lack of technical expertise in the commission, the GAC did not restrict itself to technical matters and played a large role in shaping the early policy of the AEC. By 1949 the committee perceived its influence decreasing as the AEC and its staff gained experience; until then its political presence further diffused responsibility and differed from the Manhattan Project's concentration of authority.[61] The organization departed from the preference of Truman's Bureau of the Budget for single administrators of executive agencies, which would be more democratic and efficient; instead it returned to Vannevar Bush's dispersal of responsibility for the wartime OSRD in collegial committees.[62]

The AEC also broke with the Manhattan Project in its philosophy of administration. The MED had dispersed its facilities but had administered them through a military hierarchy centralized under Nichols in Oak Ridge. The AEC adopted instead the principle of decentralization, spurred by chairman Lilienthal's belief in its importance in a democracy and his experience with the TVA. The AEC thus delegated authority to operations offices in New York, Santa Fe, Chicago, Oak Ridge, and Hanford.[63] Each operations office had responsibility for a broad geographical area and could further delegate authority to local area offices near large installations, such as those in Berkeley and Los Alamos (under Chicago) or Brookhaven (under New York). Administrative decentralization allowed the fledgling AEC to operate with a minimal staff in Washington, which could concentrate on policy and leave operational decisions to the regional offices, which in turn could rely on contractors to carry out their programs.

The AEC delegated line administrative functions to the field offices, which reported to Carroll Wilson, the general manager in Washington. To assist the general manager, the commission set up four staff divisions along broad functional categories: military application, production of fissionable material, engineering, and research.[64] The AEC thus followed the classic form of decentralized organization developed by American railroad corporations in the 1850s: decentralized units with authority and responsibility to direct work in the field and functional staff divisions to advise a general manager at a central office.[65] The AEC shifted the MED's Research Division from Oak Ridge to Washington and placed it under the direction of James B. Fisk, a solid-state physicist and former director of research at Bell Labs. The national labs in theory would lie in Fisk's bailiwick, though in practice this was not clear-cut. The Division of Production had directed field operations for Groves in the interim postwar period and continued to oversee the operations offices and hence the labs for the AEC.[66] The early Production Division in effect combined line and staff functions and, in addition to its nominal responsibilities, helped to administer lab programs, a precedent for the later organization of the national lab system.

In 1948 a GAC subcommittee drafted a scathing criticism of the AEC organization, especially decentralization. The report took Los Alamos as a model, on the assumption that weapons development was the one successful area for the AEC, and noted that the geographic area of Los Alamos and its field offices coincided with the functional responsibility of the Division of Military Application. The lesson of Los Alamos: "The original decision of the Commission to decentralize their operations on a geographic basis was a mistake." The advisers proposed to replace the current staff organization at AEC headquarters with a line organization based on functional instead of geographic areas. "The fear that was expressed by the Chairman of the Commission at our first meeting, namely, that a functional organization run from Washington would be over-centralized, is in our opinion unjustified." They recommended that each laboratory report to a single division in Washington.[67] It was not the last time scientists would support centralization against the AEC's policy of decentralization.

The AEC followed the advice when it reorganized in August and September 1948. Wilson delegated some of his operational authority to the program managers of the various divisions and assigned to each division oversight of particular installations, including the laboratories. But Wilson shied away from complete centralization—"the principle of decentralized operations must be maintained"—and kept "administrative support" functions with the regional operations offices.[68] The result: a confusing line of authority from the labs through local area and operations offices to various divisions in Washington, thence to the general manager and the commission itself.

The reorganization tangled further the lines of administration by adding new divisions for reactors and biomedicine. Both fields initially fell under the Division of Research. The early AEC reactor effort lagged; to jumpstart it Wilson created the Division of Reactor Development in August 1948.[69] Biomedical research wandered initially in the wilderness, with no representatives on the GAC or in Fisk's division to lead it. To give guidance the AEC appointed an Advisory Committee for Biology and Medicine, which recognized a need for staff support in Washington for the biomedical program. Wilson accepted the advice and formed a Division of Biology and Medicine in September 1948.[70]

The creation of divisions of Reactor Development and Biology and Medicine spread responsibility for programs at the labs among four divisions. The attempt to assign line authority for individual labs to particular divisions overlooked the pursuit of multiple programs at each of the labs that overflowed neat functional barriers. Wilson's reorganization entailed the administration of Argonne by the Division of Reactor Development, Los Alamos by the Division of Military Application, and the other labs by the Research Division. But Brookhaven, Oak Ridge, and Los Alamos had or were building reactors, Argonne supported basic physical and chemical research, and all of the labs, including Los Alamos, maintained biomedical programs. The AEC would struggle for many years to solve the resultant Laboratory Problem, called "the least palatable feature of the organization," which stemmed from the multiplicity of multiprogram labs.[71]

In an attempt to untangle the lines of authority, the AEC in 1950 assigned responsibility for coordination of the programs of each national lab to a director of one of the staff divisions. Each laboratory coordinator would serve as "a Washington 'spokesman' for each laboratory," arrange that lab's research programs every year to meet its long-range mission, and coordinate its activities with the programs at the other labs. The scheme met "practically unanimous endorsement throughout the organization," including the labs and operations offices. Some of the assignments of labs to program managers were easy: Argonne fell to the Division of Reactor Development and Berkeley to the Research Division. Others appeared arbitrary: Oak Ridge under Research and Brookhaven as a bone thrown to Biology and Medicine.[72]

The plan aimed to lay a single line from each lab to Washington, but there were needles to thread along the way, namely, the local area and operations offices. The shift from geographical to functional decentralization and the delegation of line authority to program divisions left a knot. Each lab initially reported to its regional operations office—Berkeley and Argonne to Chicago, Oak Ridge to Oak Ridge, Los Alamos to Santa Fe, Brookhaven to New York—which would then consult the appropriate division in Washington. The new plan assigned not only labs but also operations offices to particular divisions: Chicago to the Division of Reactor Development, New York and Oak Ridge to Production, Santa Fe to Military Application (figs. 1 and 2). Hence Brookhaven, for example, reported through New York to the Production Division (with Biology and Medicine as its coordinator), and Berkeley through Chicago to Reactor Development (with Research as its coordinator). The organizational lines would not stand still: Brookhaven shifted from Production to Biology and Medicine and thence to Reactor Development; Berkeley went from Reactor Development to Military Application after the formation of Livermore.

With such solutions, the Laboratory Problem persisted. As the GAC perceived, it derived from the AEC's policy of decentralization. If the AEC wished to delegate line authority to its field offices, it would have to staff each office with the broad range of technical managers necessary to implement that authority. Instead the AEC took the advice of the GAC and assigned line authority to staff divisions in Washington. Even if the AEC had abandoned decentralization, the functional delegation of authority overlooked the multiprogram status of the labs; the GAC based its recommendations on the model of Los Alamos, which it perceived as a single-

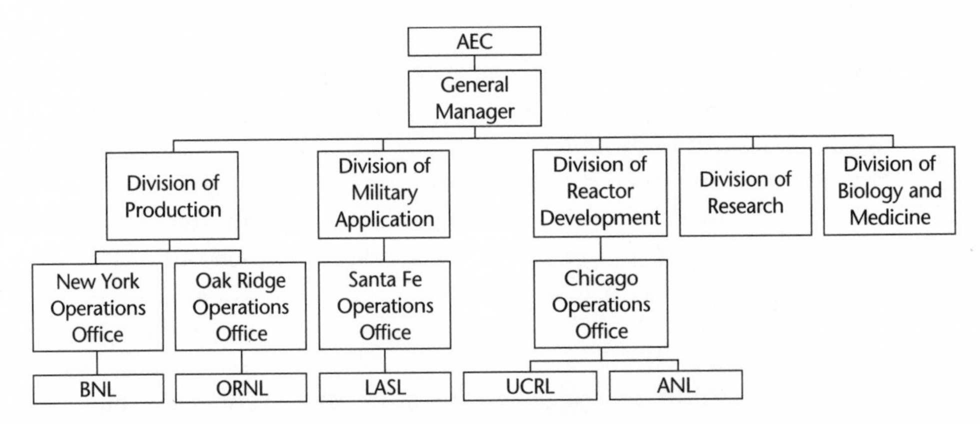

Figure 1. Line authority for the national labs circa 1950. Author chart from AEC organization chart, June 1950 (Sec'y 47-51, 23/sec. 4).

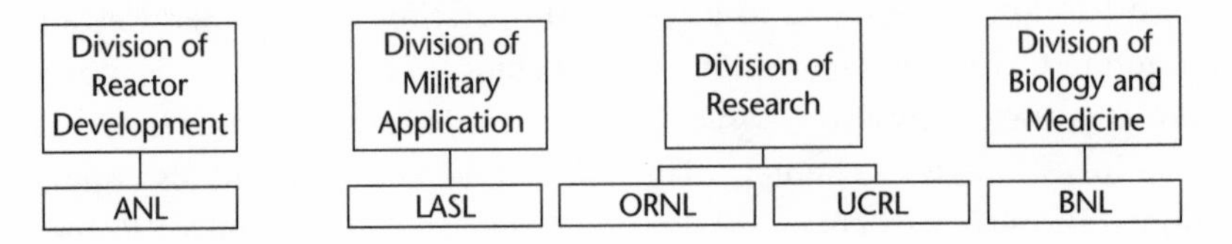

Figure 2. Laboratory coordinators circa 1950. Author chart.

purpose lab despite the lab's pursuit of several programs besides weapons. At the time, however, the AEC had just decided to centralize reactor development at Argonne, on the advice of the GAC. The proposed functional organization was perhaps planned to capitalize on geographical centralization of research programs.

The GAC's arguments for a central lab for reactor development and the retreat from administrative decentralization stemmed from its "despair of progress in the reactor program."[73] In the first years after the war the AEC faced, and sometimes encouraged, public expectations of peacetime benefits of atomic energy, especially nuclear power. Although the GAC tried to dampen overoptimistic estimates of the imminence of nuclear power, it too felt the pressure for progress in reactor development.[74] High expectations for nuclear reactors may have encouraged impatience; the GAC recommended broad changes in the AEC's organization after only a year on the job, instead of giving the initial organization a chance to build up staff and demonstrate its worth. The AEC's acceptance of this advice illustrates its reliance on scientific experts even in matters of administration. The advisers admitted that they were dispensing advice "on a subject not within their jurisdiction."[75] The problems that followed implementation of the plan suggest that scientists were not experts in every subject.

To correct the consequences of the reorganization of 1948, the GAC offered more organizational advice, which continued to push centralization. In 1951, only months after the assignment of the laboratory coordinators, the GAC suggested that the AEC appoint an individual in Washington who could bridge divisional lines in the identification of research problems. By the end of the year the suggestion had developed into a planned reorganization, including the appointment of an assistant general manager for laboratories to provide "a direct channel of communications for the national laboratories."[76]

The idea developed in embryo until 1 April 1954, when the AEC established the position of assistant general manager for research and industrial development and appointed Alfonso Tammaro, previously director of the Chicago operations office, to fill it.[77] The labs would report through their respective operations offices to Tammaro, who would coordinate the programs at the various labs.[78] In practice, however, line authority to the labs continued to run through the division directors, and the plan collapsed under the weight of the second part of Tammaro's title, which the initial proposals had not included. The AEC had begun to push the transfer of reactor development to private industry. Tammaro—a contract officer, not a scientist—became preoccupied with problems of industrial development and delegated coordination of research programs back to the division directors.[79] The Laboratory Problem continued to percolate, heated from below by the agitation of the laboratory directors and their scientists; in the future it would boil over.

THE PATH OF PROPOSALS

The pressure on the reactor program that spurred the reorganization of 1948 indicates that the national laboratories did not exist in isolation. They were tied via the AEC to the rest of the executive branch, Congress, and the American public, and these external linkages transmitted diverse pressures that shaped the evolution of the system. The main means the federal government could use to alter the behavior of the labs was fiscal. The labs did not enjoy limitless access to resources, and they competed or colluded to obtain them. In order to find out what the labs were doing and why they were doing it, we must follow the money.

Lab budgets traced the convoluted path from the labs to the AEC, the budgets of the various programs within each lab becoming separated along the way in the thicket of operations offices and program divisions. Once reunited in the AEC budget, they traveled through the Bureau of the Budget and into the president's budget, which then had to make its way through Congress. Along the way various budget examiners poked and prodded lab programs to find any fat that they might trim. This tortuous path, and the attitude toward fat, changed with time. By examining its evolution we may understand how the external environment—that is, the historical context—shaped the development of the national laboratory system and how the system influenced the context. We may also learn to what degree the framework limited the actions of lab scientists and, in turn, how they created some room to maneuver within those constraints.

The Lab Level

As with other aspects of the labs, the emergence of the AEC marked a break with previous practice in the presentation of budgets. In the interim period after the war and for the first year under the AEC, the labs had operated on a cash reimbursement basis: the contractors met the cost of work from their own funds and then applied to the government for reimbursement; where costs outran the capital of the contractor, the government provided lump-sum cash advances. The labs submitted their budgets in terms of general categories of expenses, such as salaries or equipment, along with a report describing in broad terms the proposed research program. The AEC's controller recognized the inefficiency and potential abuse of the system, which invited criticism of the new agency from Congress.[80]

His fears were well founded. With only general justifications required for lab programs, lab directors felt free to think big. No one thought bigger than Ernest Lawrence, who in the interim period multiplied men and machines at his lab with little restraint; one scientist would recall of the postwar Rad Lab, "We ran it with a big barrel of greenbacks."[81] Brookhaven settled for a budget of $2 million for its first year and debated "whether we should go all out on the budget figure which the staff thinks necessary for full scale expansion . . . [or] go ahead much more slowly." The consensus: full speed ahead. Morse noted "a few raised eyebrows at our talk of budgets of the order of twenty to thirty million a year"; but "unless the Commission thinks of *at least* this budget, it will not have a Laboratory worth having. . . . [T]he AEC must realize that it is not running a penny ante game."[82]

To rein in its high rollers, the AEC in July 1948 adopted a cost-budget arrangement: the contractor would submit a detailed budget broken down to standard categories along with monthly financial reports, and the AEC would advance the money for costs.[83] The AEC established several broad program classes corresponding to staff divisions in Washington, including weapons (the so-called 300 class), reactors (400), physical research (500), and biology and medicine (600). At the time each program class included only a few subcategories, but the range of numbers available for each gave the AEC ample room for further detail in the future.[84] Standardization requires compromises. Although the program classes divided along the lines of the program divisions of the AEC, they did not necessarily align with the organization or research program of the labs; indeed, the individuality of each lab guaranteed they would not. A single laboratory division working on, say, high-energy physics could find its work divided into two budget categories, particle accelerators and basic physical research; conversely, a single budgetary item could be spread across several lab divisions. The budget process did not easily accommodate interdisciplinary research: in the AEC's eyes, a research program "by definition may not cover work in more than one classification."[85]

Standardization can also entail paperwork. The dreaded government forms, in this case AEC Form 189 and Schedule 92, started appearing on the desks of lab scientists. The first asked for information on each program: the responsible scientist, number of people working on it, a brief description, the requested budget and costs for any previous years, and any related work going on in the same lab or elsewhere. Schedule 92 covered construction of new facilities, asking for justification, a timetable, and a budget. Each lab compiled copies of all of these forms into the so-called gray books, or total lab budgets, each year. Before they filled out the forms, lab administrators informally sounded out the local operations office and program managers in Washington, and then met with them more formally once the gray book was compiled. In this annual "program review," usually held at the lab over a few days, AEC staff reviewed the individual proposals, assessed their relation to similar work elsewhere and the rest of the AEC program, and recommended adjustments before sending the budget up through the operations office to Washington.

The involvement of AEC staff was not confined to the program review, nor did the various offices and program divisions display the same level of interest. The inconsistent activism of AEC staff brought to light another aspect of the Laboratory Problem, which stemmed from the AEC's policy of decentralization. With the reorganization of 1948, concomitant with the new accounting system, the AEC intended "scientific and technical direction" to fall to the program divisions in Washington, with the operations offices responsible only for "administrative support."[86] The distinction between administrative support and scientific direction was blurry at best. If administration "may to some ears sound devoid of technical and scientific implications," and if "none of the [local] Managers contacted has expressed a desire to 'direct research,'" neither did the operations offices wish to give budgetary carte blanche to the labs: "A lot of money is involved; they need to know what they're buying." It was a two-way street: while they encroached on research direction, some of the local managers complained that "the Program Divisions have invaded the 'funding business.'"[87]

The local AEC offices varied in their interpretation of their mandate. Los Alamos enjoyed a measure of independence from the Sante Fe operations office, which tried not to interfere with the work of its contractors.[88] The local manager, Carroll Tyler, was preoccupied by administration of the community of Los Alamos and weakened by congressional criticism of it.[89] The manager of the Brookhaven area office, E. L. Van Horn, who worked under the New York operations office, got on well with Brookhaven administrators and cut the lab some slack. "He is not a snooper. He expects us to manage." But if Van Horn was "less 'thinskinned' than some of the New York staff," who resented that the lab sometimes went over their heads to the divisions in Washington, he still asked the lab to go through the normal channels, that is, through his office and New York.[90] The thin-skinned staff in New York noticed no improvement; they complained to examiners of the Bureau of the Budget that "working relationships with Brookhaven have until recently been superficial" and proclaimed their intention "to move in on AUI and stiffen the supervision of their contract as much as circumstances would permit."[91]

At Berkeley, Lawrence and Underhill treated the Berkeley area manager, A. P. Pollman, as "an errand boy"; Pollman complained "bitterly that anything which he was directed to do on behalf of the Commission was subject to a complete turn-down by Underhill or Dr. Lawrence, and . . . that these gentlemen chose to deal directly with Washington, over [his] head."[92] The Oak Ridge operations office exercised more control over the lab at Oak Ridge, perhaps owing to its legacy as the administrative center of the Manhattan Project. Its staff by 1949 nevertheless was planning "to press pretty hard to do a more thorough job of review and evaluation."[93]

The reorganization of 1948 assigned line responsibilities to program divisions in Washington and put them in charge of operations offices. The activism of each operations office thenceforth reflected the attitude of the division to which it answered, and thus the latitude allowed the labs. The Division of Production exercised firm control over its offices in Oak Ridge and New York and staffed them with technical experts, who transmitted the pressure from the division to the respective labs. The Division of Military Application, by contrast, gave Santa Fe freer rein and did not provide it technical staff, so Los Alamos had more independence.[94] Los Alamos benefited doubly from the tolerance of the Division of Military Application, since it had both line authority for the lab and staff authority over most of the lab's program at the time.

Lab Budgets at the AEC

The staff authority of the program divisions introduced the other aspect of the Laboratory Problem. Once a laboratory's budget had progressed through the operations office and program division with line responsibility, its various sections (300, 400, 500, 600, and so on) were dispersed to the relevant division (Military Application, Reactor Development, Research, and Biology and Medicine, respectively). The budget of each lab was not considered in Washington as an integral unit, but was folded into the budgets of the several divisions. The multiplicity of multiple-program labs resulted in the "peculiarly difficult" problem of coordinating all of the programs.[95] The programs required coordination in two ways: first, of the various programs within a single lab, which could be easily separable, like the physical research and biomedical programs at Berkeley, or hopelessly entangled, like the reactor, physical research, and biomedical work at Oak Ridge; and second, of a single program among the several labs, since each of the labs pursued work for each of the four divisions.[96]

The appointment of laboratory coordinators in 1950 was intended to overcome the disintegration of lab budgets in Washington and to coordinate programs across the system. In the long run, the solution proved ineffectual, since the coordinators lacked the time to defend their labs, owing to their primary responsibility as division directors, and more important, they lacked the budget authority of their primary role. But while the dispersal of lab budgets to the various divisions cost the labs a sense of integration, they gained protection from budget scrutiny from higher levels as distinct institutions. That diminished the possibility of institutional biases and pork-barrel politics, at least in the consideration of operating budgets. The process also left individual programs to the authority of divisional staff, who had the functional expertise to evaluate them and with whom lab scientists often developed an informal rapport, and who then served as advocates.[97]

The divisional budgets next advanced through the general manager to the AEC, which had to balance the budgets of the various divisions within the overall AEC budget, according to its general policy goals. The commission did not examine individual projects, trusting the program managers and their staffs to adjust these to divisional budgets. It did, however, consider the construction budget item by item; hence new accelerators, research reactors, chemistry hot labs, or medical research clinics had to run the gauntlet of the AEC. In its consideration of new facilities, and for the optimum level of effort of each of the labs in general, the AEC could appeal to one of its advisory committees. The GAC and the Advisory Committee on Biology and Medicine did not approve detailed annual budgets, but they did occasionally review general lab programs and specific facilities and hence could interpose another checkpoint on the path from proposal to program.

The AEC questioned each item in the construction budget and the level of operations for each division because it knew that others would apply the same scrutiny to them. The AEC worked under the watchful eyes of budget examiners in other stems of the executive branch, which in turn felt the pressure from Congress. The attitudes of the program divisions were shaped by the interest taken by these outside parties, as well as by the AEC itself, in the work of the division. The divisions of Research and Biology and Medicine sought less detail in their budgets and less control over their contractors; as the AEC put it to Congress, "Because of the greater difficulty in describing goals in basic research, and because responsible scientists plan their own basic research projects, such programs are not budgeted or controlled in the detail found in applied research planning."[98] The justification did not attempt to define the terms "basic" and "applied," nor did it explain the latitude granted the Division of Military Application. A look up the ladder will help resolve these questions.

The Executive Branch

The proposals of the national labs, once incorporated into the AEC's budget, were not yet out of the woods of the executive branch. In the original debates over the Atomic Energy Act in 1946, Harold Smith, the director of the Bureau of the Budget, had insisted that the new AEC fall under the executive control exercised by his bureau through its disbursement of funds.[99] The budget examiners of the Bureau of the Budget served as guardians at the executive gate and controlled the passage of programs. Responsibility for the AEC's budget, which reached into the billions of dollars by the early 1950s, rested with the three examiners in the Atomic Energy Unit (later a section) of the bureau's Military Division; the assignment to the Military Division reflected the early emphasis of the AEC. The head of the unit, Fred Schuldt, had trained as a scientist, but the budget examiners were not supposed to evaluate the AEC's program from a technical standpoint. On the contrary, Smith argued that the bureau's policy control depended on legislative and administrative experience. "I would say the technical people, in the main, are the worst people to deal with large policy issues."[100] His examiners would agree: "The highly technical nature of the AEC's program causes the Commission to believe that it alone has competence to reach conclusions on policy matters. . . . [But] most policy problems can be posed in non-technical terms."[101]

These examiners enforced executive policy through a lengthy give-and-take with the AEC. The formal procedure usually entailed some form of discussion by the summer before the start of the fiscal year, which ran from July to June of the year in question (so the budget process for, say, fiscal year 1955 would start by the summer of 1953). The Budget Bureau furnished budget estimates for each program division; the AEC would attempt to adjust its budget accordingly, though not exactly: the submitted budget would often fail to meet the targets. The bureau's examiners then went through each activity of the operations budget and each item of the construction budget and returned the budget to the AEC with recommended cuts and their reasons for them. Each budget classification, down to the subcategory within programs (such as basic physics research within the Division of Research budget) or particular construction items (such as new accelerators or research reactors), had to pass through the review of the bureau's staff. The examiners could cut a project with sweeping judgments: "not considered essential," "adequate facilities are now in existence," or "not sufficiently justified." The AEC and its staff would then decide which items to appeal, if any. For appealed items again rejected by the Budget Bureau, the AEC could make a final appeal to the president, which it usually reserved only for the largest items.[102]

The annual negotiations over the budget forced painful decisions on AEC staff and helped to shape the programs of the labs.[103] Budget examiners maintained that some sort of target was necessary to force program managers to prioritize and eliminate items and to save bureau staff work. Not that budget examiners were lazy; one could note that "the mass of detail is so great that it is impractical for a central review group to make major adjustments in the level of the program on an item-by-item basis."[104] The AEC operating budget for fiscal year 1955 was four inches thick. But budgeteers had only themselves to blame. When Nichols, the AEC general manager, complained of the amount of detail required, Schuldt of the Budget Bureau allowed that the AEC might cut the "narrative chit chat" but insisted on tabular budgets at least to current levels of detail. Nichols, Schuldt complained, wanted "to show practically nothing in the way of specifics for research programs."[105] Schuldt similarly perceived a proposal to simplify the AEC budget as "a serious backward step."[106] Lab scientists came to anticipate budget reviews with an air of resignation: "As usual, it is expected that questions will be forthcoming for additional justifications of the project programs and/or construction projects."[107]

The Budget Bureau needed special zeal to overcome problems posed by the AEC, including its status as an independent agency. Further, the AEC as an operating, as opposed to regulatory, agency tended to multiply programs instead of controlling them (although the regulatory role began to increase after 1954).[108] The independence of the AEC increased in the early 1950s with the ascendance of Strauss to the chair of the commission at the same time that he served as Eisenhower's special adviser on atomic energy, which gave him a direct line to the president around the Budget Bureau. Finally, bureau staff had to contend with the "*magic* that still surrounds the atomic bomb and the new magic that clothes 'Atoms-for-Peace.'"[109]

At the same time Budget Bureau staff complained of the independence of the AEC, they also cited its ties to another federal agency, the Department of Defense. The military took a strong interest in the work of the national labs, in particular the development of atomic weapons and reactors for propulsion. The Defense Department, created soon after the AEC in 1947, transmitted its interest to the commission through the Military Liaison Committee, a byproduct of the congressional debate over civilian control of atomic energy.[110] Representatives of the three military services composed the Military Liaison Committee, which conveyed their desires, or "requirements," for military applications to the AEC. The Department of Defense, however, did not have to pay for its wants. Examiners at the Bureau of the Budget noted that the military thus had little incentive to exercise restraint; nor did the AEC, which "accepted uncritically" the requests by the military as long as the Department of Defense supported the commission in budget negotiations. The mutual reinforcement foiled the fiscal controls of the Budget Bureau and tended to expand programs.[111]

If lab scientists protested the cuts and detail imposed by Budget Bureau staff, they all could agree in their perception of the Laboratory Problem. Bureau staff noted that AEC research budgets "seem to be ad hoc and based largely on the needs of a current or proposed facility, without any clear reference to any general plan for the laboratory's long-term mission." The labs lacked clear missions, the AEC distributed their budgets to several divisions, and the laboratory coordinators had proved ineffectual; so bureau staff approved the AEC's plan to appoint an assistant general manager for research and development to administer the national labs.[112] But lab scientists and budgeteers agreed for different reasons: the former sought to cut through the tangled lines of authority and answer to a single individual in Washington, to acquire more freedom of action; the latter, to achieve greater definition of lab missions and coordination of programs by the AEC. Both would be disappointed by the solution of a new assistant general manager, who, when finally appointed in 1954, could devote little time to the national labs. Both looked to Congress for support.

The Legislative Branch

Atomic energy in the legislative branch occupied a similar position to that in the executive branch. It suffered the scrutiny of its budgets by the appropriations committees of both houses, but enjoyed some independence thanks to the support of the Joint Committee on Atomic Energy. The path from proposal to program ran through these committees.

The AEC budget passed in turn through appropriations committees of the House and Senate. Like the Bureau of the Budget, the committees consistently pressured the AEC for fiscal control and more budget detail from the commission's first appearance before Congress in the spring of 1947: congressmen demanded "a clear idea of how this money will be spent"; "in this agency the sentiment seems to prevail that the sky is the limit." A three-hundred-page budget provided "little or no information as far as break-down and details are concerned."[113] The appropriations committees followed their usual pattern: the House cut the budget with abandon, and the Senate served as an appeal board to restore the more drastic cuts by the House. But even senators, who felt that their "committee has been rather friendly to the Atomic Energy Commission," complained that it "always wants to talk in terms of glittering generalities."[114]

The technical nature of the AEC program limited the action of congressional appropriators. Congressmen admitted that "when it comes to talking in scientific terms, I do not want any answers of that kind."[115] "None of us are scientists"; "anyone who is not a nuclear expert has to take the word of the Atomic Energy Commission and their staff as to their requests for money."[116] Like the Budget Bureau, congressional appropriators complained as well about military requirements and their uncritical acceptance by the AEC.[117] Congress had also to contend with classification. Laboratory and AEC budgets were segregated into classified and unclassified portions, with only the latter obtaining a public hearing and strict scrutiny. The classification of weapons budgets and consequent protection from political scrutiny enhanced the relative autonomy of the weapons labs. The budget for the weapons program was not broken down at all; in the official record for the budget for 1952 it consisted of one item, for $161 million, with a one-page justification. The committee did receive a classified briefing on the weapons program, but Representative Albert Gore pointed out, "The committee will have to know more about the justifications for the $161,000,000. . . . That part of the American people who read this hearing would like to see some additional justification rather than this one page."[118]

The appropriations committees were stymied in their attempts at fiscal control by some of their own colleagues. The Atomic Energy Act of 1946 set up a congressional Joint Committee on Atomic Energy, consisting of eighteen members, half from each house, to have jurisdiction over atomic energy legislation in Congress. The Joint Committee did not initially have authority over the AEC budget and served instead to protect it from the budget knives wielded in the executive branch by the Bureau of the Budget and in the legislative branch by the appropriations committees. "More often than not . . . [members of the Joint Committee] find ourselves saying to the Executive Branch *not:* 'Do less; do it less rapidly,' but 'Do more; do it more boldly.'" Representative Gore hoped for some help to keep the AEC in check but "never had one suggestion from the Joint Committee on Atomic Energy as to where we might save $1. . . . The only thing I have heard out of that committee is what a grand, great thing it is and that we ought to spend more money on it."[119] The Joint Committee encouraged the independence of the AEC, which failed to support the president's budget and instead criticized it as inadequate in congressional hearings. In other words, an executive agency turned to the legislative branch for support against the executive. The Bureau of the Budget noted the betrayal and bemoaned "the extent to which the [Joint] Committee is concerning itself with the detailed operations of AEC's research program."[120] The Joint Committee in turn expressed its dissatisfaction with the "dominant role assumed by the Bureau of the Budget in the atomic energy construction program."[121]

Lab scientists did not fail to avail themselves of such an ally. Ernest Lawrence in particular cultivated committee staff and impressed them with his enthusiasm, which matched their own. By contrast, the same staff could complain after a briefing by Bradbury that he was not "sensational in his presentation."[122] In 1954 the Joint Committee increased its influence by including in the revised Atomic Energy Act the requirement that it authorize all construction appropriations, thereby adding itself as another formal step in the budget process. The Joint Committee thence could restore projects eliminated in previous steps. For instance, for the 1959 budget, AEC staff learned that the committee might increase the research budget and recommended the modest sum of $100 million as an appropriate target; the Joint Committee obligingly restored fourteen physical research projects and several reactors eliminated by the Budget Bureau and the AEC itself.[123]

Back Down the Ladder

The approval of the president's budget by Congress was not the end of the budget process for the national labs. The money still had to trace the path of proposals in reverse to reach the scientists in the labs. Each step in the path offered some possibility for diversion. The Bureau of the Budget could, as it did in 1958, withhold the funds authorized by Congress for particular projects as a way to reclaim budget control wrested from it by the Joint Committee.[124] Most years, however, the Budget Bureau released the funds to the AEC, which then had to figure out how to apportion budget cuts. The AEC could cover cuts with savings from the previous year or, more often, by shifting money from flush programs, as it did for fiscal year 1953 to cover a large cut in physical research.[125] But it usually just let the divisions deal with cuts themselves. Each division stayed true to character in divvying up its budget: the Division of Reactor Development specified in detail which activities would be cut, while the Research Division delegated the decisions to the lab directors.

Once the labs received their budgets, the directors had some discretion in shifting funds to cover particular projects, depending on the program and the attitude of the local operations office. Lab directors could generally shift funds within a program, such as the Division of Research, up to 10 or 15 percent of the budget. Larger shifts or shifts between divisions required the approval of AEC staff.[126] Ten percent could amount to significant discretion in a physical research program of $5 million, the level at several of the labs in the early 1950s. The lab directors might also choose to hold back a small amount of the lab's budget as a discretionary fund, as Oak Ridge did with 1 percent of the operating budget in 1955, to provide flexibility and avoid overruns—again, six-figure discretion for a $30 million budget.[127] The lab directors had a final form of discretion in the distribution of indirect costs. For example, Berkeley foisted a disproportionate share of indirect costs onto its biomedical work, leaving more money for physical research; it found this technique especially useful as the lab ramped up its effort on the Materials Testing Accelerator. The Division of Biology and Medicine protested that "their program was, in fact, subsidizing the physical research program and the M.T.A. program" and sought a fairer allocation of indirect costs.[128]

The Evolution of the Process

Although the general course of the budgetary process stayed the same, the path of proposals became more difficult to negotiate. The expansive early budgets of Brookhaven buttressed an image of the early postwar days of the labs as a "golden age," when money grew on trees and lab scientists cavorted in barrels of greenbacks. One encounters laments for the late 1940s in many recollections of lab veterans.[129] The image is exaggerated. The labs labored under fiscal controls from the earliest days of the AEC. And even earlier: Clinton Lab administrators complained in 1946 of the attempts of the Manhattan Project's local representative to "interfere and assume responsibilities which are reserved only for Monsanto under the present contract."[130] The advent of the AEC and its integration into the federal budget process only increased interference by local representatives of the government. Brookhaven budgeteers in June 1947 reacted with annoyance to

> the continual presence in our midst of personnel whose attitude appears to be one of perpetual questioning and carries with it the inference a) that we are not to be trusted, b) that we will not follow sound business practice unless forced to do so by continual item by item supervision, c) that we are prone to be impractical and wasteful of Government funds, and d) that we do not exercise good judgment in our procurement program.
>
> This is particularly irksome when the questioning initiates from personnel far less qualified than are those they question.[131]

In Berkeley, if Lawrence treated Pollman, the AEC area manager, as an "errand boy" and tried to go over his head, it was because Pollman demonstrated his intent to keep the lab within its budget. By May 1947 Rad Lab administrators were already complaining that Pollman was comparing their monthly cost reports to the cost estimates in the annual budget; Kenneth Priestley insisted that, since the estimates were by definition provisional, the lab should not be held to them.[132] That same spring the lab directors were protesting the number of administrative reports required and the limits imposed on their discretionary budgets by the AEC.[133]

Some of the nostalgia may be due to the fact that many of today's elders were members of junior staff back then and hence sheltered from the paperwork and budget hassles afflicting lab administrators.[134] Many memories of the golden age, however, come from former lab directors and senior staff. Most myths contain some truth. If the "golden age" was not so free and easy as lab veterans remember, it could nevertheless seem so relative to what followed. The Manhattan Project had been sheltered from the American political process; the formation of the AEC exposed the lab system to political pressures. By 1949 Brookhaven trustees had identified a "growing tendency on the part of the Commission to tighten its operating controls." Administrators at Berkeley and Argonne shared similar complaints about requests for flash budget estimates on three days' notice and increasing budget detail.[135] AEC staff professed sympathy but pointed out that the details were necessary to defend the budget before the Budget Bureau and Congress. If the bureau suspected padded cost estimates, it would make cuts across the board, and without detailed justifications it would excise programs altogether.[136]

The AEC transmitted these fiscal pressures through its program divisions and operations offices. The role of the latter highlights the blurry boundary between technical program direction and administration. Lab scientists and administrators complained that the operations offices tampered with programs despite their lack of technical competence.[137] Their program managers in Washington agreed. After an AEC memo of 1953 proposed to formalize the philosophy of technical direction by local offices, the director of Biology and Medicine protested that few offices had the necessary technical competence, and to hire scientists for the wide range of disciplines seemed "impracticable and undesirable."[138] Budget hawks in Congress, a source of pressure for technical oversight, could also cast a cold eye on the multiplication of staff at the operations offices.[139]

The operations offices were not the only meddlers. Oak Ridge administrators complained "that much too much attention is paid in Washington to the comparison of costs and to detailed direction of work comprising small portions of projects." They cited the Division of Reactor Development as the most meddlesome. The division administered large construction projects of new and hazardous technology by industrial contractors, which required active and increasing oversight as the AEC pushed to establish a reactor industry in the 1950s. A member of the Reactor Division admitted to "a predilection for making technical suggestions, partly as a relief from the tedium of paper shuffling."[140] Paper shufflers multiplied in Washington to handle the job of detailed reviews; Congress tried to limit the growth while encouraging the reviews.[141] The AEC headquarters building expressed the trend architecturally. The first AEC headquarters had room for only 350 staff and no way to expand, which Carroll Wilson, the first general manager, hoped would limit growth. In 1957 the 1,600 employees of the AEC moved into a new headquarters in Germantown, Maryland, with plenty of room to grow.[142]

The AEC's budget classifications also reflected the trend. The initial budget scheme of 1948 (e.g., class 500 for research and 600 for biomedicine) had included only a few subcategories: biology and medicine had ones for production of isotopes (610), research and development (620), and plant and equipment (630). Less than two years later the AEC had attached another level of zeroes and subdivided the scheme to provide more detail: the biomedicine program, now designated 6000, included categories for cancer research (6200), medicine (6300), biology (6400), and biophysics (6500), some of which had several subcategories and even, for biophysics, "activities" below those that made full use of all the available digits in the scheme.[143] The Research Division distinguished between basic and applied physics (5210 and 5260, respectively), chemistry (5310 and 5360), and metallurgy (5410 and 5460), with additional categories for isotope production and special training.

The distinction between basic and applied research, with the latter referring to work relevant to weapons or reactor programs, troubled lab scientists and their overseers as well as their historians. Congressman Gore tried to pin down Pitzer, the director of the Research Division, on the difference between basic and applied research. Pitzer punted: "The difference is largely one of point of view."[144] Lab budgeteers found the categories "not very meaningful" and "frequently misinterpreted by individual scientists." Thomas Johnson, Pitzer's successor, agreed and proposed to revise the titles in 1953, in addition because "when part of the work is called applied there is a suggestion that the rest of it is irrelevant," although "some scientists find the adjective 'applied' distasteful."[145] The ensuing revisions, announced in January 1955, did not please lab scientists, however, since they split physics research into five classifications, chemistry into four, and metallurgy into three. An Oak Ridge administrator protested "that the breakdown of the classifications is entirely too detailed and complex."[146] The complex classifications for physical research were child's play compared to those for reactor development, which, true to form, broke down the budget in such detail as to exhaust the number of digits available in the 4000 series and force it into the decimals.[147]

The pressure of all this apparatus had some effect. Brookhaven administrators complained in 1950 that "the last traces of wartime liberality in Government expenditures have disappeared." The lab now aimed at an annual operating budget of $10 million, and Pitzer in the Research Division was proposing to halve that—not a penny ante game, but a far cry from the contemplated $30 million budgets of the heady days of early 1947.[148] Lab operating budgets, after recovering to wartime amounts, leveled off in the late 1940s, although construction budgets continued to expand as permanent facilities replaced temporary ones from wartime and as Brookhaven built itself up from scratch.

Lab scientists did not passively accept budget dictates from above. Entrepreneurial lab directors, like Lawrence and Weinberg, learned to cultivate alliances in obvious places, such as the Department of Defense or Joint Committee on Atomic Energy, and less obvious ones, such as the Budget Bureau.[149] Some of these alliances proved strong enough to counteract the fiscal constraints and, after an initial equilibrium, lab budgets started rising again. Budget Bureau staff noted the "seemingly continuous program growth" in the national labs evident by the mid-1950s and sought to stop it.[150] Even the Joint Committee in 1955 entertained doubts about the growth of the labs in 1955.[151]

In response, the AEC stepped up its own efforts. The increasingly intricate and formal budget process began to replace some of the less formal mechanisms and accelerate the splintering of lab programs. The program reviews for each lab, where lab scientists sat down with AEC staff to consider the overall program, began to die out by the mid-1950s, since formal assumptions and flash estimates from each division guided preparation of the budget in advance.[152] New paperwork and classification schemes appeared in the late 1950s, requiring ever finer levels of detail in budgets and patience in lab directors. The hassles reached the weapons labs: Darol Froman quit as associate director of Los Alamos in 1960, fed up with "negotiations of one sort or another with the Washington bureaucracy—budgets, programs, data sheets, reports, forms, and visits." The same year

Haworth predicted that "if this trend continues, there will have to be a head-on collision with the AEC."[153]

The trend itself was bigger than the AEC, and reflected the growth of the administrative state in the United States. As the scope of government activity increased in the mid-twentieth century, proliferating agencies (such as the Bureau of the Budget, a creation of the Roosevelt administration) and congressional committees spawned large staffs to keep pace with the work (and with the opposing branch of government). High-level horse-trading and informal, particular decisions gave way to bureaucratic process and formal rules and procedures. Red tape and micromanagement by both the executive and legislative branches thus increased through the postwar period, and extended to the national lab system. These developments reflected increasing demands for accountability in the federal government, which derived ultimately from the American public.[154]

Implications for the System

The multiplicity of the national laboratories and the potential for wasteful duplication invited consistent scrutiny from budget examiners at all levels. Congressmen on the Appropriations Committee recognized requests in 1948 for similar high-energy accelerators and asked, "Why do you want a $3,000,000 set-up in Brookhaven and at the same time a $4,000,000 set-up in Berkeley?"[155] The House Appropriations Committee in 1953 thought it "possible to reduce the number of scientists required for atomic energy work if overlapping and duplication of projects were to be eliminated."[156] The Bureau of the Budget criticized the "multiplicity of approaches" on both power reactors and Sherwood.[157] Even the Joint Committee on Atomic Energy could question the duplication in the programs of the two weapons labs.[158]

The AEC's budget process highlighted the consequences of maintaining several multiprogram laboratories. Because lab programs were lumped together in division budgets, duplication was more apparent; research items for one lab appeared right alongside those for another, instead of being separated by institution. Concurrent debates over the activism of the operations offices struck to the heart of the policy of decentralization. The AEC sought to delegate administrative authority to its local offices, but never resolved whether this included technical direction of programs. If the operations offices did not have it, then no one in the AEC had line responsibility for each lab as a whole, since their budgets were dispersed in Washington and the various mechanisms proposed to provide some coordinated authority all failed. The dispersal of lab budgets encouraged the perception of the labs not as integrated institutions but as a collection of projects, or, as Haworth put it, "a kind of job shop."[159] The budgetary atomization of the atomic energy labs and the increasing interference of meddlesome managers put lab directors on the defensive. They found strength in collective action.

THE LAB DIRECTORS' CLUB

As with the labs themselves, the origins of what came to be called the Lab Directors' Club dated back to the Manhattan Project. The directors of the main research centers convened at a series of meetings starting in 1946 to plan for the postwar period. In May 1947 the AEC gave the group formal status as a "Research Council," to assist the fledgling Division of Research "in the evolution and integration of its research program." For the first meeting of the council, in August 1947, Lawrence enticed the commission itself to Bohemian Grove in the redwoods north of Berkeley. For four days the AEC joined the lab directors for frank, informal discussions of the future of atomic energy research and development.[160] Despite the success of the first meeting, Fisk and his successors left the Research Council on the shelf.

The lab directors' meetings revived in the mid-1950s and would provide an important forum for complaints over the Laboratory Problem and nitpicking AEC budgeteers. In 1953 Thomas Johnson, the former Brookhaven physicist now directing the Division of Research, began holding periodic program reviews at individual labs, to which he invited the directors of the multiprogram labs: Zinn from Argonne, Lawrence from Berkeley, Haworth from Brookhaven, Bradbury from Los Alamos, and Weinberg from Oak Ridge; Frank Spedding of Ames also merited membership, perhaps owing to his long association with the others dating back to the Manhattan Project, and the director of Livermore would join in the future. The meetings gave the directors a chance to play a little tennis, maybe talk over a beer with old friends, but also to coordinate their programs and discuss their common problems.[161]

In December 1955 the group met in Washington with sympathetic colleagues on the General Advisory Committee. The discussion raised many of the difficulties the directors faced: the dispersal of lab budgets among

AEC divisions; the lack of discretionary funds; budgeting by fiscal year on long-term projects; the distribution of indirect costs; and the interference by legal and fiscal types in lab programs. They griped mainly about "the multiplicity of Washington contacts" and the lack of "one responsible god-father" for the labs. No answers emerged, despite much discussion; the problems "were easily equal in difficulty to those which the Queen of Sheba prepared for Solomon."[162]

The next September the same crew of directors congregated in Carmel, just down the California coast from Berkeley, this time with AEC staff. The discussion turned again to the diffusion of responsibility for the labs at the AEC. They considered but rejected vesting official responsibility in a committee of the lab directors—that is, in a revival of the Research Council, which they all seem to have forgotten. Another option, a single laboratory coordinator (another forgotten former solution), would just interpose another bureaucratic layer between the labs and their ultimate sponsors in the program divisions. There was, they conceded, "no ready remedy."[163]

The Yosemite Revolution

The revival of the lab directors' meetings stemmed from the program reviews, which in the meantime had fizzled with the increasing budget formality. The directors now continued the club on their own initiative. Lawrence, ever the gracious host, organized a meeting at the Ahwahnee Lodge in Yosemite, scheduled for March 1957 but postponed until January 1958. Their thoughts on the Laboratory Problem germinated during the delay and reached full flower in what came to be known as the laboratory directors' "revolution at Yosemite."[164]

In preparation for the Yosemite meeting, the lab directors decided to put some of their complaints in writing. To this end Weinberg and Haworth drafted a "White Paper" elaborating on some of their earlier discussions, which they planned to go over in Yosemite and then send to the AEC.[165] The final draft before the meeting bore the strong imprint of Weinberg's initial draft, with hints of Haworth's "job shops." Weinberg had identified two general approaches to the national labs:

> The first, which may be called the "institutional concept[,]" holds that the Laboratories should be strongly integrated institutions with strong central managements and that the success of programs they undertake is expected to derive from the essential strength of the Laboratories them-

> selves. This concept is held by the managements of all the National Laboratories and by the higher levels within the Atomic Energy Commission.
>
> In contrast to the institutional concept is a view that the Laboratories are merely instruments, to be closely manipulated and controlled in all respects. This view incorporates the "project concept" according to which the Laboratories are collections of projects, each of which is controlled by the staff member in Washington who disburses money for that project. . . . The project concept is, at least implicitly, held by much of the lower level AEC staff, although it is not fashionable to state one's belief in it.[166]

In terms of the budget, the institutional view sought integrated lab budgets, with the distribution of funds within the lab left to the discretion of the director; the project view entailed separate financing for each activity, "without regard to mutual interactions" and with little discretion. In other words, for the former, budgets paid for the presence of people at the lab; in the latter, the budget paid for projects that put people to work. Administratively, the institutional view devolved responsibility to lab management; under the project view, "salaries, sub-contracts, purchasing, record keeping and a host of other administrative details must be in accordance with uniform instructions from, and detailed authorizations by, 'experts' on the Commission staff." The difference boiled down to the question, "How much authority does Laboratory management have and how much authority do the program and administrative divisions of the AEC have?" The lab directors, naturally, answered in favor of the former.[167]

The old Laboratory Problem and its manifestation in the budget process reared its ugly head. No single person or group in Washington had responsibility for a laboratory. "There is no such thing as a 'Laboratory' budget; indeed, it never reaches Washington. . . . What is missing is an integration along vertical organizational lines." The laboratory coordinators had failed to effect this integration, and the field offices likewise lacked ultimate fiscal authority and served only as an additional bureaucratic channel. The problem, however, seemed not to be the existence of "*too many* echelons" but rather that the higher echelons persisted in picking programs to pieces instead of considering broader matters of policy. Again, the divisions differed in their pettifoggery. The directors lauded the enlightened oversight of the weapons labs by the Division of Military Application, which generally adopted the institutional view; by contrast, the long-maligned Division of Reactor Development took the project view.[168]

The attitude in the Reactor Division annoyed the lab directors to the point that one of their two concrete suggestions focused on it. As a short-term measure, they urged, the work of the labs for the Reactor Division should be split off in a separate division. This would free them from the hassles that spilled into their work from the large construction projects undertaken by industrial contractors for the division. Reactor development at the labs could hence resemble weapons development, with funds appropriated as lump sums instead of for specific projects. The second, longer-term suggestion elaborated the first. The new division for the reactor research of the labs would combine with the divisions of Research and Biology and Medicine into a new "Research Department," which would have responsibility for lab programs and also for the AEC's smaller off-site contracts for physical and biomedical research. The new department in effect would restore the arrangement of early 1948, before the formation of the divisions of Reactor Development and Biology and Medicine. The division directors would resume staff positions under a new deputy general manager in charge of the new department. The Division of Military Applications, however, would not join the department, which left the weapons labs to enjoy their current freedom but in isolation from the rest of the system. The labs would report straight to the deputy general manager instead of to the various divisions, and the AEC budget would incorporate their budgets intact instead of dispersing them. In short, the lab directors proposed to "simultaneously increase the responsibility and authority of the National Laboratories and centralize commensurate authority within the AEC."[169]

After the Yosemite meeting the club sent the white paper to the AEC "in a constructive rather than complaining spirit."[170] The AEC had just launched an internal study of its organization; the resultant report borrowed the directors' main ideas but shied away from their bolder suggestions. While it agreed with the "institutional" view of the labs, it endorsed the policy of decentralization to the field offices and did not propose a major reorganization. The AEC, after two days spent discussing the benefits of decentralization, took no action.[171]

The proposal by the lab directors for simultaneous devolution and centralization suggests some ironies of their "revolution." Centralization, they hoped, might free them from the meddling of AEC staff. Their embrace of centralization ran against the grain of the AEC's first guiding policy and marked a shift in attitudes. Scientists and lab administrators had initially supported the delegation of authority to local representatives in order to free the labs from Washington bureaucracy.[172] Lab scientists learned early that decentralization freed them from neither accountability nor bureaucracy and could instead entail the direction of their work by managers in local offices and in Washington. Lab directors thereafter argued for *centralization* of responsibility for the labs, which the AEC ineffectively implemented in the laboratory coordinators and then the assistant general manager for research and industrial development; the failure of these devices led to the lab directors' calls for a new deputy general manager for research.

The lab directors thought that centralization would advance the institutional view of the labs. But however much they blamed the Division of Reactor Development, they might have looked in the mirror for the prevalence of the project view. The very success of the labs at pursuing particular projects—that is, sources of support—encouraged the perception of them as collections of projects, or budget activities. Weinberg, who coined the institutional and project concepts, was perhaps the worst offender, or best entrepreneur. Budget Bureau staff noted in 1952 that Oak Ridge's "performance has attracted a steady flow of 'new business'" in reactors and chemical process development.[173]

Although the lab directors recognized that reactors "attract great public and political interest, placing great demands upon division management," they otherwise underestimated the demands of budget examiners in the executive and legislative branches for accountability. They admitted that with added budget discretion, lab management "must be subjected to a more thorough accounting and scrutiny than is now the case," but they proposed to have the General Advisory Committee perform such reviews, supplemented by the reports of the labs' visiting committees. One can see why the lab directors would acquiesce to such an audit, and why political representatives would never allow it.

The Reorganization of 1961

The lack of action by the AEC led Weinberg to lament the "somewhat abortive Laboratory Directors' revolution." A year had gone by and he did not think "the questions raised in our Yosemite White Paper have been satisfactorily resolved."[174] The failure to achieve concrete results took some of the steam out of the Lab Directors' Club. The death of Ernest Lawrence in 1958 and the illness of Leland Haworth the next year deprived the club of two of its stalwarts. A congressional investigation into the missions of the

AEC labs in 1959 forced each director to defend his own lab and diverted the group from common problems of organization. When the distraction of the hearings had died down, the directors again considered from afar the issues raised at Yosemite. In the spring of 1961 their views found a sympathetic audience in two new AEC commissioners, especially since one of them had helped to frame them.

On 16 January 1961, President Kennedy appointed Glenn Seaborg, a chemist in the Berkeley Radiation Laboratory and a veteran of the wartime Met Lab, to the chair of the AEC. Before Seaborg assumed the chair in March, he planned to meet with the lab directors; Haworth suggested that the directors draft a statement of their concerns to give Seaborg at the meeting. Weinberg, as usual, took the lead on a first draft, which proposed familiar solutions: delegation of authority; more budgetary flexibility, including 10 percent of the lab's budget left to the discretion of the director; and separation of reactor research. For solving the "central organizational problem," however—the lack of centralized authority in Washington—Weinberg proposed a new step: the designation of the one of the commissioners as "trustee" for the national labs. The AEC had traditionally included one or two scientists on the commission; one of these "scientific Commissioners," under Weinberg's plan, would assume authority for the labs at the highest level of the AEC. The proposal had no precedent: although commissioners in the past had taken an active interest in particular programs, none had ever had official individual responsibility for any aspect of the AEC's work. Another part of Weinberg's proposal did have some precedent, although he did not mention it, in the long-dormant Research Council: the Lab Director's Club should have "semi-official status as an advisory agent of the Commission" and meet with the AEC and GAC (whose functions it seemed to duplicate) on problems concerning the national labs.[175]

Not all of the directors agreed with Weinberg's proposals. Bradbury recognized "the unpleasant fact" that discretionary funds were limited not by the AEC but by Congress. He also thought it more politic to approach Seaborg informally, without any position papers and separately from the other commissioners, whose long association with the AEC would sensitize them to criticism of it: "The only hope of change is through the (hopefully) good influence of Seaborg and whoever turns up as a fifth Commissioner."[176] The lab directors soon learned that they would have a more direct pipeline to the AEC: Haworth, the director of Brookhaven and coauthor of the Yosemite paper, turned up on the vacant seat on the commission in April 1961.

They had another ally in the General Advisory Committee, whose members did not worry that the directors sought to encroach on their advisory turf. The GAC also noted the interposition of nontechnical staff at the field offices between lab scientists and program managers, and followed the lab directors in recommending the appointment of a new assistant general manager for the labs.[177] The new commissioners did not waste any time. In August 1961 the AEC announced a major reorganization, which implemented many of the proposals from the Yosemite paper. The AEC consolidated authority for the national labs in a new assistant general manager for research and development, to whom the labs would henceforth report directly. The reorganization attempted to correct the organization introduced in 1948, which assigned line responsibility to the staff divisions in Washington; the divisions now reverted to staff functions and the operations offices reported straight to the overall general manager. As the lab directors had proposed, the new organization did not affect the weapons program, which the Division of Military Application would continue to direct in its esteemed fashion.[178]

The reorganization of 1961 thus aimed to solve the long-standing Laboratory Problem by forsaking the policy of decentralization to local offices and centralizing authority in Washington for the multiprogram labs. As the new assistant general manager, the AEC appointed Spofford English, deputy director of the Division of Research and a chemist who had worked at Oak Ridge during and after the war. With sympathetic central authority and an apparent solution to the Laboratory Problem, the lab directors could relax. Brookhaven administrators immediately recognized the reorganization as "decidedly favorable from the Laboratory's point of view," and Weinberg perceived that the troublesome relationship between the labs and the AEC "has largely disappeared with the new AEC policies."[179]

But the triumph of the Yosemite principles was a hollow victory. It removed one leg from the path of proposals but added another; it did not simplify the rat's nest of the AEC organization chart nor resolve the distribution of lab budgets when they reached Washington. English had no budget authority; the AEC continued to compose its budget from those submitted by the divisions, into which the lab budgets were folded. And no reorganization within the AEC could remove the ultimate source of budgetary pressure on the labs: the fiscal authority of the executive and legislative branches of the federal government.

II

THE ENVIRONMENT

4

COLD WAR WINTER, 1947–1954

The framework of the national laboratory system—the internal organization of each lab and its contractor, the tangled lines of authority through the AEC, the path of lab budgets through the executive and legislative branches of the federal government—transmitted forces from the external environment to the labs and molded their programs. Lab programs reflected the evolution of the external environment from the focus on national security at the onset of the Cold War and the national emergency of the early 1950s to the shift, post-Korea and post-Stalin, to military standoff and sociopolitical confrontation. By the end of the decade the changing context brought a congressional investigation into the missions of the labs, in parallel with the AEC's internal examination of the Laboratory Problem.

The evolution of the environment encompassed the allocation of resources, institutional interactions, and cultural influences. In the first period, from 1947 to 1954, resources provided by national security overcame centralizing impulses within the system. Secrecy isolated the labs from institutions outside the national security apparatus. And in the winter of the Cold War, the rhetoric of competition provided a significant impetus behind the centrifugal forces and ensured the maintenance of a competitive system among the labs.

DRIVERS

The first formulation of the national laboratory concept included two primary goals: the development of applications by permanent, security-cleared staff for national defense, and the provision of large facilities for basic research by visiting academic scientists. The dual identity of the labs did not imply a simple balance between basic and applied research. The weapons and reactor programs drove research programs seemingly far removed from immediate military applications, such as biomedicine and meteorology. National security also justified the training and retention of scientists as a reserve of manpower to be mobilized in case of crisis; hence the inclusion in the system of Berkeley, which initially lacked programs in either weapons or reactors. The weapons and reactor programs demonstrate the centrifugal effects of national security on the national lab system.

Weapons

In the first years after World War II, the AEC took as its primary mission the production of nuclear weapons. "All lines flow to Los Alamos," which the AEC and GAC decided to maintain as the central lab for weapons research.[1] The impetus behind the weapons program intensified with increasing international tensions—the failure of attempts at nuclear arms control with the Soviet Union, communist coups in Hungary and Czechoslovakia, the Berlin blockade, and the fall of China—that nourished the Red Scare at home. It also derived from the sorry state of the American stockpile, which, contrary to popular perception, contained zero assembled weapons in early 1947.[2] Los Alamos multiplied weapons and their yields by developing new designs that it tested in the Sandstone series of 1948 at Eniwetok atoll.

External pressures on the weapons program multiplied faster than new designs and culminated in late 1949 with the Soviet Union's first explosion of an atomic bomb, Joe 1. The Soviet shot galvanized a debate in the United States over the desirability of a crash program to pursue thermonuclear weapons, the so-called super or hydrogen bomb, on which Los Alamos had maintained a small effort since the war. Despite the opposition of most GAC and AEC members, primarily on moral and strategic grounds but also because it would divert scarce scientific talent and materials, President Truman on 31 January 1950 committed the United States to just such a crash program.[3] Los Alamos adopted a six-day workweek that would remain in effect at least through mid-1951.[4] Five months later North Korean troops poured across the thirty-eighth parallel on the Korean peninsula to complete the context of crisis.

The response in the labs was not confined to Los Alamos. Indeed, the designation of Los Alamos as the weapons lab obscures the contributions of the other labs to the weapons program. Argonne had maintained a theoretical group working on weapons problems and helped analyze data from weapons tests, and would design the new production reactors at Savannah River, for which Oak Ridge provided the chemical process to separate plutonium and uranium for weapons.[5] Even before the crisis Weinberg had indicated to the AEC the willingness of Oak Ridge to take on weapons work, so long as it did not cut into existing projects.[6] The dire context of Joe 1 and the Korean conflict mobilized the other labs. Brookhaven's trustees perceived Joe 1 as "a national emergency which will place grave new responsibilities upon the Laboratory" and assured the AEC "that Brookhaven is prepared to undertake programmatic work of special interest to the Commission."[7]

Los Alamos meanwhile labored under its increased workload and struggled to find the scientists, especially theorists, necessary to accelerate the H-bomb program. Los Alamos took Brookhaven at its word and asked it to undertake the investigation of neutron-scattering cross-sections at fission energies, to aid in the design of tampers for lightweight fission weapons.[8] Brookhaven proposed an eighteen-inch cyclotron to provide the neutrons; the AEC's Research Division recommended immediate approval "to capitalize on Brookhaven's enthusiasm for an applied project."[9] The AEC not only approved the program but also soon expanded it. The data on cross-sections were useful too for the reactor program, also "in a state of war," and Haworth anticipated that the cross-section work "will improve our position with respect to being able to provide a greater fund of information for the Reactor and Weapons Development Programs."[10]

Brookhaven found other ways to contribute during the national emergency. It lent health physicists to weapons tests and discussed helping Sandia Laboratory with weapons evaluation work. Several Brookhaven scientists took part in Project Vista, on the tactical uses of nuclear weapons, in the summer of 1951. At the request of Argonne and Oak Ridge, chemists at Brookhaven began a secret program on separation processes for the production of heavy water. The lab's budget for what it classified as applied research doubled between 1950 and 1951 to 15 percent of the lab's research effort (though only 5 percent of the total lab budget). Hence Haworth, in April 1951, could judge that Brookhaven was "playing an accelerating role in problems directly concerned with national defense."[11]

Brookhaven was not the only lab to volunteer its scientists during the emergency. As he had done in World War II, Ernest Lawrence mobilized his laboratory after the detonation of Joe 1, to the encouragement of I. I. Rabi: "You fellows have been playing with your cyclotron and nuclei for four years and it is certainly time you got back to work!" Luis Alvarez of Berkeley gave up "any pretense of being interested in fundamental physics" and resolved "to concentrate on national defense."[12] Lawrence, Alvarez, and others on the Berkeley staff lined up in support of the H-bomb program and thought of ways to advance it. Lawrence first proposed the construction of a heavy-water reactor on San Francisco Bay and designated Alvarez to direct it. But, Alvarez admitted, the Berkeley lab "had no technical qualification in that field. We had never been in the reactor business." To make up their deficiency, Berkeley prevailed on Walter Zinn, Argonne's director, to send out some "pile experts"—an example of a lab finding necessary expertise at another site within the system.[13]

In the meantime, however, despite his assurances of cooperation, Zinn maneuvered behind the scenes at the AEC to protect his turf in the reactor field. As Alvarez would later concede, Zinn "had been designing piles for 4 years since the end of the war, and he had seen none of these being reproduced in hardware. Now if a lot of money was to be made available to build piles . . . he would like to see some of his ideas get into the piles." Lawrence and Alvarez changed their strategy and decided to pursue reactor expertise outside the system, in a consulting firm in the private sector; "if we were to make too much use of the Argonne Laboratory and the Oak Ridge Laboratory in the design of our piles . . . people could criticize us for taking effort away from those laboratories." Their strategy failed, Zinn's succeeded, and the AEC scrapped the plans for a reactor at Berkeley. Argonne retained its lead role in reactor development and provided the plans for the heavy-water reactors eventually constructed at Savannah River.[14] In this case, despite lip service to cooperation, an individual lab could resist systemic considerations in favor of defending its own program, and could thus block the diversification of another lab.

Lawrence, undaunted but wiser, returned with another plan but stuck to his specialty. He now proposed a linear accelerator, based on an earlier design by Alvarez, to produce neutrons from an intense beam of deuterons. The neutrons from the Materials Testing Accelerator (MTA) could then produce uranium 233, plutonium, or tritium. Lawrence, typically, thought big: a Mark I prototype, with a vacuum tank 60 feet long and 60 feet in diameter, to produce 50 milliamp of 30 MeV deuterons; and the full-scale

Mark II production linac, which stretched in final plans to 1,500 feet. Mark II would give 500 milliamp of 350 MeV deuterons, or a gram of neutrons a day, which could then produce 3 kg of plutonium from uranium tailings or 3 g of tritium per day. The plan aimed to provide "a big source of free neutrons"—free only in the physical sense: Lawrence at first priced the Mark II machine in the neighborhood of $100 to $150 million and eventually doubled his estimate. To support his plan Lawrence fired the enthusiasm of his friends on the congressional Joint Committee and suggested that the project's "Number One priority" was not high enough: "If real speed were desired . . . the project should be put on a crash basis." The AEC, on the GAC's advice, decided to wait on Mark II until Mark I was working successfully.[15]

The scale of even the smaller Mark I machine proved too large for the hilltop Rad Lab, and a new site was selected forty-five miles southeast of Berkeley at the Livermore Auxiliary Naval Air Station. Early estimates predicted the diversion of half of the Berkeley staff to the project. Berkeley also borrowed several experienced metallurgists from Oak Ridge to help with the preparation of uranium targets. The Mark I MTA fired its first beam in May 1952, but the development of cheaper domestic sources of uranium ore in the meantime had obviated the need for the machines. The AEC canceled the Mark II that August and maintained the Mark I as a small project under the Division of Research.[16]

The response of the AEC labs to the context of crisis in the early 1950s imparts several general lessons. The initiation of weapons work at Berkeley and Brookhaven, the two labs perceived to lie at the farthest remove from applied research, demonstrates the integration of basic and programmatic research effected by the context of national emergency in the early 1950s.[17] Remobilization thus blurred the early differentiation between labs with a primarily programmatic mission at Argonne, Los Alamos, and Oak Ridge, and those at Berkeley and Brookhaven with a more fundamental, academic orientation. To ease the transition, labs with less experience in particular fields could appeal to others within the system. It also provides evidence for the general tendency of centralized programs to diffuse to the other sites in the laboratory system. The centrifugal impulse in the lab system soon reached the weapon design program, where it spun off a new weapons lab from Berkeley's annex for the Mark I at Livermore.

Under the pressures of the crash program for the H-bomb, Los Alamos had delegated work to other labs: theoretical problems to Argonne, cross-section research to Brookhaven, and a program for diagnostic measurements of the Greenhouse-George weapons shot in the spring of 1951, which would test principles of thermonuclear explosions, to about forty Berkeley scientists, some of whom set up shop at Livermore.[18] The diffusion of weapons work extended outside the lab system and included a team under John Wheeler at Princeton on theory and a cryogenics program at a National Bureau of Standards facility in Colorado. The H-bomb program, however, continued to suffer for lack of theorists, at least in the eyes of Edward Teller, who consistently urged a stronger effort.

By September 1950 the GAC was entertaining the idea of a new weapons laboratory to take the load off Los Alamos. There were no concrete plans at the time, however, and Teller continued to chafe within the Los Alamos organization and agitate for acceleration of the program. Taking a page from Lawrence's playbook, Teller circumvented the channels of the AEC and cultivated allies on the Joint Committee and elsewhere in Washington.[19] In April 1951 Teller presented AEC chairman Gordon Dean with a proposal to establish a second laboratory for weapon design at the National Bureau of Standards in Boulder, Colorado, apparently to take advantage of cryogenic apparatus necessary for the extant H-bomb designs using liquid deuterium. The AEC considered the proposal over the summer of 1951 but reached no decision.[20]

The debate over the second weapons laboratory, sparked by Teller and fanned by the Joint Committee, soon engulfed the GAC and AEC. It continued to center on the H-bomb program, despite the technical breakthrough by Teller and Stanislaw Ulam in March 1951 that appeared to ensure the feasibility of a fusion weapon; but it also raised deeper issues about the structure of the laboratory system and the implications of systemicity. Teller continued to advocate a second lab and in September 1951 suggested a new site at Rocky Flats, Colorado, where Dow Chemical was building a weapons production plant. Teller's latest proposal argued that a new lab should ensure "competition" with Los Alamos.[21]

The GAC took up the theme at its meeting in October and asked, "Would inter-laboratory competition be desirable in the weapons development field?" The committee considered that it might entail "duplication of certain at present unique facilities" but also that some "sort of division of responsibilities might be made between two such laboratories" to avoid duplication. For instance, Willard Libby, a proponent of a second lab, proposed that the new lab take over thermonuclear development and leave Los Alamos to work on fission designs. Libby "urged that a spirit of competition be brought into the weapons development field," especially "because of the conditions of secrecy under which the work is carried on." The secretary recorded that "there was general lack of agreement with this thesis." Rabi noted that the progress made by Los Alamos, including the Teller-Ulam breakthrough, provided evidence for keeping thermonuclear work there. Oliver Buckley observed "that the existing examples of competition in the Armed Service laboratories do not bear out Dr. Libby's thesis of the benefits of competition." The GAC had more familiar examples at hand: in the same meeting it considered, as it had done for several years, the difficulties in the AEC's reactor program and found that they demonstrated "the disadvantages of having several competing development laboratories rather than a single central laboratory with clear responsibilities and incentives for a whole field."[22] We see again the support by scientists for centralization.

The GAC concluded that a second weapons lab was "neither necessary, nor in any real sense feasible." A second lab would only compound the problem, namely, the shortage of scientists. Instead, the committee urged that Los Alamos delegate routine development work to Sandia and thus free some staff to work on new, advanced designs. Libby maintained his support of competition in a minority statement, although he allowed that the new lab should develop "close and friendly cooperation" with Los Alamos. He recognized the problem of recruitment and suggested that a second lab could find sufficient staff "mainly from outside the AEC laboratories, though a considerable number would come from the Commission laboratories including Los Alamos itself."[23]

The GAC continued to oppose a second lab at its next meeting, two months later. Bradbury likewise kept up his opposition, resenting the "rather thinly veiled criticism" of Los Alamos. The AEC's Division of Military Application supported Bradbury: centralized responsibility for the program would better ensure progress than competition. The AEC took the collective advice and decided in December against a second weapons lab.[24] But the lone advocates on the GAC and AEC—Libby and commissioner Thomas Murray—had powerful allies. Senator Brien McMahon of the Joint Committee had provoked the interest of the military services in the issue, and the chairman of the AEC's Military Liaison Committee in

November 1951 spoke up in favor of "competition and new ideas in weapons development" and a second lab. At hearings before the Joint Committee in February 1952, Senator Bourke Hickenlooper joined McMahon and Murray in invoking the merits of competition against the monopoly of Los Alamos: "Where you have one outfit like that, they can be leisurely or not as they choose. . . . The competitive spirit is not existent." The AEC should not put "all [its] . . . eggs in one basket."[25]

The agitation of the military, especially the air force, increased through the spring of 1952, to the point where the air force threatened, despite the legal authority of the AEC, to sponsor its own nuclear weapons lab and even talked to AUI, Brookhaven's contractor, about managing it.[26] An alternative accommodation had meanwhile emerged. In late 1951 Murray had discussed the issue of a second lab with Lawrence, who had already mobilized his scientists for the MTA and weapons test diagnostics. In January 1952 Lawrence sent Herbert York, one of the young scientists working on diagnostics, to Chicago, Princeton, and Washington, where he sounded out opinions on the second lab, and to Los Alamos, where he offered to establish a permanent program at Livermore to support weapons diagnostics. The waning fortunes of the MTA encouraged Lawrence's search, as a Rad Lab report put it, for "new fields of work [that] can fully utilize the skills and abilities of the group presently engaged in MTA work." The next month Teller visited Livermore, where Lawrence broached the possibility of establishing the second lab at the site. Teller supported the plan and agreed to come to California to implement it, provided the lab included thermonuclear work; but he also kept his options open in case his earlier proposals found favor.[27]

The concrete alternative offered by Berkeley and its annex at Livermore and the administrative experience and enthusiasm of Lawrence gave the AEC a way to appease advocates of a second lab while limiting any inroads against Los Alamos. The GAC in April 1952 agreed that Berkeley "should be encouraged to go ahead on a broader front as long as it could do so without pirating Los Alamos people." But, as Oppenheimer summarized the discussion, "whether this will lead to a second thermonuclear laboratory is not clear."[28] The AEC agreed to continue Berkeley's program of diagnostics in support of Los Alamos and to encourage its interest in thermonuclear research—but nothing more. "To draw upon the skills of people outside Los Alamos," said AEC chair Gordon Dean, "was not a departure from our policy in any way."[29] The new program remained an adjunct of the Radiation Laboratory and its director, York, reported to Lawrence. In September 1952, even as he planned a more ambitious program beyond diagnostics, Lawrence was still assuring the AEC that "it was not intended that Livermore should become the second major weapons laboratory."[30]

Teller, unsatisfied, announced that summer at a Berkeley banquet celebrating the new program that he would have nothing to do with it. The deputy director of the AEC's Division of Military Application, Admiral John T. Hayward, warned that the AEC had to satisfy the Department of Defense "whether we meant this step or [whether] it was an expedient brought about by political considerations." It had also to face the implications: "Is it our intention to have it grow to the ultimate size of LASL without getting into facilities that are peculiar to that place . . . ? What will be the relative effort budgetwise between the two places? . . . What system will be set up for the meshing of the second program?"[31]

The AEC would struggle to answer these questions for years afterward. It did quickly commit to the growth of the new lab: by early 1953 Livermore had 350 staff on site, with another 200 or so joining the effort in Berkeley; by mid-1954 it anticipated total staff of 1,400, of whom 600 would be scientists.[32] The prevalence of Berkeley staff highlights one of several ironies in the early history of Livermore. The impetus for a second lab came from the drain on the theoretical effort on thermonuclear weapons at Los Alamos. Most of the staff at Livermore, however, were young scientists from Berkeley, grounded in the experimental tradition of the Rad Lab.[33] Berkeley had lost what theorists it had during the loyalty oath controversy at the university. Hence the second lab, whose opponents had warned that it would depopulate Los Alamos, instead recruited from Berkeley and as a consequence did not fill the need for theorists that helped create it.

The University of California agreed to the expansion of the Rad Lab to include weapons work at Livermore, after Lawrence assured the Regents of "cooperation" with Los Alamos and Bradbury, who lacked Lawrence's strong relationship with the Regents, agreed that "there is every reason for the two laboratories to work together continually."[34] In other forums Bradbury resented the intrusion on Los Alamos that a second lab entailed; he found his lab undermined by competition supported by its own contractor. The betrayals extended to the military services. The advocacy of the air force helped ensure the foundation of a second lab, but in the meantime Los Alamos had developed the high-yield strategic weapons the air force had sought. As the air force warmed to Los Alamos, the new lab at Livermore turned to the army and then the navy for weapons development programs.[35]

A final irony emerges in the evolution of attitudes toward competition. The AEC and GAC supported centralization in the debate over a second lab, while the Joint Committee, especially its Democratic chairman (joined by Republicans abandoning fiscal restraint for the cause of national security), invoked the merits of competition. In late 1958, at the outset of the weapons test moratorium, the Democrat who chaired the Joint Committee questioned the redundancy inherent in the two-laboratory setup, compelling the AEC and GAC to defend the two weapons labs and the duplication required "to promote complementary competition."[36] Laboratories, once established, acquired institutional inertia.

The Joint Committee may have abandoned its earlier rhetoric about the virtues of competition because it perceived that the labs tried to circumvent competition in practice. But if the weapons labs cooperated in the distribution of new projects, they continued to confront the consequences of competition. Hence the sensitivity of Livermore scientists to their test fizzles in the Nevada desert, which not only subjected them to "a heavy dose of ribbing from [their] colleagues at the original weapons laboratory" but also undermined their pursuit of new programs.[37] Hence also the reaction of Los Alamos scientists to the publication in 1954 of a book by two journalists for *Time* assigning credit for development of the hydrogen bomb to Livermore. The book, along with the Oppenheimer security hearings, which attacked the former director of the lab and raised anew the insinuations that the H-bomb program had stalled at Los Alamos, aroused resentment there. Los Alamos sympathizers produced hostile reviews of the book, Bradbury called a press conference to defend his lab, and five hundred scientists from Los Alamos signed a protest of the Oppenheimer decision, addressed to Eisenhower. To smooth ruffled feathers, the president dispatched Lewis Strauss to the lab with a presidential citation for its role in the development of the hydrogen bomb. "Operation Butter-up," as the local paper dubbed the exercise, would not have been so necessary without the presence of the upstart in Livermore.[38]

Reactors

The spread of centralized programs through the lab system is also evident in reactor development. The program in the immediate postwar period remained divided between Argonne and Clinton, but the same impulse to centralization that had spurred the establishment of the weapons lab at Los Alamos soon came to the reactor program. It would confront the centrifugal forces that created Livermore, with similar results.

The pessimistic postwar view of the long-range potential of Clinton Labs prevailed in the GAC and AEC, which felt no obligation to sustain all the institutions inherited from the Manhattan Project. An AEC staff paper of April 1947 proposed "that for effective concentration on urgent problems and for security" the AEC conduct its primary program "as completely as possible with Atomic Energy Commission facilities, essentially disentangled from nonprogrammatic, fundamental research."[39] The GAC, at its third meeting later that month, interpreted the proposal to imply a centralized AEC laboratory. At the same meeting the GAC's subcommittee on research urged the concentration of all programmatic activities—that is, weapon and reactor development—"into a single central laboratory." The subcommittee consented "for practical reasons pertaining to military security and the necessity for close military liaison to keep weapon development in a separate laboratory at least for the present." But for the rest of the programmatic research, "we have in mind a new laboratory" operated directly by the AEC.[40]

The subcommittee's report listed several advantages to a single, central lab, which perhaps reflected wartime experience in Los Alamos and the MIT Rad Lab for radar research. A central lab would have a clear purpose, with "no divided responsibility." Security restrictions, particularly prevalent in programmatic work, prevent "adequate communication and contact between separate laboratories." A single laboratory under one director could concentrate effort on the most important problems: "Within such a laboratory the entire sweep of primary A.E.C. activities come into view and into focus." Hence the lab director and staff could formulate their program "with a minimum of detailed supervision from Washington." A central lab would make best use of scarce scientific talent, and a "laboratory of such size, importance, and clear-cut function would be a challenge to the best men." In addition, "it may be just possible to find one man fully qualified as director; it is very uncertain that several can be found for several effective laboratories." In short, a single, central lab operated by the AEC would avoid the problems of contractors, security, manpower, and administration of multiple labs.[41]

The subcommittee recognized that the transition to a central lab would be difficult and that it might interfere with short-term projects. It did not recommend a site for the lab but implied that it would not be one of the existing labs; the new lab would draw personnel from Clinton and Argonne in particular.[42] When the GAC considered the report, however, it soon seized on the idea of using the Argonne site, which was near a large city but could accommodate both reactors then under design at the two labs, Argonne's fast breeder and Clinton's high-flux reactor. Clinton was not an option: "Most of us think that the evidence is in that Clinton will not live even if it is built up." The AEC received the recommendation, but Fisk, the director of research, and Wilson, the general manager, had too many other pressing problems to solve at the time.[43]

The GAC revived the question at its next meeting at the end of May, attended by the commissioners. It had "no good news" and no easy answers, but it still recognized the downsides of duplication. For instance, "the chemical and metallurgical facilities at Argonne and Clinton are almost identical to each other. This . . . indicates that something is wrong." The committee again raised the possibility of committing the high-flux reactor to Argonne instead of Clinton; against this, Lilienthal "felt a bit apprehensive about carrying on research in this mysterious field near a city." Rabi used the forum to advance the cause of Brookhaven as an alternative site, at the expense of Clinton. The main advantage of Brookhaven, as Oppenheimer put it, was that "there is a brand new, fresh contractor full of beans and interested in the subject"; at Clinton, by contrast, Monsanto displayed no such vigor. Fermi felt that Clinton was not so bad; there was also the possibility that the staff there would refuse to move to another AEC site. Conant doubted the value of Clinton but thought that moving its program to Brookhaven would not gain any "unification of the effort." There would still be "too many shows trying to do too many things."[44]

As Conant's objection indicated, Argonne and Oak Ridge were not the only sites in the reactor program: Los Alamos had one reactor and was building another, Brookhaven was building a research reactor, and the AEC was sponsoring the specialized reactor lab at Knolls. The far-flung reactor work undermined Fermi's proposal from the April meeting to keep Argonne and Clinton but put the two labs under common direction by a joint planning board—that is, to centralize administration of reactor research but not location. Oppenheimer took the opposite tack. He perceived two main objections to a centralized lab: it would create too large a lab, and centralization in itself was undesirable. The first he did not think an intractable problem. To the second, more theoretical objection, he countered that "centralization means that no place knows enough to run its own affairs without directions from Washington and the purpose of making a strong reactor laboratory was to decentralize, to give local autonomy to that laboratory as far as the long range field of development went." In other words, institutional centralization would foster administrative decentralization; the canny proposal thus jibed with Lilienthal's philosophy of administration. In the long run, the AEC would attain neither; in the short run, it would keep trying.[45]

Two considerations conspired against the proposal to centralize: the delay in the program that would result and the desire of Clinton scientists, despite the disadvantages of Dogpatch, to stay where they were to work on the high-flux reactor. The reappearance that summer of the University of Chicago as a candidate to replace Monsanto as contractor strengthened the position of Clinton, although some on the GAC, notably Conant, feared that such an arrangement would make a central lab even less likely. Fisk warned that "if the negotiations with Chicago should fail then the only alternative would be to liquidate Clinton."[46] After negotiations with Chicago did stall in the fall and Union Carbide emerged as an alternative contractor for Clinton, the possibility of centralization reemerged. In several days of hectic meetings and phone calls over the Christmas holiday in 1947, Wilson and the AEC decided that Carbide would take over Clinton and Argonne would take over all reactor development. Wilson dispatched Fisk with the bad news to Clinton, where scientists greeted him without holiday cheer, "not only among the chronic complainers but quite generally." A profane ditty, to the tune of "Deck the Halls," concluded: "Fisk considered many factors / Then he stole all our reactors. / Now the New Year's here to greet us / Can the bastards really beat us?"[47]

The AEC tried to soften the blow by assuring that Clinton would continue to operate its graphite research reactor and maintain the related chemistry and physics research, the isotope development program, and biomedical work; it would also "undertake a strong program of research on chemical process problems" and chemical engineering. But all work on the high-flux reactor and another reactor for power production, and associated metallurgy, chemistry, and physics programs, would transfer to Argonne. The area manager appealed to Clinton scientists with the GAC's argument: "In a central reactor laboratory, the opportunity to plan and execute important reactor developments is far greater than in separate locations."[48]

Clinton scientists did not accept the decision without protest, but neither did they question the premise behind it. Weinberg fired off a letter to Oppenheimer to present the case for Clinton to the GAC. Weinberg professed "no quarrel" with the decision to centralize. He protested only "the decision to concentrate the reactor work at Chicago rather than concentrate it at Oak Ridge." He recognized that the main strike against Clinton was its rural location. But the GAC, he said, underestimated the appeal of Oak Ridge; perhaps it would not attract scientists with the status and sensibilities of GAC members—what programmatic laboratory could?—but it would attract scientists of reasonable competence, and those already present had developed a sense of community. He pointed out that Argonne was itself over an hour from Chicago. Finally, he noted the "inconsistency" in the decision: the AEC had not included the reactor work at Knolls and Brookhaven in the centralization.[49]

Weinberg found an unlikely ally in Zinn, the director of Argonne. At an emergency meeting of the GAC, called to consider the fallout from the decision to centralize, Zinn surprised the assembled advisers with his "grave doubts about the success of the proposal." Like Weinberg, Zinn accepted the premise of centralization. On a practical level, he dreaded the personnel and administrative problems that centralization would entail at Argonne. More fundamentally and surprisingly, and also like Weinberg, he asked: "Granting that the present reactor situation did not add up to what is wanted, would it do so if one meddles with the place (Clinton) which potentially has the greatest strength? Some people will say, 'Why should one take a strong place and move it to one relatively weaker?'" Zinn thought that Argonne and Clinton were about even in experimental physics, chemistry, and metallurgy, but that Clinton had the edge in engineering and theoretical physics. And though he supported the decision to centralize, he asked the assembled advisers, in what was becoming a popular image for the labs, whether they "worried about placing 'all the eggs in one basket.'" The GAC responded with the wisdom of Mark Twain—"it would be simpler to watch 'one basket'"—and refused to believe that Clinton might be the stronger lab. In any case, the decision was a done deal.[50] So they thought.

Zinn's magnanimity, although motivated in part by self-interest, indicates that the labs did not always engage in cutthroat competition with one another for programs. It also illustrates that cooperation at the level of the labs could circumvent decisions made at higher levels of management.

Zinn's stance encouraged the resistance of scientists at Clinton (soon to be renamed Oak Ridge), who seized any opportunity to retain reactor work. AEC staff unintentionally provided one. Wilson, the general manager, attempted to "ease some of the disquiet" at Clinton and reassure its scientists of the long-term future of the lab. He emphasized in a letter that Argonne would take over the high-flux reactor and that "the development of reactor designs will not form a part of the Clinton Laboratory program beyond the period of orderly transfer of these activities to the Argonne reactor laboratory." But he added, "Any decision regarding future research reactors at Clinton will depend both on the evolving research program at Clinton and on progress in reactor development at the Argonne Laboratory." A handwritten note in the margin, apparently by Weinberg, perceived that this "leaves the door open."[51]

Weinberg walked through it. In March 1948, scant months after the decision to centralize reactor work, Oak Ridge submitted its program for review by AEC staff. The program included items for the "possible location of [the] High Flux Reactor at Oak Ridge" and "possibility of a reconsideration of the decision to transfer the Power Pile development to Argonne because of the international situation," which might require immediate construction of the naval propulsion reactor at Oak Ridge. The AEC stuck to centralization and warned the lab to leave reactor work to Argonne.[52] Weinberg returned a month later with another proposal to build "a 'poor man's' high flux pile at Oak Ridge." The poor man's pile would use the same design as the high-flux pile but would operate at 3 MW instead of 30 MW and cost less than $5 million. The pile would provide ten to twenty times the neutron flux of the current research reactor and hence sustain the neutron and solid-state physics and metallurgy groups, which, Weinberg claimed, were pursuing not "'reactor' studies in the sense that they are directed towards design of specific reactors" but rather basic research, which the AEC had agreed to support at Oak Ridge. It would also aid the isotope production program. Weinberg saw the poor man's pile as a stepping-stone to bigger and better reactors: "This has always been the pattern in the development of large scientific equipment—the 37″ cyclotron at Berkeley was followed by a 60″ one which was followed by the 180″ machine; and it should be the pattern at a laboratory like ORNL."[53]

Weinberg pressed his case in person in Washington, where Wilson and Fisk allowed that Oak Ridge might have a new research reactor. Before reaching a final decision, Fisk deferred the question to Zinn at Argonne, since he ostensibly led the AEC's reactor effort in the absence of an AEC staff division with responsibility for reactors.[54] Fisk also suggested that Zinn and Weinberg meet to coordinate their plans, which they did in June. Weinberg now tried to have it both ways: a high-flux reactor was a research tool, not a design prototype, and hence one might be built at Oak Ridge; but it would contribute to the reactor development program and hence deserved support. For his part, Zinn felt his staff had plenty to do without adding the burden of the high-flux design, especially as they ramped up that spring to build a reactor for the navy; he also had evidence that most of the Oak Ridge reactor scientists who were supposed to transfer to Argonne would not come. The two directors arrived at an agreeable compromise: the AEC could authorize two high-flux reactors, one for each lab. The two labs could cooperate on the final designs and perhaps use the same contractor to build them; the two reactors might differ only in their power of operation.[55] Like the Manhattan Project before it, the AEC need not run a zero-sum game.

A year earlier, in the debate over the central lab, commissioner Sumner Pike had raised such a possibility: "There is no reason why we shouldn't have two high flux reactors and that would mean perhaps a third duplication of chemistry and metallurgy."[56] The Bureau of the Budget and the appropriations committees would inject one reason. Zinn anticipated their objection in his reply to Fisk. He noted first that the Oak Ridge estimate of $5 million was optimistic: the cost would probably be twice that, at least from the experience of Brookhaven, whose own research reactor had cost (at that point) $9 million. Assuming a comparable cost, the AEC had to decide if it was willing to support three such research reactors, for Brookhaven, Argonne, and Oak Ridge, in addition to the power reactors under development at Argonne and Knolls.[57]

The AEC would have to support a research reactor at Argonne, according to Zinn, because Zinn and the AEC had come to the conclusion that the high-flux reactor was too hazardous to locate at the Argonne site. According to AEC guidelines, a reactor operating at the proposed power level of the high-flux design could not be so close to a metropolitan area like Chicago nor to a "a national vital installation" like the production plants at Oak Ridge or, for that matter, "any present AEC site." Hence the price tag for the AEC's reactor program would have to include $30 million for the high-flux machine at a new remote site, plus the cost of a research reactor for Argonne; Zinn was not willing to support a reactor for Oak Ridge but forgo one for his own lab.[58]

Although he offered these observations on the practical aspects of the reactor program, Zinn objected to the premise behind Fisk's request for them: "It was not my understanding when Argonne took responsibility for reactor development that this in any way gave it control over the type and quantity of research reactors which might be acquired by any installation other than Argonne. . . . [T]he research reactor for the Oak Ridge National Laboratory is a matter of negotiation between the Oak Ridge contractor and Washington."[59] Zinn's refusal to decide the program of another lab may have contributed to the creation of the Division of Reactor Development at the AEC in Washington. It definitely doomed centralization, along with a few other factors: the AEC's sensitivity to morale at Oak Ridge after Black Christmas; Zinn's agreement to cooperate rather than compete with Oak Ridge, made necessary by the plethora of reactor projects coming down the pipeline; the decision that the high-flux pile was too dangerous for the Argonne site; and above all the refusal of Oak Ridge scientists, especially Weinberg, to accept centralization at Argonne.[60]

Buoyed by its reentry into the reactor business, Oak Ridge pushed for more. Centralization required the shift of the high-flux reactor (soon renamed the Materials Testing Reactor) to Argonne; but since safety concerns precluded Argonne as a possible site, Oak Ridge offered itself as an alternative, raising its sights from a poor man's pile to the deluxe version. The AEC's Reactor Safeguard Committee refused to consider the Oak Ridge alternative and instead endorsed a remote proving ground. Undeterred, Weinberg returned with a proposal for a research reactor to run at 15 MW, half the design power of the high-flux but five times that of the poor man's pile. Weinberg warned that a remote reactor station would turn into another laboratory and further drain scientific manpower. He could have pointed out that similar considerations had produced his own lab at Oak Ridge.[61]

The AEC rejected Weinberg's arguments and decided in October 1948 to build the Materials Testing Reactor at a new remote site, which it would acquire in Idaho. Argonne and Oak Ridge would cooperate in the design and construction of the reactor under a rough division of labor: Argonne would handle the cooling system and auxiliary plant, and Oak Ridge would engineer the reactor core and shielding. Argonne got relief for its overtaxed staff, and Oak Ridge kept its reactor team employed and intact. Instead of centralization the AEC had come around to interlaboratory collaboration.[62] The attempt to establish a central lab for reactor research resulted in the construction of the high-flux reactor, the centerpiece of centralization, at neither Argonne nor Oak Ridge, although both labs helped build it and had access to it in Idaho for research. The labs got their research reactors a few years later: the AEC approved the so-called CP-5 pile at Argonne in May 1951; Oak Ridge dusted off the idea of the poor man's pile in August 1950 and, not surprisingly, had developed the proposal into quite a regal reactor by the time it was approved in 1953.[63]

Research reactors alone did not ensure the reentry of Oak Ridge into the reactor business. Weinberg had taken pains to emphasize that a research reactor would not involve the lab in power reactor design. But its gradual resumption of reactor work encouraged the lab to seek more. Weinberg began to agitate for support of his pet project, a homogeneous reactor that would circulate fissionable material in an aqueous slurry through the reactor. He again exploited, or created, a loophole in the approved program to justify new reactor work: "The laboratory has interpreted its directive to stress chemical technology to include a serious investigation of those reactor types, notably the homogeneous reactors, whose development primarily involves chemistry and chemical engineering." In the fall of 1948 the AEC approved research on the homogeneous system at Oak Ridge.[64]

The homogeneous reactor would become a major part of the Oak Ridge program in the 1950s, but for its first few years it struggled with long-term problems of materials and dynamics of the system. AEC staff perceived in October 1949 that Oak Ridge "possessed needed facilities, that it had a competent staff which liked to live in the region, and that ANL [Argonne] cannot handle the full workload." Oak Ridge, however, seemed to lack "a large long-range central problem to knit the laboratory to a common purpose."[65] It was already in the process of acquiring one. The navy's interest in the potential of nuclear reactors for propulsion predated the Manhattan Project and now supported a substantial project under the AEC at Argonne, Knolls, and Westinghouse.[66] The air force had also developed an interest in nuclear propulsion during World War II and in 1946 had contracted with the Fairchild Engine and Airplane Corporation for a project for "Nuclear Energy for the Propulsion of Aircraft." Fairchild set up shop in some of the abandoned production facilities at Oak Ridge, but the small staff focused on what they knew, aircraft engines, and neglected what they did not, atomic energy. The project languished until the summer of 1948, when the AEC commissioned a review by a group of consultants. The so-called Lexington report predicted that development of a nuclear-powered airplane would take fifteen years and cost $1 billion, but the potential importance for long-range bombers justified spending $200 million over three to five years for research and development. Easy to say for the consultants, who would not have to pay for it. The AEC agreed, in December 1948, only to support a program at the level of $3 million over the next few years.[67]

Before the AEC had a chance to digest the Lexington report, it approved in October a small effort on aircraft propulsion at Oak Ridge so long as it did not interfere with the high-flux reactor work. After the AEC decided to support the larger program, Oak Ridge scientists overcame their earlier skepticism about the project and sought a larger piece of the action. In April 1949 they met with representatives of the air force, Fairchild, and AEC reactor staff to consider how Oak Ridge could contribute. After several months of negotiation, Oak Ridge and the AEC agreed in September to establish an Aircraft Nuclear Propulsion program at the lab, "with a priority second only to that of the MTR [Materials Testing Reactor] project." The work involved research on problems of materials, radiation damage, shielding, and also reactor design.[68] Although Lawrence Hafstad, the AEC's director of reactor development, at the time pronounced himself "open-minded but skeptical" about the prospects of the aircraft propulsion program, and other AEC staff considered it "something of a joke," the program had support from an important quarter: the Joint Committee on Atomic Energy, which first took an interest in the program in 1948.[69] In March 1949 William L. Borden, executive director of the Joint Committee staff and a driving force behind the creation of Livermore, pressed Hafstad on the possibility of a crash program for aircraft propulsion. The Joint Committee would continue to prod the AEC on the subject throughout the 1950s.[70]

The Aircraft Nuclear Propulsion project, along with the homogeneous reactor, gave Oak Ridge reactor projects to sustain it when the MTR effort dwindled. The AEC in mid-1950 officially approved an operating policy for Oak Ridge that included reactor development.[71] The resumption of reactor work helped turn the dismay of Black Christmas to optimism at Oak Ridge; the new attitude and encouraging technical results converted the AEC, which now pointed to Oak Ridge and its industrial contractor as a model for the other labs. A couple of years later an AEC report would note that reactors "thrive better at Oak Ridge than they do at Argonne."[72] The emergent context of crisis in 1950 further fired the enthusiasm of the air force for aircraft reactors and led the AEC to reorient the project from research to development. By the start of 1951 Oak Ridge had over 250 people engaged in the project, by far the largest at the lab, and another 60 working on the homogeneous reactor.[73] The context, and budgetary climate, of national emergency also fostered the rhetoric of competition evident in the creation of Livermore: Weinberg admitted to some duplication in the reactor work of Oak Ridge and Argonne but thought it "desirable," since "competition was healthy."[74]

The failed attempt at centralization underscores the contingent status of the labs after the war; the AEC did not have to maintain the status quo and in this case tried to change it. Centralization could not overcome the institutional momentum acquired by the labs and the already well-developed enterprise of lab scientists that propelled it. We also see again the spread of programs through the system as a result of external pressures. In this episode, as with weapons research during the national emergency of the early 1950s, the Cold War drove the diffusion of reactor research. The navy's desire for submarine reactors overtaxed the Argonne staff and led them to seek help from Oak Ridge. The interest of the air force in reactors for long-range bombers then sealed the reentry of Oak Ridge into reactor work. For both the weapons and reactor programs of the national labs, pressure from the external environment—that is, the historical context—provided a centrifugal force.

BIG EQUIPMENT

The Addition and Multiplication of Accelerators

One of the reasons why Oak Ridge pressed ahead with plans for the poor man's pile so soon after the decision to centralize reactors was that the AEC also was questioning the presence of accelerators in their program. Weinberg may have realized that, despite the commission's expressed intent to maintain "a strong permanent regional Laboratory" at Oak Ridge, the absence of both reactors and accelerators would undermine its status as a national lab.[75] The national labs were not intended only to serve as homes for programmatic, often classified work; they were also supposed to provide big, expensive facilities for visitors. Labs that lacked the second aspect of this dual purpose might be considered incomplete. As Brookhaven and Berkeley consciously sought a programmatic role to complement their research facilities, the programmatic labs sought facilities to bridge the gap from the opposite direction. We can anticipate the result: a blurring of the distinction between the programmatic and basic research laboratories.

The attempt by Oak Ridge to develop an accelerator program raised a general question for the AEC implicit in the second function of the national labs: What sort of facilities should they provide? The AEC's response in this case applied more generally. The original conception of the labs involved the provision of nuclear reactors for research. Accelerators did not fall under the legal monopoly on nuclear material maintained by the AEC and had remained within the reach of universities, and hence did not seem to lie within the mission of the labs.

The hybrid case of Berkeley could support either the relegation of accelerators to university research units or the addition of accelerators to the mission of the national labs. When Ernest Lawrence first proposed in September 1945 that the Manhattan Project pay for his postwar plans for accelerators, in addition to a nuclear reactor, Groves had refused and asked why the government should pay for them. That December, however, in the absence of an atomic energy agency, Groves agreed to subsidize accelerators at the Rad Lab. The fledgling AEC faced a fait accompli; after the new commissioners visited the lab in November 1946, they assured Lawrence of their "vigorous support for . . . the development of high-energy accelerating equipment."[76]

The Berkeley precedent need not have applied elsewhere. The founders of Brookhaven sought only a nuclear reactor; at one of their first meetings, in February 1946, they deferred discussion of accelerators since several northeastern universities were already pursuing them separately. A month later, however, they had appointed a committee to plan for "electronuclear machines," and their proposed program of June 1946 included a 1-BeV electron synchrotron, a 500-MeV synchrocyclotron, and for the future a multi-BeV proton synchrotron.[77]

Despite its earlier assurances and the big plans of Berkeley and Brookhaven, the AEC was still not convinced that it should sponsor accelerators. In particular, James Fisk, a solid-state physicist from Bell Labs newly appointed director of research for the AEC, did not sympathize with academic accelerator builders. In the spring of 1947 he questioned the relevance of high-energy accelerators to nuclear weapons and reactors and whether the AEC should support them. His skepticism stemmed in part from the activity of the navy's Office of Naval Research (ONR), which had set up an aggressive program to fund basic scientific research after the war.

The ONR was sponsoring the construction at universities of a dozen accelerators up to the size of the 184-inch cyclotron at Berkeley. Accelerators seemed to lie closer to the domain of the AEC than the navy; but if the ONR was willing to support accelerators at universities, the AEC could cede the field to them.[78]

The GAC provided a more hospitable forum for accelerator aficionados. Fisk's deputy, Ralph Johnson, asked the advisers "whether a large-scale support of accelerators was appropriate to the interest of the AEC, compared with the need for efforts in metallurgical and chemical research," and noted the possibility of ONR involvement. The committee urged that the AEC as a civilian agency was a better sponsor for unclassified accelerator programs than the navy. Fisk and the AEC continued to waffle until the meeting arranged by Lawrence at Bohemian Grove near Berkeley in August 1947. Two weeks earlier President Truman had vetoed a bill for a National Science Foundation, which might have assumed responsibility for accelerators. In California the GAC reiterated its support for basic nuclear research and Lawrence lobbied Lilienthal in person. Fisk held out against AEC sponsorship of accelerators, but in October the commission appropriated $15 million for them.[79]

In the meantime Berkeley and Brookhaven had proceeded with their plans. At Berkeley, William Brobeck in 1946 had designed a proton synchrotron to reach 10 BeV—in the Berkeley tradition, far beyond the 400 MeV of the 184-inch machine, just then coming on-line. Lawrence thought that might be too ambitious for the AEC and suggested half that energy; he finally settled on 6 BeV after Edwin McMillan and Wolfgang Panofsky pointed out that it might thus produce antiprotons. Berkeley passed its proposal to the AEC in November 1947; at nearly $10 million, it would take most of the money appropriated by the AEC for accelerators. That did not sit well with other accelerator aspirants. Brookhaven scientists had endured an internal debate over how best to proceed. M. Stanley Livingston, who with Lawrence had built the first Berkeley cyclotron and now headed the Brookhaven accelerator effort, suggested an intermediate, 750-MeV synchrocyclotron before proceeding to higher energy. Rabi instead urged that Brookhaven not "take the safe path or little steps, but be bold" and aim right away for 10 BeV. Rabi tried to educate Livingston: "We would have to live with this competition with Berkeley." Rabi and Brookhaven could afford to be bold and compete with Berkeley in part because the two labs also collaborated; Rabi had visited Berkeley in the fall of 1946 and returned with blueprints of Brobeck's design, and Berkeley engineers continued to aid the Brookhaven team.[80]

In October 1947 the AEC decided the matter for Brookhaven: Fisk and commissioner Robert Bacher informed the lab that Livingston's smaller machine did not go far enough beyond existing energies. It did not often happen that program managers criticized a plan offered by lab scientists for its lack of ambition. Nevertheless, Livingston wrote a protest, which did not display much political acumen: "It now becomes evident that the program must be based primarily upon the policies of the Atomic Energy Commission and that the Laboratory staff and the Universities' advisory groups have relatively little significance. As such, it is not the 'free' laboratory for fundamental research which had been visualized, but is now directly controlled by the national interests of the Atomic Energy Commission."[81]

The GAC and AEC were indeed aware of national interests, and the multiplication and size of accelerator plans alarmed them. The GAC in November 1947 could not reach a consensus on the desirability of the Berkeley design. How did it relate to the Brookhaven work? Would it compete with other programs for money or manpower, within the Rad Lab or throughout the country? And should the government even pay for big accelerators?[82] The proposals renewed Fisk's skepticism: the AEC "would like, at some time, to see built a multi-billion volt proton accelerator and . . . would like also to have some part in the undertaking. But there is not yet agreement on the Commission's proper role in such an undertaking nor on its timeliness . . . in view of other commitments." The AEC had higher priorities than accelerators and could only support modest paper studies.[83] The AEC's inconsistency, after encouraging the ambition of Brookhaven, underscores the lack of consensus on the addition of accelerators to its mission.

The chastened but persistent labs returned in a few months with smaller proposals, considered by the GAC at its meeting in February 1948. Both labs now aimed for 2.5 BeV and both emphasized that their machines would make efficient use of money and manpower. The committee registered "considerable divergence of opinion" on the two plans. Some members approved identical machines; others sought some differentiation but could not find sufficient scientific justification for particular energies; some continued to oppose federal support for accelerators as a matter of principle. Each lab had a champion on the GAC, neither of whom recused himself from the debate owing to conflict of interest. Rabi pointed to the existing range of accelerators and insisted that it was Brookhaven's turn: Berkeley should get nothing. Glenn Seaborg agreed to support only one machine—at Berkeley. The committee also considered what Lee DuBridge termed the "Clinton effect," which just then confronted the AEC after the decision to centralize reactors: the "psychological reaction" (in Fermi's words) of scientists at a lab to rejection of their program, and the wider perception in the labs that the AEC would emphasize applications at the expense of basic research.[84]

The GAC did not want to disappoint either group, both of which were qualified. It offered instead the same non-zero-sum compromise that resolved the high-flux reactor situation: an accelerator for each lab. To avoid duplication, however, the two machines would aim "at substantially different maximum energies," which they should negotiate with each other and with Fisk.[85] Berkeley revived and slightly enlarged its earlier proposal to push for 6–7 BeV. Brookhaven scientists conceded the higher energy to their more experienced competitors in Berkeley—"up against an outfit like that, we really didn't have much of a chance"—and stayed with 2.5 BeV, despite the fact that they had chosen that energy to produce pions, which Berkeley scientists had just discovered in their 184-inch machine.[86] Still, Brookhaven staff believed that the smaller machine would come on-line sooner, giving them "the advantage of two to four years of operation with the biggest machine in existence."[87] In addition, they were planning to build a high-flux pile of their own to go beyond the graphite reactor then under construction, which in combination with an accelerator would give the lab an unparalleled combination of facilities. Finally, they still viewed 10 BeV as the ultimate goal; they conceded 6 BeV to Berkeley with the tacit understanding that their turn would come for the higher energy. Fisk agreed to the division of energies at a meeting with the two groups in March; he acknowledged that the AEC had no plan to limit the size of accelerators but noted that the commission could not guarantee a future machine for Brookhaven. He also discussed the "need for close cooperation and rapid exchange of information between the two groups," but scientists from the two labs felt that the existing informal arrangements were sufficient. The AEC approved the settlement the next month.[88]

The two machines survived the budget examiners in the executive branch but still faced the scrutiny of Congress. A member of the House Appropriations Committee acknowledged the division of energies but wondered whether the Berkeley machine would put the one at Brookhaven out of business: "So how do you justify building one at Brookhaven when you are building another larger one [at Berkeley] . . .? Why don't you take less money and send your scientists to Berkeley and let them use that one?" Kenneth Pitzer, a Berkeley chemist who had replaced Fisk as director of the Research Division, replied that smaller machines were not rendered obsolete: "Just because a better microscope was invented does not mean that all the poorer microscopes have been thrown away."[89] But in this case the microscopes had yet to be built, at large public expense. Pitzer's evasion avoided basic questions about the lab system—the supply of scientists, the desire to keep in-house staff occupied, the duplication of equipment, and the availability of facilities to outside visitors—which the AEC and the labs would eventually have to face.

The settlement of early 1948 sealed the acceptance by the AEC of high-energy accelerators within its program. The AEC also conveyed the impression that it would not limit the number of machines at any one lab—Berkeley now boasted four large accelerators—nor in the system as a whole, and that it would support future steps to higher energy.[90] The commission presented its program to Congress with a martial metaphor: "The postwar assault of the physicists upon the citadel of the atomic nucleus is being mounted with new and heavier artillery."[91] The military uses of accelerators were not just rhetorical. Accelerators developed, or reacquired, a military component in the deepening Cold War. In addition to their application to production of bomb material in the Materials Testing Accelerator of Berkeley, in late 1949, after the first Soviet atomic test, the GAC addressed the possibility that accelerator beams might be used as a defense against nuclear bombs. It urged the AEC to take responsibility for the problem and assign groups from Berkeley and Los Alamos to study it.[92]

In January 1950 a group of GAC members, lab directors and scientists, and AEC staff met at Caltech to discuss possible countermeasures against atomic weapons. The main topic, which had "recently been considered in England," was the idea to aim gamma rays from an accelerator at an incoming bomb by radar and computer-guided servomechanism; gamma-neutron reactions in the heavy metal of the bomb would bathe the fissionable material in neutrons and preinitiate the weapon. The range for a 100-MeV beam might be 1 kilometer; for a 1-BeV beam, 2 kilometers. The slant range provided by existing accelerators could only limit damage, not prevent it altogether; 1 kilometer would not provide much protection from an atomic bomb. But the group agreed that the idea deserved exploration.[93]

Plans to use accelerators to preinitiate weapons languished for two years under the limits imposed by current accelerator energies, although the Division of Military Application appointed Edwin McMillan of Berkeley to supervise work on countermeasures.[94] The discovery of the principle of strong focusing in the summer of 1952 promised an easier route to high energies. Soon after the discovery the AEC's director of research included in an outline of the high-energy accelerator program the possible application of the principle to the countermeasures program. The Division of Research set up a program with the Department of Defense in which the AEC would support unclassified work on the accelerator end of the countermeasures system.[95] A staff report noted that a $5 million item in the research budget stricken by the Bureau of the Budget had "considerable relevance to the entire problem" of countermeasures and urged that "the military usefulness of particle accelerators should be included as an additional justification in further defense of ultra–high energy particle accelerator construction items before the Bureau of the Budget and Congress."[96]

Although the AEC judged the scheme "rather improbable of success," and a subsequent study for the Department of Defense found it "neither militarily nor economically feasible at this time" owing to the high cost, size, and power requirements and limited range, both agencies thought that the potential for future developments justified support of research on high-energy accelerators. The AEC appropriated $500,000 for further studies of the countermeasures problem in the budget for the 1954 fiscal year and probably followed the advice of the staff report in defending accelerators before budget examiners with appeals to their military uses.[97]

Smaller Stepping-Stones

The military countermeasures studied in the early 1950s depended on accelerator beams of very high energies. But accelerators were not restricted to high-energy physics nor to Berkeley and Brookhaven. Accelerators of lower energies had long formed an important part of programmatic research. All of the labs either had or quickly acquired Van de Graaff generators, at energies from 2 to 12 MeV; the Van de Graaffs covered most of the spectrum of fission energies and thus served both weapons and reactor programs. An assortment of small (less than 60-inch) cyclotrons and betatrons also populated the labs and gave them access to the 20–30 MeV range, which got them into interesting nuclear physics problems and also chemical exploration of radionuclides and the transuranic elements, all of which were part of the AEC's mission.

These small accelerators did not fall into the category of large facilities for use by visiting researchers. First, they had their programmatic uses, and hence access to them was restricted. Second, Van de Graaffs, cyclotrons, and betatrons, although big and expensive by prewar standards, were not beyond the reach of individual universities and were not unique to the labs. A number of firms—General Electric, Collins Radio, High Voltage Engineering, and others—had begun to market the machines commercially; a university could acquire a 92-inch cyclotron or a 100-MeV betatron for under $400,000, a bargain compared to the $10 million price tag attached to the Berkeley Bevatron.[98] The lab system, too, provided a healthy market for accelerator commerce; sometimes, as in the case of Brookhaven, at the insistence of the AEC, which sought to maintain a role for private enterprise instead of letting labs build devices themselves.[99]

The AEC did not intend its support of these low-energy accelerators to provide a stepping-stone to bigger machines. In 1947 Argonne had contemplated entering the high-energy sweepstakes, but Fisk and Bacher recommended that the lab concentrate on reactors. Argonne also had to steer clear of the program of Enrico Fermi and other Met Lab veterans at the University of Chicago, who at the time were planning a 170-inch, 450-MeV synchrocyclotron under the primary sponsorship of the Office of Naval Research.[100] Like Brookhaven, Oak Ridge in early 1946 formed a committee to consider accelerators and adopted plans for a 100-MeV betatron, but the AEC questioned the need for it and ended up lending it to the National Bureau of Standards.[101] With typical persistence, Weinberg, when he returned with his proposals for the poor man's pile, allowed that although accelerators were not the "natural direction for physics" at Oak Ridge, they had "an important subsidiary function." He added: "It is not the intent to use the 'poor man's' pile simply as a substitution for the accelerators. . . . Even with the installation of the 'poor man's' pile, accelerators may be in demand if sufficient competent personnel are in the Laboratory whose interest[s] lie in the field of accelerators."[102]

Oak Ridge found such personnel next door, in the Y-12 facilities of Union Carbide, where a group of scientists under Robert Livingston had continued research on calutrons after the war to explore new methods of separation of weapons material. In 1950 Oak Ridge acquired the Y-12 facilities and Livingston's group joined the lab as the Electronuclear Division. While the lab's Physics Division ran the high-voltage Van de Graaffs, the Electronuclear Division took the circular route and pursued cyclotrons. The inherited program included an 86-inch, fixed-frequency cyclotron that came on-line in November 1950 and provided a more intense beam than the higher-energy synchrocyclotrons. The 86-inch cyclotron produced polonium 208, an isotope of interest for radiological warfare, and also performed radiation damage studies for the Aircraft Nuclear Propulsion project.[103]

The addition of the Electronuclear group encouraged lab administrators to ask the AEC in 1950 to add accelerators to the lab's official mission. The Division of Research pointed to the preeminence of Berkeley and refused, despite the precedent of Brookhaven, to admit challengers. The next year GAC also noted the overlap with Berkeley and suggested that Oak Ridge scientists leave accelerators to their more experienced colleagues in California. Weinberg and Clarence Larson responded that the Y-12 group had long experience in high beam currents (which, they did not add, derived from Berkeley's efforts during the war), and that the program would keep together "an able group interested in this field."[104] They also had a vote of confidence from the AEC, which, a month after it discouraged official recognition of accelerator work at Oak Ridge, had asked the lab to build a machine to accelerate heavy ions—specifically, nitrogen. The hydrogen bomb program had raised the possibility that a thermonuclear reaction might ignite the nitrogen in the atmosphere. Oak Ridge built a 63-inch cyclotron and a nitrogen ion source and, in 1952, measured the cross-section of the nitrogen-nitrogen reaction. The world would not explode from the hydrogen bomb, at least not from a test of it. Thus reassured, Oak Ridge scientists turned the cyclotron to basic research.[105]

Through the legacy of the Manhattan Project, a convenient organizational rearrangement, and the requirements of the AEC, Oak Ridge found itself invested in the accelerator business. Although the lab's program aimed at lower energies than Berkeley or Brookhaven, it occupied a niche in high-intensity and heavy-ion acceleration. The lab aimed to capitalize on its experience and in 1953 proposed the construction of a 114-inch, continuous-wave cyclotron. The cyclotron would accelerate heavy ions, from beryllium to neon, to energies of 10 MeV per nucleon; the heavy projectiles could thus penetrate the potential barriers of the heaviest nuclei in the periodic table, create new radioelements, and serve as a probe of nuclear physics. The proposal meshed with the AEC's program and extended that of Oak Ridge.[106]

The Oak Ridge cyclotron faced competition from two quarters, one within the system and one without. In August 1953, around the same time as the Oak Ridge proposal, Seaborg and other Berkeley chemists suggested that the Rad Lab build a linac to accelerate heavy ions.[107] Another group of nuclear physicists at Yale University were rallying behind a similar proposal. In 1948 the AEC had joined forces with the Office of Naval Research to grant research contracts to university groups for basic research, including the construction of accelerators. The AEC continued the program after the navy withdrew from it and even after the establishment in 1950 of the National Science Foundation, whose paltry budgets precluded support of expensive equipment. Although the national labs had for the time being a monopoly on high energies, there was no guarantee the AEC would continue it nor any indication of where the high-energy region began. The Yale proposal thus enjoyed some precedent in the AEC program.[108]

All of the proposals aimed at the same energy of 10 MeV per nucleon, but their means and ends differed. All of the machines banked on the design experience of the particular lab: Oak Ridge planned to scale up its 63-inch and 86-inch cyclotrons; Berkeley's linac would use a single-cavity drift tube similar to the one on its existing proton linac; Yale designed a two-stage linac, the first stage with a single-cavity drift tube and the second a multicavity section developed by their scientists. The linacs at Berkeley and Yale would each cost $1.2 million; Oak Ridge estimated its cyclotron to cost $2 million. Each machine advanced the interests of a particular group. Livingston and his Electronuclear Division pushed the Oak Ridge cyclotron. The impetus for the Berkeley linac came not from accelerator builders but from Seaborg and the chemists. The Yale machine aimed more broadly at nuclear physics. Each group also offered its own justification. Oak Ridge, the center of isotope production for the AEC, proposed to use its cyclotron to produce radioisotopes throughout the periodic table. Yale cited the desirability of training graduate students. The Berkeley machine would give Seaborg and his group, who had discovered many of the elements beyond uranium in the periodic table, the means to make elements beyond number 98 (named californium, just after 97, berkelium).[109]

The GAC considered the merits of the proposals at its meeting in November 1953. The committee agreed that at least one of the designs deserved to be built, but that "three heavy particle accelerators might be unwarranted duplication." Both Oak Ridge and Berkeley "already have a great abundance of nuclear machines"; Yale, meanwhile, had none, and would contribute university funds. Yale and Berkeley had agreed to pool their engineering facilities, giving Yale the benefit of Berkeley's experience. Fisk, now a member of the GAC after retiring from the Division of Research, stayed true to form and asked why the AEC should build yet another cyclotron. But the GAC agreed with Rabi "to get Yale back into nuclear physics," and the AEC gave Yale the go-ahead.[110]

In January 1954 Seaborg again pitched the Berkeley proposal to the GAC and AEC with a new justification: in November it was to produce transcalifornium elements; now it was nuclear physics and chemistry. Chemists at Berkeley, Argonne, and Los Alamos had found evidence for elements 99 and 100 in December 1952 in radioactive debris from the Ivy-Mike test of the first thermonuclear device. Thermonuclear bombs were not the most convenient way to produce the new elements; in late 1953 Berkeley chemists managed to produce element 99 in the lab's 60-inch cyclotron, and both Berkeley and Argonne were irradiating samples in the Materials Testing Reactor and would soon find element 100. Hence Seaborg, at the January meeting, emphasized not the production of new elements but instead general knowledge of nuclear structure and processes, closer to the Yale approach; he may also have wanted to divert attention from an ugly priority dispute, conducted in secret, then dividing Berkeley and Argonne over the discoveries.[111]

Seaborg sold his former colleagues on the GAC and the AEC on the merits of the Berkeley machine. The Berkeley accelerator closely resembled the Yale design, despite Seaborg's claim that "the machine would be unique, not like any other now at Berkeley or elsewhere"; but Thomas H. Johnson, the director of research, noted that Berkeley, unlike Yale, intended to accelerate ions heavier than neon. The two groups would pool their resources in the design of the machines and hence save money; the AEC could obtain linacs at both Berkeley and Yale for a little more than the cost of the Oak Ridge cyclotron alone. The Oak Ridge plan suffered also from the perception that its main supporters were its designers, not its users. The AEC approved the Berkeley heavy-ion linac, called the HILAC, and it came into operation in 1957.[112]

Oak Ridge tried again for its cyclotron later in 1954 and in the following years. The AEC deferred Oak Ridge not because of a policy to limit the spread of accelerators but for more practical concerns over budgets and reactor workload.[113] It thus did not discourage the diffusion of bigger accelerators through the lab system. The approval of the Yale linac indicates that accelerators for heavy ions still seemed small enough, relatively, for the AEC to place them at individual universities. The labs justified smaller accelerators like Van de Graaffs on programmatic grounds; the heavy-ion accelerators could go to universities because they aimed their beams toward research, albeit in nuclear science, a field closer to the AEC's interests than high-energy particle physics (though the AEC would soon extend the support of accelerators at universities to higher energies). The programmatic labs such as Oak Ridge, soon to be joined by Argonne and Los Alamos, could meanwhile use the ambiguous status of medium-energy accelerators to gain a role as providers of research facilities—that is, as national labs. But the ambiguous status of these machines did not ensure that these facilities would be available to visitors.

The labs expanded the definition of big, expensive facilities beyond reactors and accelerators. Jerrold Zacharias of MIT suggested that Brookhaven equip some air force planes for cosmic ray research. John Wheeler at Princeton encouraged the plan: a cosmic ray group at Princeton wanted to send up some detectors, and a B-29 would fall right in line with the concept of a national lab as provider of facilities no university could handle individually. The air force obliged and by 1950 Brookhaven had use of a B-29, based at Rome, New York, for cosmic ray studies to thirty thousand feet.[114] Expensive, elaborate facilities were not restricted to the physical sciences, nor to inanimate objects. Oak Ridge set up a program to study the long-term genetic effects of radiation in mice. Lab staff estimated that the project, eventually dubbed Megamouse, would require ten thousand mouse cages and $1 million a year; their proposal of 1947 noted, "It is solely the expense that has prevented the accumulation of data on mammals."[115]

In principle, the labs intended to provide large facilities for use by visiting academic scientists. In practice, outside users did not enjoy easy access to them. Classified programmatic research, like polonium production or the Megamouse project, put some facilities off-limits. Access to research reactors, the original justification for big facilities, was limited to cleared scientists for several years under the AEC's policy of classifying reactor information. Although high-energy accelerator research was unclassified, the machines might lie, as at Berkeley, in restricted areas. Besides classification, visitors had to contend with the lab's own scientists for access. In the debate in early 1948 over the location of high-energy accelerators at Berkeley and Brookhaven, a strike against the former was its lack of access for visitors: "Perhaps," posited Fermi, "some method of making Berkeley more of a national laboratory would ease the feeling with respect to use of government funds for this accelerator."[116] When the Bevatron came on-line, Bureau of the Budget staff expressed surprise that 25 percent of its beam time went to "foreigners." Luis Alvarez explained that at Berkeley "the word 'foreigner' in this case means anyone not employed at the Laboratory."[117] Even Brookhaven's commitment to the national lab ideal proved paltry in practice. Hence Hafstad, head of reactor development for the AEC, "questioned the accuracy of the term 'national' laboratory." And "'regional' would be an equally incorrect description." Instead, Hafstad said, the AEC should banish the adjective and, for instance, call it "Oak Ridge Laboratory."[118]

Despite their indifference to the national lab concept, lab scientists and their overseers in the AEC would increasingly recognize its importance and implications, and would appeal to "nationalness" in various ways. As Hafstad recognized, "national" meant regional, in that the labs would make their facilities available to nearby academic scientists. But regional labs implied duplication of facilities: the AEC would have to supply an accelerator and reactor to each lab for that region's scientists. The AEC, however, never enjoyed budgets that might have supported such a duplication of equipment, nor did all of the labs have the personnel to build it. The shortage of resources forced the labs to compete and collaborate—to compete for the money for machines while sharing their scientific and engineering talent to build them. The combination is evident in the early encounter on high energies between Berkeley and Brookhaven and in the later one over heavy ions between Berkeley and Yale.

Both of these episodes illustrate another consequence of the avoidance of duplication among the labs: specialization. Even though the AEC might allow only a little duplication, the duplicated machines differed. The differentiation of facilities could come about at the insistence of the AEC and its advisers, as in the Berkeley-Brookhaven deal, or at the initiative of lab scientists, as in the heavy-ion accelerator proposals. The latter case still reflects the individuality of the labs: the different programs and traditions at Berkeley and Oak Ridge produced different designs for different purposes. Specialization extended to the other types of big facilities. Although

Brookhaven's first research reactor copied the design of the Oak Ridge graphite reactor, the lab's planners intended it only as a cheap stopgap (they were wrong on both counts) on the way to a higher-flux reactor "of unique design."[119] Weinberg and Zinn's proposed settlement to the reactor centralization debacle, which recommended a research reactor for both Argonne and Oak Ridge, eventually produced different parameters for the two approved designs.

The Cold War context and national security concerns affected even so esoteric a field as high-energy accelerators, and to boost the budget the AEC and its labs could appeal to programmatic uses. But military applications, whether realized or potential, brought restrictions on access and stymied the purpose of provision of equipment. Military programs also involved other agencies and thus illustrate the permeability of the boundaries of the lab system. The Department of Defense not only interacted with the labs in the weapons and reactor propulsion programs but also intersected the accelerator field, first through the Office of Naval Research and then the countermeasures program. Institutional relationships extended to industry, as with the nascent commerce in accelerators, and universities, evident in Argonne's early avoidance of the University of Chicago's accelerator plans or the appearance of Yale in the heavy-ion derby.

Accelerators provide an early example of another characteristic of the lab system: diversification. The AEC agreed to support work on the margins of its mission; these paste-ons then grew to become a major component of its program. By the late 1950s high-energy physics would dominate the research budget, at the expense of nuclear physics and to the alarm of program managers. The AEC at first resisted the addition, then tried to limit it to just one or two labs; but the presence of lower-energy accelerators for nuclear physics and chemistry provided a path for the diffusion of more elaborate machines throughout the system. Again, the path could be cleared by both the activism of lab scientists and the efforts, with unintended consequences, of the AEC. A good example was the AEC's request to Oak Ridge (in an example of the job shop approach) to build a small cyclotron to accelerate nitrogen ions, as an adjunct to the weapons program; lab scientists then used the machine as a springboard to a larger heavy-ion cyclotron.

In the case of research facilities, diffusion did not counteract an official policy of centralization as it did in the weapon and reactor programs. The diffusion of programs, driven by national needs and entrepreneurial scientists, brought the labs closer to the original ideal of regional facilities. Specialization, effected also by national priorities (imposed through the budget) and shrewd scientists, took the labs away from the regional ideal and closer to a national one, in which each lab would have a facility unique within the United States. But to lab scientists and the AEC, the concept of "national" labs remained, at this point, not national but regional. The definition still entailed access for visiting researchers to the facilities of the labs, a point increasingly recognized by the labs. But most of the users of lab facilities were not visitors but in-house scientists pursuing their own programs in basic research. What were these scientists doing, and why were the labs paying them to do it?

SMALL SCIENCE

The AEC did not have to support basic research. Its initial emphasis on weapons production put research on the back burner. Accumulating proposals from scientists in the labs and in universities and the advocacy of the GAC quickly forced the attention of the AEC to its research policy. The immediate justification offered by the GAC was the perceived shortage of scientists: support of basic research by the AEC would train new scientists, some of whom might then staff AEC programs. The AEC found another reason in the notion that current atomic bombs and reactors "had their beginnings in the fundamental studies of many years ago"; "The research program is our first line of defense for 10 or 20 years hence." Research would "provide a fund of new theoretical and experimental knowledge for later adaptation to practical use"; or, to modify the metaphor, against the oncoming winter of the Cold War the AEC had to add "to our store of basic knowledge."[120]

Neither of these ends necessitated research by scientists within the national labs. On the contrary: the GAC, despite its membership of current or former lab representatives, opposed the support of basic research in the labs on both grounds. For the first, the committee noted that the support of resident research staff in the labs would deplete university faculties and hence stunt the supply of new scientists. As DuBridge argued, "Trying to create ten Berkeley radiation laboratories around the country . . . would have a great devastating effect on our universities and the Atomic Energy Commission activities." Supporting university scientists and their students "will eventually create more people rather than lose people." The committee considered the dilemma of staffing the labs for programmatic work at the expense of training future scientists and agreed on the need for balance between short-term and long-term considerations. But basic research at the labs did not enhance either side of the equation. Oppenheimer suggested instead a sharper distinction between academic research and the "special kind of semi-industrial laboratory which is not academic, should not be academic, and which will derive part of its people from this training ground." The latter category appears to have included the national labs.[121]

Oppenheimer's distinction also ruled out the labs as a source for the storehouse of basic knowledge. It stemmed from the GAC's support of centralization, which it had stated at the same meeting. DuBridge made the connection:

> For doing a job, a programmatic applied job, of getting industrial reactors, let's say, or making radar sets, or atomic bombs, a large central coordinated facility is ideal and is the most effective and efficient way to do a specific job. But if you want to learn new things, you want to increase your knowledge [through] basic research, then this is not the best way of doing it. . . . [B]y and large new knowledge comes best from a large group of people working individually, alone and free with their own facilities and groups. Therefore, we simultaneously suggest that a central laboratory is a good thing for the main, applied, programmatic job of the Commission. But also, that the best thing for increasing knowledge in basic physics is a wide distribution of a small group of people working in universities or around the country.

Centralization applied only to reactor work, and DuBridge's statement could have implied the support of research groups in the regional labs. The committee did seem to leave that door open when it allowed that Berkeley, because of its "special history," could continue its basic research program. But the GAC refused Rabi's plea for a similar dispensation for Brookhaven, and Oppenheimer shut the door: "It would surely not be something we would want to write down, but if we said we don't support [basic research] except at universities it would come very close to what we mean."[122]

The GAC continued to oppose basic research in the labs as it lobbied for it in universities, but its position eroded over time. Later in 1947 it reiterated that its support for academic research did not imply "a large increase in the research facilities within the Commission's regional laboratories," but rather "limiting the scope of the regional laboratories to work for which Commission-owned facilities are unique and necessary to the healthy growth of the basic sciences." The committee did not define these terms, however, and it further recognized Brookhaven, in addition now to Berkeley, as a special case.[123] The AEC also failed to discourage the labs: in its decision to centralize reactors, it softened the blow at Clinton by assuring the lab of "a vigorous program of basic research." The AEC added that "training of personnel is a necessary adjunct" to basic research.[124] The "Clinton effect," which persuaded the GAC to support accelerators at both Berkeley and Brookhaven in 1948, also brought the AEC to demonstrate support of basic research to sustain morale at the labs.

The persistent agitation of the labs for basic research helped to overcome opposition to it in the AEC and GAC. All of them included basic research in their programs from the outset. Their justifications for it differed. Argonne, Los Alamos, and Oak Ridge claimed that they needed basic research "to attract and hold research personnel of the highest qualifications," since "first-class men will not remain on the laboratory staff unless they have reasonable freedom to carry on research in fields of their own choosing."[125] The labs here presented an early variant of the scientists-on-tap argument: basic research would keep qualified scientists on the staff, who might then be tapped for programmatic work as necessary. AEC staff came to agree: "Maintenance of quality in the national laboratories requires supporting a reasonable number of desirable basic research projects in them, even though more can usually be accomplished per dollar expended at universities than at the national laboratories."[126] Hence, for example, the Physics Division at Los Alamos, according to one of its members, "traditionally served as a source of people (often highly placed people) for other divisions of the laboratory and as a reservoir of uncommitted people who are ever-willing to interrupt their researches in order to contribute to phases of the programmatic activities of the laboratory."[127]

Berkeley and Brookhaven would demonstrate the concept in their assumption of programmatic work, but at the outset they relied on their contributions to training and the storehouse of knowledge. The AEC learned to play up the training angle for all of the labs: it was "an integral part" of Argonne's research, a "major purpose" at Brookhaven, and a growing function at Oak Ridge.[128] It could also point to impressive research results: the production of mesons at Berkeley; transuranic and heavy-element chemistry at Berkeley, Argonne, and Oak Ridge; neutron physics at Oak Ridge and Argonne; the development of the Monte Carlo calculational method at Los Alamos; pioneering work on helium 3 at Argonne and Los Alamos.[129]

Much of this research utilized unique capabilities or was related to the applied program, but the labs tested the boundaries of acceptable research programs. Brookhaven early proposed to create a Psychology and Sociology Department, which might capitalize on the growing interest in the social sciences after the war. The Atomic Energy Act had included a provision for "study of the social, political, and economic effects of the utilization of atomic energy," and the GAC at its first meeting considered the establishment of a counterpart advisory committee for the social sciences. Brookhaven did not implement the plan, however, and the labs and the AEC stuck to the natural sciences.[130]

The appeals to training were and would remain mostly lip service, as resident scientists far outnumbered visitors and students and as tenure policies kept a static, aging staff in the labs. Nevertheless, the GAC came to support research in the labs, but gradually and grudgingly. When Congress cut the AEC's research budget in 1949, the GAC recommended that the labs bear the burden: their size made them better able than universities to absorb the blow, and their large overhead made them twice as costly per scientist as research contracts in the universities. A reduction might impose some efficiency.[131] Two years later the labs had developed satisfactory staff, morale, and programs and the committee changed its tune: "Within the Commission's laboratories basic research is often not adequate in scope nor its contribution adequately recognized." By 1952 the GAC reversed its earlier stance in the prospect of cuts in the research budget, and Johnson in the Research Division agreed to distribute a larger reduction to off-site contract research.[132]

The GAC may have come to support research in the labs out of realism: the AEC had demonstrated that the labs would form the backbone of its basic research effort. In fiscal 1949 only 15 percent of the physical research budget went to off-site contract research. The percentage had doubled by fiscal 1951, and it would hover between one-fourth and one-third of the total after that.[133] Despite the early advice of the GAC, most of the physical research of the AEC was done in the labs. Not all of this work classified as "basic" research: the Research Division budget, as we know, was broken down into categories for "basic" and "applied" until this distinction proved too problematic, and the applied part included work relevant to weapons and reactors. Nor was basic research the sole province of the Division of Research. The divisions of Biology and Medicine, Reactor Development, and Military Application also sponsored research toward the basic end of the spectrum, although it was, and is, difficult to distinguish between basic and applied work in many areas. For instance, Oak Ridge pursued radiation metallurgy under both the divisions of Research and Reactor Development, which it attempted to allocate based on a vague distinction between short- and long-term problems.[134] Los Alamos received no support through the early 1950s from the Division of Research: "All of the research work at Los Alamos is under the Military Application Division."[135]

To take an example of how programmatic work could foster basic research under a programmatic AEC division, consider the detection of the neutrino by two Los Alamos physicists. In the summer of 1951 Frederick Reines asked J. Carson Mark, the head of the theoretical division, for a break from the H-bomb work in order "to do some fundamental physics." At the time the crash program for the H-bomb had been under way for a year and a half and now had an end in sight. That spring the Teller-Ulam design, followed by the Greenhouse test series, in which Reines participated, had shown the feasibility of a fusion weapon. Mark granted Reines the time to turn to basic physics, with no particular problem in mind. Reines soon thought of one: the use of a nuclear explosion as a source of neutrinos, which had not been detected since their theoretical postulation by Wolfgang Pauli in 1930. Reines obtained the support of Clyde Cowan of Los Alamos and designed an experiment to place a detector underground fifty meters from a nuclear test. Reines pointed out, "The idea that such a sensitive detector could be operated in the close proximity . . . of the most violent explosion produced by man was somewhat bizarre, but we had worked with bombs." Bradbury approved the experiment and the Division of Military Application funded it. Reines and Cowan subsequently decided they could get by with the neutron flux from a fission reactor and set up detectors first at Hanford, then at the more powerful Savannah River reactor, which produced the tritium fuel for hydrogen bombs. In 1956 they isolated the signal from the passage of a neutrino through their detector, a discovery for which Reines received the Nobel Prize in 1995. Los Alamos maintained the neutrino research for the remainder of the decade in groups under Reines in the Theoretical Division and Cowan in the Physics Division. The episode demonstrates how the labs rewarded talented staff with time for basic research, which often used the unique capabilities of the labs and could grow to become large, long-lasting programs. The policy kept scientists on tap for programmatic work, and important research results redounded to the reputation of the lab.[136]

Research programs outside the Division of Research only increased the fraction of AEC research carried out in the labs. Although some developmental work in reactors and weapons went to other AEC facilities, such as Sandia or Knolls, and to industrial contractors, little went to universities. The national emergency of the early 1950s shifted the emphasis in the labs toward programmatic work; but often not, lab scientists insisted, at the expense of basic research, and instead applied research came as additions to lab programs.[137] In 1950, at the outset of the national emergency, the AEC spent $31 million in operating expenses for physical research but another $50 million on research under the reactor and biomedical programs (both in the labs and off-site; a majority went to the labs).[138] One may assume that the weapons program provided even more support of research, even if these figures were but a fraction of its overall budget. If research formed only a small part of the general AEC budget, much of which went instead to weapons production, it was not insignificant—$80 million was not a trifling sum—and a larger fraction than would appear from a glance only at the physical research budget.[139]

The size of the AEC's research budget dwarfed that of the National Science Foundation (NSF). The act creating the NSF in 1950 limited its annual budget to $15 million, half that of the AEC's Division of Research. In practice NSF got far less: its first appropriation was for $350,000, which would barely buy a small betatron, let alone the scientists to use it.[140] Although the NSF was intended to serve as the main federal sponsor of basic research, the AEC and other agencies continued to provide for most of it. Congressional watchdogs growled at the presence of basic research in the AEC's program: "One of the most forceful and telling arguments in Congress when the basic law was passed creating this [National Science] Foundation was that you were going to be a great force to prevent duplication in research effort. . . . [Yet] you seem to be doing the same thing they [the AEC] are doing." AEC commissioner Smyth pointed to the penury of the NSF: "The National Science Foundation in the past has . . . been able to support so little actual research—that while we are in touch with them, we do not run into this question of overlap."[141] The same year, 1953, the AEC likewise repelled an attempt by the Bureau of the Budget to assign basic research to the NSF.[142]

Although the AEC defended its prerogative to direct basic research, the commission did not itself attempt to exercise this responsibility for some parts of its program. The labs broke down their operating expenses for the research budget into projects under one or two individuals, which received short descriptions on the standardized forms in the annual lab budget. The annual program review by the local operations office and program managers in Washington would go over these project descriptions, but that level of detail was lost in the incorporation of lab budgets into divisional budgets in the AEC. Nor would budget examiners outside the AEC have desired to deal with them: if they quailed at the four-inch-thick overall AEC budget, they would have despaired at the detailed budgets of the labs, each of which was almost as long.

Basic research programs enjoyed more independence from budget scrutiny than applied projects. Pitzer, as director of the Research Division, required only one project description form for each "basic" activity, such as physics, chemistry, or metallurgy, thus allowing the labs to lump several research projects under one broad heading. Applied projects, however, required specific, detailed forms.[143] The leeway varied from lab to lab, with Berkeley and Brookhaven enjoying more than the others. A visitor to Brookhaven, "slightly skeptical of National Laboratories" after a stint at Oak Ridge and "wondering to what extent problems of particular interest to the laboratory would be 'suggested' to me," was "delighted by the entirely free atmosphere I found there, with everyone working on problems of his own choice."[144] Oak Ridge suffered from its larger portion of work for the meddlesome Division of Reactor Development, but it too could cut basic research more slack; the AEC "recognized that in fundamental research the selection of the subject of investigation (within reasonable limits) and of the methods of attack are functions of the Laboratory."[145] Hence Thomas H. Johnson, head of the physics department at Brookhaven and soon to become director of the Division of Research, could claim:

> In a laboratory supported by a government agency which has definite responsibilities to the taxpayer, one might think that physicists would have their work planned by some central authority in an attempt to slant it towards the advancement of the general program. At Brookhaven, that is not the case. I would not say that none of the work is influenced by central planning, nor would I say that we dislike all central planning, especially if it is competently done, but surely most of our research has been undertaken on the responsibility and initiative of the individuals doing the work.[146]

Some of this research flexibility derived from the scientists-on-tap model. The AEC attempted to follow the philosophy long espoused by philanthropies and expressed by Oppenheimer at an early GAC meeting: support not particular research projects but talented scientists.[147] At Oak Ridge the AEC agreed to judge "the budget proposed for fundamental research in a given field primarily in terms of the number of scientists (individually named) of established productivity to be engaged in the work. The intended topics of investigation will be secondary evidence."[148] In the case of Los Alamos, the AEC "committed itself to the long-term support of a cadre of selected senior scientists capable of serving either as project leaders in developmental work or as individual research workers."[149] The approach conflicted with the desires of federal budget examiners to whittle down research programs to specific projects. The AEC could succeed in it to some degree because the individual programs in operating budgets for research disappeared in divisional budgets, which appeared as line items in the budget presented to the Budget Bureau and Congress.

Small science thus flew under the budget radar, which gave it room to maneuver. The GAC and AEC were surprised to learn in early 1954 that Livermore devoted 15 percent of its program, sponsored by the Division of Military Application, to fundamental research, which included work in nuclear physics, radiochemistry, magnetohydrodynamics, cryogeny, and computing. "Was it justified," asked commissioner Smyth, "to devote 15% of this weapon laboratory's budget to activities classed as basic research?"[150] The AEC could justify basic research at the national labs for both of the main purposes of the labs: to keep scientists on tap who might work on applications of atomic energy, and to support the facilities provided by the labs to visiting scientists. In addition, although the labs might divert scientists from universities, basic research would train new scientists in the art of nuclear science. The presence at the labs of low-flying basic research would help them to diversify in the future.

5

FALSE SPRING, 1954–1962

NATIONALISM AND INTERNATIONALISM

Through the early 1950s the two superpowers assembled their atomic arsenals, which soon included thermonuclear weapons, and settled into a standoff in the Cold War. After the hot war in Korea and the death of Stalin, the battleground in the Cold War shifted to include the social sphere, exemplified by Nixon's kitchen debate with Khrushchev in 1959. Victory in the Cold War might come not in a quick military action but in a long sociopolitical struggle. Khrushchev in February 1956 asserted that "there is no fatal inevitability of wars"; instead, "our faith in the victory of Communism is based on the fact that the socialist way of production has decisive advantages over the capitalist." Two years later Eisenhower, in his State of the Union address in the wake of Sputnik, perceived the "all-inclusiveness" of the current crisis: "The Soviets are, in short, waging total Cold War."[1]

Eisenhower named science as a contested field. Cold War competition was not new to science and technology in general nor to the national labs in particular; international scientific competition stemmed not only from the immediate contributions of the labs to national security but also from the sense that research formed the basis for future military applications. The reaction to Sputnik would highlight the importance attached to science in the new phase of the Cold War, and the shift from the relevance of programs for military applications to the status of programs themselves relative to those of international rivals. Scientific prestige and its propaganda value now provided the goal of competition and justified support of research. But atomic energy scientists and policy makers perceived the shift long before Sputnik brought it to the attention of the American public.[2]

A series of Soviet technical achievements—from a thermonuclear weapon test in August 1953, a few months after the Ivy-Mike shot, to the revelation in 1955 of a high-energy accelerator under construction at Dubna that would outstrip the energies of American accelerators—convinced the AEC and its scientists and overseers that the Soviets were nipping at their heels. Three lab programs became bellwethers of American scientific standing: nuclear power reactors, high-energy accelerators, and fusion. As such, all three saw their budgets rise through the mid- to late 1950s: research on civilian power reactors far surpassed efforts on military reactors at the labs; high-energy physics and fusion would each consume about a quarter of the operating budget of the Division of Research, most of which went to the labs, by the end of the decade.

The constitution of lab programs as instruments of national foreign policy clashed with the ideals of an international scientific community. The internationalism espoused by some scientists therefore acquired added importance in the post-Stalinist era of peaceful coexistence: cooperation among scientists might provide a model for cooperation among nations. The same three lab programs exemplify the internationalist ideal that coexisted uneasily with nationalist policies. All of them, but especially power reactors and fusion, benefited from Eisenhower's Atoms for Peace plan, which attempted to turn the ideal into official policy. The three fields found a forum in the International Conferences on the Peaceful Uses of Atomic Energy, organized in Geneva in 1955 and 1958 as a consequence of Atoms for Peace. The Geneva conferences combined the rhetoric of international cooperation with the reality of Cold War competition.

Atoms for Peace and Power

On 8 December 1953 Eisenhower captivated the General Assembly of the United Nations with his proposal that nuclear nations contribute uranium and fissionable materials to an International Atomic Energy Agency, which would utilize the material to develop peaceful uses such as nuclear power. Eisenhower's plan for Atoms for Peace had elements of both competition and cooperation. Its high-minded appeal to cooperation reflected Eisenhower's genuine belief that diversion of fissionable material from weapons to power reactors would lead the world toward peace instead of war. At the same time, he and his advisers recognized the plan's propaganda benefits and the relative advantage to the United States, which possessed at the time a larger stockpile than the Soviet Union.[3]

The emphasis on the redeeming features of nuclear energy was not new and had figured in many public pronouncements since Hiroshima promoting the peaceful side of the atom, to the point where the GAC sought to correct overly optimistic predictions of nonmilitary applications. David Lilienthal in particular played up potential peaceful uses, and anticipated Eisenhower, during his tenure at the AEC: "My theme of Atoms for Peace is just what the country needs." The rosy rhetoric did not match official policy, which emphasized weapons at the expense of nuclear power and focused reactor research on military propulsion; breeder reactors for civilian atomic power were last on the AEC's list of priorities.[4]

The national emergency of the early 1950s reinforced the military orientation of the AEC but also attracted attention to the contest for peaceful uses. Lilienthal warned in 1950 that the Soviets might "beat us at developing the peaceful side of the atom." The congressional Joint Committee on Atomic Energy resolved the next year that "never again" shall it "be truthfully said that the reactor of the most advanced design and performance operates anywhere but in the United States." The AEC responded to the growing chorus by allowing industrial access to classified information on reactor technology, in the hopes of inciting some interest; the response was underwhelming.[5]

The competition was real, and it was not confined to the Soviet Union; both Canada and Great Britain had experimental reactor programs under way. The chorus rose in a crescendo: the United States must "consider entering what we [the Joint Committee] choose to term the atomic power race"; "It would be a major setback to the position of this country in the world to allow its present leadership in nuclear power development to pass out of its hands." The impetus for atomic power was political, not economic; the utility industry stuck with much cheaper and still plentiful coal. Nevertheless, Congress revised the Atomic Energy Act in 1954 to encourage industrial development of nuclear power. To demonstrate the superiority of democratic capitalism, the AEC adopted a five-year plan for power reactors.[6] Pressure to develop nuclear power would increase after the Suez crisis of 1956, which threatened the stable supply of oil from the Middle East; in the midst of the crisis, Eisenhower's emissary to Egypt tried to use the potential of nuclear power to undermine Egypt's position.[7]

The interest in nuclear power also fostered Eisenhower's Atoms for Peace proposal. In the eyes of lab directors, Atoms for Peace would counteract the push for industrial development, to the advantage of the laboratories. Weinberg, with Bradbury's concurrence, argued that "if the development of nuclear energy . . . is to be an instrument of national policy, then there is no alternative to government responsibility." In other words, the labs should retain the initiative for reactor development instead of relinquishing it to industry.[8] The AEC obliged: most of the five-year plan, intended to create a role for industry, focused instead on reactors under design at Argonne and Oak Ridge; and budgets for power reactor work increased at most of the labs throughout the decade, more or less doubling from 1956 to 1958 (table 2). Some of the additional work derived directly from international competition: after the British built several gas-cooled reactors in the early 1950s, the congressional Joint Committee asked the AEC in 1956 to design one too. The AEC turned to Oak Ridge, whose work on a gas-cooled design it had canceled amid the uncertainty at Clinton in late 1947; by 1960 the revived program had grown to represent almost a third of the lab's reactor work. Oak Ridge would spend a decade on the gas-cooled project until the AEC again pulled the plug on it in 1966.[9]

The labs would play a more immediate role in Atoms for Peace as cooperative institutions. After Eisenhower's original plan to pool material bogged down in Soviet resistance and diplomatic maneuvering, Atoms for Peace came to consist of bilateral agreements with individual countries, under which the United States helped to establish research and power reactors abroad. By mid-1955 the United States had negotiated research bilaterals with two dozen nations.[10] These countries turned to the national labs as sources of expertise for export. Brookhaven expected to play a large role in Atoms for Peace, particularly as a training ground for foreign scientists, owing to perceptions that "much of what Mr. Eisenhower has proposed is already being done at Brookhaven." Perceptions did not always match reality: foreign visitors grumbled that "it was not always so easy to

Table 2. Reactor development program costs by lab, FY1956–60

Year	ANL	BNL	LASL	ORNL	Total
1956	11.8	2.5	1.5	10.6	28.1
1957	14.4	3.7	2.2	14.7	38.6
1958	17.9	5.1	3.0	20.5	49.8
1959	19.6	4.4	3.2	24.9	59.6
1960	22.3	3.4	5.5	24.6	70.6

Source: Material for lab directors' meeting, May 1966 (GM, 5625/15).

Note: Excluding reactors for missile, aircraft, and naval propulsion; in millions of current dollars. Berkeley and Livermore had no civilian reactor development work in this period.

work" at Brookhaven, especially since its research reactor remained a classified facility. In 1955 the AEC declassified the reactor and allowed the lab to become "a true scientific Mecca."[11] Brookhaven sponsored courses in radiation physics and biomedicine for foreign students, hosted visiting scientists, and sent its own scientists abroad, often bearing blueprints. Exchange programs developed with nuclear research labs in Britain, France, and Scandinavia and extended even beyond the iron curtain: Brookhaven and the Joint Institute for Nuclear Research in Moscow exchanged reprints and annual reports, and Brookhaven entertained visitors from eastern bloc countries.[12] Similarly, Argonne, at the request of the AEC, established an International School of Nuclear Science and Engineering. The first class of students, thirty-one from abroad and ten from the United States, entered in March 1955 and spent seven months studying reactor physics and engineering, including a visit to Oak Ridge; by 1959 the school had trained 325 foreign students from over forty countries.[13]

The labs served not only as a source of expertise but also as models. Several Brookhaven scientists and one from Oak Ridge took a ten-week field trip to Asia in 1956 as part of the Colombo plan, a State Department program under which a coalition of Asian nations aimed to develop a nuclear research center. Not surprisingly, the lab representatives recommended the establishment of a regional center modeled on Brookhaven, with cooperating nations taking the place of AUI's universities.[14] In turn, international labs could serve as models for the development of the national labs. The President's Science Advisory Committee in 1961 discussed the possibility, suggested by Rabi, of converting Brookhaven to "an Inter-American laboratory" modeled on CERN for countries in the Western Hemisphere.[15] CERN, an accelerator lab established in Switzerland in 1954 by a dozen European nations, itself derived impetus from a suggestion by Rabi and took Brookhaven as a model of collaborative laboratory administration.[16]

Commitment to internationalism was not consistent throughout the system. The AEC would not have expected Los Alamos or Livermore to open its doors to foreign scientists. The other labs had to limit their accessibility to foreigners because of AEC policies, which required adjustments to accommodate the presidential pronouncement of Atoms for Peace. The AEC permitted each lab to employ a limited number of alien scientists at any one time for unclassified research: Brookhaven's quota was six, Berkeley's three. It was a matter not of security but of money: the AEC would clear any number of visitors but would pay for only a few. The AEC also required a statement from the lab testifying to the unique ability of each visitor, to ensure that a foreigner would not take a job away from a U.S. citizen.[17] The old problem of lack of housing also limited the number of visitors the labs could accommodate. Last, although lab scientists were willing to travel to foreign labs and conferences to share their knowledge, the AEC was sometimes reluctant to pay for it.[18]

For two conferences, however, the AEC spared no expense. The International Conferences on the Peaceful Uses of Atomic Energy, under the auspices of the United Nations, gave scientists a chance to discuss their programs with international colleagues and gave nations an opportunity to maneuver for Cold War prestige and propaganda. To the first end, scientists from all of the labs presented papers on diverse aspects of nuclear energy at the first conference to nearly two thousand delegates representing seventy-two nations. To the second, the AEC made the centerpiece of the American exhibit in 1955 a swimming pool reactor designed by Oak Ridge and powered up by Eisenhower during his visit to the conference. The Oak Ridge exhibit demonstrated the fruits that the United States was willing to share in a spirit of international cooperation, but also signaled its scientific standing in the international race for nuclear power. International competition and cooperation did not remove domestic rivalries: Zinn at Argonne felt that the "notoriety" of the Oak Ridge reactor overshadowed Argonne's technical presentations. But the exhibit highlights the situation of the national labs within not only a domestic political context but also an international one. Hence Weinberg, basking in the figurative and literal glow of the Oak Ridge reactor: "1955 will be remembered as the year nuclear energy went international."[19]

Accelerators

The first Geneva conference gave American high-energy physicists the opportunity to hobnob with international colleagues. They thus learned of the Soviet accelerator under construction at Dubna, which threatened to outpace the next generation of American accelerators. Like nuclear power, accelerators provided a surrogate arena in the mid- to late 1950s for Cold War competition, to the benefit of American accelerator builders. But the esoteric field of high-energy physics and its self-proclaimed fundamental character supported internationalist ideals, especially after hopes for military uses of accelerators, in either weapons production or defense, faded in the early 1950s. The potential for peaceful cooperation on accelerators would find expression in the late 1950s.

As with nuclear energy, international competition in accelerators was not new, nor was it limited to the Soviet Union. In England the University of Birmingham's plans for a 1.3-BeV machine in 1948 goaded Berkeley and Brookhaven to design for higher energies. Accelerator builders were already learning that references to an international race helped elicit funds for new machines. "In most discussions concern has been expressed that America is lagging the British and possibly the Russians in this promising new approach to an understanding of the nucleus." The AEC responded to these concerns with two machines, both bigger than Birmingham's.[20] After the approval of the Cosmotron and Bevatron put the United States far in the lead in accelerator energies, appeals to international competition for high energies died out. Brookhaven designers planning possible next steps in early 1952 were not even sure that higher energies were desirable scientifically; higher intensities might produce more interesting physics.[21] Furthermore, the route to higher energies appeared blocked by prohibitive magnet sizes, vacuum volumes, and costs, all of which presented a daunting obstacle even to ambitious accelerator builders.

The discovery of strong focusing in 1952 (or rediscovery, given Christofilos's previous patent on the principle), which avoided the need for large magnets, opened up the accelerator field. It also provided an example of the benefits of international cooperation. The discovery stemmed from correspondence between scientists at Brookhaven and CERN, itself a concrete implementation of international cooperation and a consequence of accelerator costs beyond the means of a single nation. In 1952 CERN was planning a 10-BeV proton synchrotron and asked Brookhaven scientists for their opinion on the design. While contemplating the machine, M. Stanley Livingston, Ernest Courant, and Hartland Snyder conceived the idea of strong focusing, which used an alternating gradient around the circular path of a particle to compress the cross-section of the beam and thus freed synchrotrons from the constraints of huge magnets and vacuum volumes.[22]

By virtue of the discovery, and the tacit settlement of 1948 that gave Berkeley the Bevatron, the AEC awarded Brookhaven the right to build the next generation of high-energy accelerator. CERN also capitalized on the discovery and revised its design to aim for 30 BeV. Brookhaven, however, determined the energy of its machine not through competition with colleagues abroad nor from scientific criteria but from fiscal considerations: "The proposal will be to achieve the highest possible energy for a fixed dollar amount rather than making the energy the fixed point and leaving the cost uncertain."[23] Optimistic early plans for 75 to 125 BeV or higher at $25 million soon dropped to 25 BeV for the same cost. Although AEC staff noted CERN's plans, Brookhaven resolved to set its machine's size "independently of what CERN might do" and settled for slightly lower energy.[24]

If Brookhaven scientists did not perceive an international competition in accelerators, in the wake of Atoms for Peace and amid changing international contexts the AEC paid more attention to the possibility. The AEC approved the Brookhaven plan and hoped it would go to 35 BeV, under the assumption that "American scientists have held the lead in nuclear science since the invention of the cyclotron and they do not now wish to fall behind."[25] Western European colleagues at CERN, however, could provide no substitute for Cold War competition. That challenge issued from the Soviets at the Geneva conference in 1955, where they unveiled plans for a 10-BeV proton synchrotron, based on the older weak-focusing principles, which was already under construction and scheduled for completion in 1957. The reaction in the United States suggests that Geneva represented the equivalent of Sputnik for American accelerator builders. The revelation spurred the AEC into action: it urged advancing Brookhaven's schedule "so that the date of completion will be nearer that of the Russian 10-BEV accelerator," and expansion of Berkeley's program, especially around the Bevatron, since "the dominance of the United States in the high energy field is clearly diminishing." It also quickly proposed, with the advice of the GAC, a plan to build at Argonne a 12-BeV scale-up of the Bevatron, to be completed by 1959, to regain the lead over the Soviets.[26]

In January 1953, soon after the announcement of strong focusing, Argonne had proposed with a group of midwestern physicists the construction of a high-energy accelerator at Argonne. The plan had degenerated into a squabble between Argonne administrators and university physicists unsatisfied with their access to Argonne's facilities, many of which were still classified.[27] The AEC's proposal for Argonne to beat the Soviets conveniently co-opted the coalition of university scientists; but Argonne scientists balked at merely scaling up an existing design and instead proposed a new two-stage accelerator to reach 25 BeV by 1963. To save the situation, Lawrence discussed with commissioner Willard Libby the possibility of cannibalizing the Bevatron, then only one year old and fresh from the discovery of the antiproton, in order to reach 10–12 BeV by 1958. Instead the AEC insisted that Argonne race the Soviets and demanded a design by 1 February 1956.[28] The deadline allowed the AEC to rush the proposal into congressional budget hearings, where Libby laid out the AEC's motivation: "To get down to brass tacks, the thing we saw in Geneva, which the Russians showed us, is exactly in [this] energy range. And when their machine starts the Russians are going to take over the show for several years. . . . On this accelerator the particular aim is to get the lead back from the Russians as quickly as we can."[29] The Joint Committee promptly approved the proposal.

Two more conferences on high-energy physics in early 1956 at CERN and in Moscow substantiated the Soviet challenge and provoked testimonials to its strength: the Dubna machine, "an awe-inspiring sight," "almost twice as big as the Berkeley machine," showed that "Russian physicists were equal to the best physicists of the West in this field."[30] The challenge provoked the growth of accelerator budgets. Operating expenses for high-energy physics, most of which went to the national labs, grew from $7.3 million in 1954 to $33.2 million in 1960, and as a percentage of the AEC's expanding physical research budget, from 18 percent in 1954 to 25 percent in 1960.[31] Construction costs for accelerators, including the Alternating Gradient Synchrotron at Brookhaven and the Zero Gradient Synchrotron at Argonne, rose from $1.6 million in 1955 to $26.3 million in 1959, to which was added $105 million for a linear accelerator at Stanford.[32] In December 1958 the AEC projected that high-energy physics costs would expand to $125 million by the 1963 fiscal year.[33] The trend raised eyebrows in the Bureau of the Budget by early 1956 and, the following year, led its examiners to question the "necessity for seemingly continuous program growth, particularly in [the] area of ultra-high energy physics."[34] By 1959 the GAC and AEC were worrying that high-energy physics was attracting too much money and manpower at the expense of other fields, with no end to the growth in sight; and midwestern accelerator builders were employing "a new cost unit called a pit[t]ance, defined as $100,000,000."[35]

Americans were not the only ones appealing to international competition to support larger accelerators. According to a member of the congressional Joint Committee who talked to a Soviet physicist at Dubna: "The Dubna Laboratory asked our group when we were there 2 years ago how we got the money to build our accelerators. We told them the legislative process of getting money on our program. He said, 'That is not the way I understand.' He said, 'I understand you get it by saying the Russians have a 10 million electron volt synchrotron and we need a 20 billion electron [volt] synchrotron and that is how you get your money.' I said, 'There may be something to it.' I said, 'How do you get your money?' He said, 'The same way.'"[36] CERN scientists, despite their history of cooperation with the United States, also invoked American achievements as a spur to higher energies and budgets—indeed, they owed the origins of their laboratory to the need to meet "le défi américain."[37] The feedback loop from country to country, lab to lab, ensured spiraling accelerator costs that threatened to outgrow the willingness of national governments to pay for them. In response, national governments, even those on opposite sides of the iron curtain, proposed cooperation.

The presence of American physicists and congressmen at Dubna indicates that nationalism coexisted with internationalism. High-energy physicists upheld the ideal that science respected no national boundaries—at Geneva they found that cross-section research gave "the same values on both sides of the iron curtain"—and attempted to revive an international community.[38] Conferences on high-energy physics in the early 1950s had been national affairs: the Rochester conference in January 1952 included only a single European. By 1956 the tide had turned, with international attendance at Rochester, Moscow, and CERN; the following year physicists created the High-Energy Physics Commission, with two members each from Europe, the United States, and the Soviet Union, to coordinate the international rotation of conferences.[39]

At the Rochester conference in 1956, Senator Clinton Anderson, chair of the Joint Committee, was inspired by the internationalist ideal to propose the establishment of international atomic energy labs.[40] The nascent accelerator lab at CERN provided a concrete model for such suggestions in its combination of national resources from a dozen western European nations. In 1958, in response to the growth of high-energy physics and questions over which federal agency should sponsor it, the GAC and the President's Science Advisory Committee convened a study committee on the subject. The panel, under the chairmanship of Emmanuel Piore and including Leland Haworth from Brookhaven and Edwin McMillan from Berkeley, did not do much to restrain the growth of the field but did suggest the exploration of international cooperation.[41]

Another committee assembled a year later to follow up on the Piore panel's report. The new group, consisting of Haworth, Edward Lofgren from Berkeley, and Wolfgang Panofsky from Stanford, met with Soviet and western European delegations at a conference at CERN in September 1959 to consider collaboration in high-energy physics. The meetings of the International Accelerator Study Committee forced lab scientists into diplomatic roles: they spent much of their time haggling over semantics, with the negotiations perhaps complicated by language barriers.[42] The outcome of these deliberations was included in an agreement of November 1959, the result of parallel negotiations between AEC chairman John McCone and his Soviet counterpart Vasily Emelyanov on exchanges between the United States and the Soviet Union in nuclear science. The McCone-Emelyanov agreement included the possibility of collaboration on "the design and construction of an accelerator of large and novel type."[43]

Toward that end, in the spring of 1960 a group of American physicists visited various Soviet accelerator labs, despite the recent U-2 spy plane incident, and that summer several Soviet physicists toured American labs. In September Soviet and American physicists met at the American Institute of Physics and recommended cooperative study of a 300-BeV strong-focusing synchrotron.[44] Although the Soviets sought to exclude other countries, American representatives assured CERN scientists of wider participation. Rabi and Haworth had to repeat the reassurance to CERN after another meeting between American and Soviet physicists the next year in Vienna. The setting reflected plans to put the project under the jurisdiction of the International Atomic Energy Agency in Vienna; possible sites included a new international neutral zone or in a corridor from Berlin to West Germany.[45]

European scientists need not have worried about their exclusion. American physicists were already expressing doubts about the feasibility of cooperation. Collaboration would require either American or Soviet accelerator experts to relocate to the other country, something Americans probably would not want to do. Berkeley and Brookhaven were jockeying for position for the next American machine, joined by groups in the Midwest and at Caltech. Although the discussions continued at a conference at Dubna in 1963, with CERN representatives now in attendance, prospects appeared dim. Maurice Goldhaber, Haworth's successor as director of Brookhaven, judged that "the time is not ripe for a cooperative effort with the U.S.S.R."[46]

The plans for international collaboration foundered on the association of accelerators with national prestige and security. High-energy physicists might claim that their field was "certainly the most basic and purest part of modern physics. . . . It is almost certain that no practical applications can ever be expected from this kind of research."[47] But if possible applications receded from sight in the mid-1950s, they did not disappear altogether. The Piore panel of 1958 and the President's Science Advisory Committee suggested that the Department of Defense, not the AEC, fund the linear accelerator at Stanford, a recommendation based on both potential countermeasure uses and the development of advanced microwave technology.[48] American scientists could argue that international exchange of information and accelerator experts would provide useful intelligence to the United States: "The Russians readily find out most of what we are doing but it is very difficult for us to obtain information on their work."[49] Soviet scientists recognized the relative advantage: in the talks at CERN in 1959, Panofsky perceived that the "Russians are highly suspicious of our motives in these negotiations." Dubna was not entrusted with classified work, apparently because it hosted physicists from China, which in the late 1950s was challenging the Soviet Union for leadership of the communist world. Nevertheless, the Soviet Union still feared that American visitors might pry information of military value out of engineers.[50] American scientists also thought that access to Soviet accelerators would provide industrial intelligence on Soviet production of copper and iron, useful data in the total Cold War. Their Soviet colleagues recognized the possibility; in discussions over an international accelerator for biomedical applications, Soviet scientists hinted "that should such a project be embarked upon, they would give serious consideration to significant opening up with respect to their technology and industrial practices relating to such things as steel and copper."[51] Hence, despite their reassurances to CERN colleagues, American scientists emphasized collaboration with the Soviets: the United States already had easy access to and exchange with CERN, whose sponsoring nations were American allies in the Cold War, so formal collaboration was not worth the administrative hassle.

Internationalism also failed because of the impatience of American accelerator builders, who pressed ahead with their own plans rather than wait for uncertain diplomatic developments.[52] As with power reactors, domestic competition for accelerators among the national labs persisted amid international cooperation. Laggardly lab scientists might lose out to domestic rivals on the next generation of accelerators, which would affect an individual scientist about as much as falling behind in the international race to high energies. The competing proposals from Berkeley and Brookhaven in the early 1960s and the claims of new groups to an entry in the high-energy sweepstakes distracted American accelerator builders from the prospects for collaboration and hence doomed them to failure.

Fusion

The combination of nationalism and internationalism is also evident in the Sherwood program on controlled thermonuclear reactions. The possibility of fusion reactors for power had arisen during World War II, but first received wide attention in 1951 after the announcement, later proved false, of the achievement of controlled fusion by an expatriate German in Argentina. By 1952 the AEC had agreed to support small classified projects at Livermore, Los Alamos, Princeton, and Oak Ridge. The interest of the first three groups jibed with the crash program to build the H-bomb, which included a theoretical group at Princeton. In July 1953, when Lewis Strauss ascended to the chair of the commission, the AEC was supporting about thirty scientists with $500,000 for fusion research.[53]

Strauss brought to the AEC his strong personal interest in the fusion program, which stemmed from his earlier support of the hydrogen bomb and his determination that it also produce peaceful applications. His conviction meshed with the concurrent development of Eisenhower's plans for Atoms for Peace. If fission reactors were to provide cheap power and help unify the world, fusion reactors offered a safer, cleaner, and inexhaustible source of power. The AEC soon considered a "quantum jump" in its effort. Some of its advisers—in particular, Rabi and von Neumann—cautioned against rapid expansion of the nascent program. AEC staff were equally ambivalent: Thomas Johnson of the Research Division wanted to maintain only current levels of support, and neither Kenneth Fields of the Division of Military Application nor the directors of Livermore and Los Alamos wanted to divert effort from the thermonuclear weapons work. The AEC overruled its advisers, staff, and lab directors, to the satisfaction of the scientists working on fusion, and asked for a reorganization of the program and a budget up to three times larger than the current one.[54]

The AEC was motivated not just by international humanitarian concerns. If it were, it would not have kept Sherwood highly classified. Part of Strauss's support of Sherwood stemmed from his desire to demonstrate, through the achievement of fusion power, the superiority of the American capitalist system in the Cold War. It had less ideological sources too: the AEC knew the British were working on fusion, and predicted "real economic benefits" if fusion succeeded.[55] Fusion research did not have direct military applications, although a fusion reactor could possibly serve as a neutron source for the production of weapons material.[56] The combination of ideological, scientific, economic, and military competition among nations kept Sherwood classified. Hence the U.S. delegation to the first Geneva conference did not reveal any of the research on fusion, and Americans had to stay silent after Homi Bhabha of India, president of the conference, used his keynote address to tout the potential of fusion energy.

The Geneva conference had contradictory effects. Bhabha's speech forced the AEC to reveal the existence of its program and reconsider its classification. Its rivals also opened up. In April 1956 Igor Kurchatov, the physicist who led the Soviet weapons project, visited Harwell, Britain's atomic energy lab, and surprised his hosts with a detailed description of Soviet fusion research. Soon afterward Strauss proposed to include fusion in the cooperation agreement signed by the United States and Britain under Atoms for Peace. One of the reasons for exchange, however, was to allow the United States access to the still secret British program.[57] As fusion research came into the light, recognition of the progress of rivals intensified national competition. Knowledge of Soviet, British, and West German programs spurred the AEC "to expand our efforts if we are to enjoy the prestige of being the first to demonstrate a controlled thermonuclear reactor."[58] The British brought a new device on-line in August 1957 and quickly announced spectacular success: the first controlled production of thermonuclear neutrons. American scientists soon poked holes in the British claims; but before they did, they feared that they had lost the race and stepped up their efforts.[59]

As the British and Americans squared off, the Soviets sent up Sputnik. The satellite inspired the AEC to push for an exhibit on fusion at the second Geneva conference in 1958. Earlier suggestions for a demonstration of the American program at Geneva had received a lukewarm response, since there seemed little chance of a breakthrough that might support a spectacular exhibit. Strauss, however, feared that the Soviets might mount their own exhibit and "claim that their device was producing thermonuclear neutrons, although there would be no way to verify this claim during the Conference." With Libby's support, and against the advice of lab scientists and AEC staff, Strauss urged Sherwood scientists "to obtain 'thermoneutrons' and to demonstrate *at Geneva* a satisfactory thermonuclear device—one which would command worldwide attention."[60]

The American exhibit at Geneva in September 1958 failed to produce thermonuclear neutrons but still qualified as a success. Los Alamos, Oak Ridge, Princeton, and Livermore and Berkeley all contributed demonstrations that proved to be the most popular of the conference, attracting 100,000 visitors.[61] The United States owed its success to the resources the AEC was willing to commit to the fusion program, which in turn derived both from the potential for peaceful uses and the stimulus of international competition. The resources included scientists, whom the AEC was willing to divert from weapons research, and money. From 1953, when Strauss rejoined the AEC and began to nurture the program, operating budgets for fusion grew from $0.8 million to $18.4 million by 1958 (table 3). As one scientist put it in May 1956, after surveying the field: "Surely there is no shortage of money allocated for this work; in fact, one gets the feeling in visiting the various sites that the number of dollars available per good idea is rather uncomfortably large. There is certainly a feeling of some pressure to spend the money made available."[62] J. M. B. Kellogg, the head of the Los Alamos Physics Division, sought to spend some of it later that year: "Sherwood will almost certainly not contract, and I believe it is possible that we ought to decide on an all-out Sherwood effort starting within six weeks."[63]

Although the crash program cooled off after Geneva in 1958 and settled into a long-term effort aimed more at research than development, it continued to grow through the end of the decade. The free-flowing funds fostered several competing groups, evident in the representation in the fusion

Table 3. Operating costs for controlled thermonuclear research in national labs

Lab	1953	1954	1955	1956	1957	1958	1959	1960
Argonne	—	—	—	—	—	0.2	0.1	0.1
Berkeley	—	—	—	—	0.5	0.8	1.0	1.3
Los Alamos	0.1	0.3	1.0	1.1	1.8	2.3	3.0	3.3
Livermore	0.3	0.7	2.2	3.2	4.1	5.6	5.3	5.5
Oak Ridge	—	0.1	0.4	0.4	0.9	3.4	4.0	5.6

Source: *The Future Role,* individual lab budgets.

Note: All figures in millions of current dollars. Figures for 1960 are estimates. Most of the balance of the program went to Princeton.

exhibit in Geneva. Again, international competition did not presume a unified domestic effort. Kellogg urged, "I have an uneasy feeling that we do not want to be accused of dragging our heels on Sherwood, and that perhaps something is about to break. If s[o], let us try to make it break in Los Alamos rather than Berkeley."[64] As with accelerators and reactors, and the Manhattan Project before them, when programs became an international race the government proved willing to fund parallel approaches.

BOUNDARY DISPUTES

As the labs opened up to develop international relations, they also emerged from their relative isolation from domestic institutions. The relaxation of security concerns around 1954 exposed their wide-ranging programs to institutional friction on three sides: with industry, over the proper location of applied research and development; with universities, over prerogatives of basic research and education; and with other federal agencies, over the extension of the labs into fields beyond atomic energy. The disputes highlight the difficulty of a strict definition of the boundaries of the national laboratory system.

Labs and Industry

The development of nuclear technology by the national labs required interaction with its military and industrial users. In today's terms the interaction was a matter of technology transfer, which begged the question: when to transfer? The question applied, for example, to radiation instruments, such as dosimeters and scintillation counters, which the labs developed and built for their own use. In 1950 the AEC noted, "Virtually every atomic energy installation has an instrument laboratory."[65] The commission instead encouraged the growth of a radiation instrument industry, which by 1952 amounted to a $20 million business, over a quarter of which came from purchases by the AEC (with the military accounting for another 50 percent).[66] Lab scientists, however, found that industry could not meet all of their needs, while industry representatives complained to Congress that the labs were continuing to build their own instruments instead of buying them on the market.[67]

The question applied also to the reactor program and thus came to occupy the system as a whole as other labs joined Argonne and Oak Ridge in reactor work. The issue arose for reactors in the early 1950s, as the postwar programs of Argonne and Oak Ridge began to bear fruit in the form of specific power reactor designs. The AEC had assumed from the outset that the labs would relinquish reactors to industry when specific designs were available for construction. The model most frequently invoked was the aircraft industry and the National Advisory Committee on Aeronautics (the forerunner of NASA); NACA supported basic laboratory research on aeronautics and on aircraft components, but left full-scale aircraft projects to industry. The NACA model, however, assumed the existence of a viable industry; in the early 1950s the nascent nuclear industry, still stymied by secrecy, was not ready to take on large projects. At least Weinberg did not think so: "The nuclear energy laboratories . . . are superposed on an industry which in a real sense is yet to be established. . . . For this reason the nuclear energy laboratories cannot now take NACA as their pattern; they must make themselves responsible for large reactor or chemical projects, and carry them to completion, presumably in cooperation with concerns interested in helping establish a nuclear energy industry."[68]

AEC staff and commissioners acknowledged Weinberg's intentions, which Argonne shared, but instead of encouraging full-scale development by the labs, the AEC suggested that "increased participation by industrial groups might advance the trend towards use of the Commission's national laboratories primarily for supporting work, rather than for construction and operation of reactors."[69] The AEC intended to restrict the labs to "the NACA pattern" and leave construction of reactor mock-ups and pilot plants to industry; but its own attempts to promote nuclear power, egged on by the enthusiasm of the Joint Committee, forced its staff in Reactor Development to turn to the labs to provide the progress requested by the AEC. Zinn "assumed that the entry of the National Laboratories so deeply into project work was not at the urging of the staffs of the Laboratories. . . . By and large, projects have been set up to fulfill a need indicated by the U.S. Atomic Energy Commission."[70]

Not only were the labs pulled away from the NACA model by AEC demands and lack of industrial capability, but also lab scientists wanted to push their projects through to construction. In 1951 Lawrence Hafstad, the director of Reactor Development, complained about "ORNL's apparent interest in getting into bigger and bigger construction jobs," especially for the homogeneous reactor, but he "anticipated hurt feelings at Oak Ridge" if the AEC gave the reactor to industry.[71] The entry of Brookhaven into power reactor development exemplified the enterprise, and sensitivity, of lab scientists. The lab's first research reactor displayed the difficulty of large construction projects: it finally went critical in 1950, years behind schedule and more than $10 million over budget.[72] Amid the debacle one of Brookhaven's administrators proclaimed, "The scientific staff does not wish to undertake the responsibility for any more construction programs of pile magnitude."[73] After the completion of the research reactor, the lab's reactor engineers began to explore liquid metal fuels for power reactors and in 1951 contemplated pursuing their promising results toward development of a specific design. Some of AUI's trustees thought reactor work "highly desirable," but others sought to stick to basic research, and Haworth thought a development program "premature."[74] By 1952 the reactor proponents had won out after "fierce and heated arguments" at the lab, and the AEC established a separate project for the Liquid Metal Fueled Reactor at Brookhaven.[75]

Three years later reactor engineers had forgotten the fiasco of the first research reactor and were seeking to construct large-scale experiments of their reactor design.[76] The proposal provoked another debate among AUI's trustees, who wondered again whether the lab ought to focus on basic research. Arguments for the plan appealed to the absence of industry and to the desire of the lab's nuclear engineers to pursue it: the plan would maintain morale and demonstrate to lab staff that they would have the freedom to pursue good ideas. Though they did not name it, the trustees were here invoking the "Clinton effect," the desire to avoid hurt feelings and low morale that might result from rejected programs. Other trustees, notably Rabi, strongly opposed developmental projects, but in the end they voted to support the reactor and its engineers.[77] Hence Haworth in 1955 echoed Weinberg's identification of a basic question facing the national labs: "whether to encourage the national laboratories to continue such heavy engineering activities or whether, like NACA, to relinquish these activities entirely to the nuclear industry."[78]

The development of a reactor industry after 1954 weakened the argument that the labs had to undertake large projects lest reactors falter at the development stage; but it did not prevent scientists from continuing to push into heavy engineering. For its part, industry overlooked its initial and, to a lesser degree, continuing lack of interest in atomic energy and castigated the national labs for doing industry's work. To focus their interests, fledgling firms had joined the Atomic Industrial Forum, established by former AEC commissioner T. Keith Glennan to promote industrial participation in nuclear energy.[79] In 1959 the forum met "to consider the national laboratory problem." Although the representatives exuded an attitude of "respect" for the labs, and only a few individuals criticized them for "'competing' with industry," the forum agreed that the AEC "should have a positive program for transferring projects from national laboratories to industry." The lab directors responded with an equally tolerant stance—"industry could and should take over when the R&D concepts were fully advanced"—but could not reciprocate respect: "Industry is not yet capable of developing and maintaining the types of organizations necessary" for large projects.[80] Vague references to "positive programs" for transferring "fully advanced" technologies would not resolve the ill-defined border between the national lab system and industry.

Labs and Universities

While industry protested that the labs had strayed too far into applied research and development, the basic research at the labs butted up against academic interests. The labs had developed basic research programs over the opposition of the GAC and other academic scientists, but the AEC continued to support research in universities and, increasingly, to provide them the large facilities previously reserved for the labs. As the National Science Foundation matured, and especially after Sputnik, which provoked the National Defense Education Act of 1958 and a windfall of research contracts, universities with newfound strength challenged the right of the labs to pursue basic research. The universities had reason to be defensive: the labs not only defended their basic research programs but also trespassed on the pedagogical turf of academia.

One of the initial justifications for the labs had been as training grounds for new scientists. In the early 1950s the labs had developed more explicit educational programs to fill an academic vacuum: universities had no curricula in nuclear engineering because they lacked the engineers and the facilities to teach it and because the field remained classified. In 1950 Oak Ridge started its School of Reactor Technology, a one-year course in nuclear engineering for recent college graduates and employees of government and industry. The school admitted forty-six students in its first year (all of whom had to get security clearances) and quickly grew to meet demand.[81] A few years later the trustees of AUI at Brookhaven noted "growing sentiment" at the lab for some teaching activity, from both potential teachers (senior staff) and students (junior staff, lab technicians, and graduate students). Columbia and New York University also asked Brookhaven to offer courses in nuclear engineering to their graduate students. The trustees agreed to consider a course but asserted that the universities should administer it: "Brookhaven in no sense would assume the functions of a university."[82] Argonne displayed similar consideration when it set up its School of Nuclear Science and Engineering in 1955 to help implement Atoms for Peace: the school would be "temporary in nature until such time as American universities are equipped to assume this task. No degrees will be given and it is not expected that the course will ever be longer than one year."[83]

The trickle of trained nuclear engineers provided by the labs could not fill university departments, especially since the growing reactor industry diverted much of the flow. The shortage of nuclear scientists and engineers, evident in the competition for personnel, continued through the 1950s and, as the Cold War grew to encompass nuclear science and technology, led the AEC and the Joint Committee to encourage educational initiatives. The declassification of reactor information in the Atomic Energy Act of 1954 allowed university professors, in addition to industrial firms, access to the field. Congress amended the act in 1957 to emphasize training as a function of the AEC.[84] The previous year commissioner Sumner Pike had questioned the relation of Brookhaven to northeastern universities—not the presence of academic visitors at the lab, but the role of the lab at the universities. Pike seemed to suggest that Brookhaven could help universities design curricula and offer its own formal courses. Haworth objected: "We at Brookhaven have never felt that it was the mission of this Laboratory to carry out this . . . role in the northeastern area." But Brookhaven did offer several ad hoc courses for nuclear engineers and a summer school in radiological physics, activities that the lab thought "should continue and, indeed, be intensified."[85]

Haworth could not afford to encroach on the prerogatives of northeastern universities, as he had to answer to their representatives on AUI's Board of Trustees. Other labs were not so constrained. Pike had taken as a model the role of the Oak Ridge Institute of Nuclear Studies, a consortium of twenty-four southern universities formed on the AUI template but independent of the lab at Oak Ridge. The AEC supported the institute, which brought university professors and their students to the lab for research and training, offered courses, and served as a liaison between the expertise available at Oak Ridge and southern universities. The lab also boosted its own efforts: in 1957 the School of Reactor Technology began to offer a standard two-year curriculum in collaboration with six universities.[86]

The AEC had another precedent in Berkeley, whose hybrid status combined a national lab with a university campus. The Rad Lab provided space for over a hundred graduate students from the University of California at Berkeley each year, and those professors with half-time appointments on campus brought their knowledge down the hill to undergraduates. Lab staff portrayed "training of scientists and engineers" as "of comparable importance to the actual research accomplishments."[87] The number of grad students jumped from 130 in 1957 to 180 in 1958, reflecting the availability of the Bevatron but also presaging a renewed emphasis on scientific training in the wake of Sputnik.[88] The general push for education to beat the Soviets, codified in the National Defense Education Act, led some of the labs to consider extending their role: AUI's trustees now approved the possibility of courses for academic credit at Brookhaven, which extended beyond nuclear engineering to radiobiology and nuclear medicine.[89]

While the labs took on some of the traditional role of universities, universities also encroached on the domain of the labs. Their sallies are not immediately apparent in relative levels of operating budgets: off-site contracts still constituted around 30 percent of the operating budget of the Division of Research in 1958. Instead the incursions came in the realm of facilities: the total ratio of on-site to off-site research spending went from 4:1 in 1950 to 2:1 in 1958.[90] In order to retain their educational function the universities had to obtain the big equipment seen as central to contemporary science; but they would thus nullify one of the original justifications for the labs. Research done at the labs helped the universities to undercut them. The swimming pool reactor developed by Oak Ridge for the first Geneva conference provided a model for research reactors at universities, and the Argonaut reactor exhibited by Argonne at the second conference was designed specifically as a low-cost training reactor. Affordable reactors and the declassification of reactor information accelerated the trend started when North Carolina State built, with AEC support, the first reactor on a college campus in 1952. Two years later Weinberg noted that unclassified, inexpensive reactors had come within reach of universities.[91]

Similarly, the discovery of strong focusing at Brookhaven again put high-energy accelerators within the grasp of universities. In the wake of the discovery, the Division of Research agreed to support studies at Princeton and MIT in addition to Brookhaven.[92] The support encouraged the MIT group to join forces with Harvard and propose that they build a 15-BeV machine in Cambridge. The GAC viewed the plan as a threat to the national laboratory concept in general and to Brookhaven in particular. The accelerator would not contribute to the normal functions of a university department but instead belonged to the category of unique, advanced equipment. Guided by Rabi, the GAC recommended that high-energy machines go to national labs and that the strong-focusing design go to Brookhaven.[93]

University scientists were as persistent as their colleagues in the labs. The next year both Princeton and the Cambridge group returned with plans for multi-BeV accelerators, with the proposal from midwestern university scientists added to the pot. The GAC again took up the debate and suggested, with some dissent, a general policy for new accelerators: top priority should go to national labs "or other large AEC laboratories," which thus included Los Alamos as well as Oak Ridge; second priority to university groups, such as MURA or the Harvard-MIT collaboration; last priority to individual universities like Princeton. The GAC did not specify where Berkeley fit in this scheme.[94] The Division of Research acceded to the pressure from academic scientists and recommended that the AEC support two accelerators for universities; although the AEC's general manager disagreed, commissioner Libby persuaded the AEC to approve university accelerators, based on the argument that the "best research is usually done by the young men," whom the graying labs lacked. The AEC selected the Cambridge proposal for a 6-BeV electron accelerator and a 3-BeV proton synchrotron for Princeton, which had observed the GAC's priorities and added the University of Pennsylvania to its plans.[95]

Although the two accelerators lagged those already approved at Argonne and Brookhaven in energy, they did seem to prepare a path to high-energy physics for university campuses, a possibility lamented by the lab directors. Weinberg in particular warned the GAC in December 1955 that the assumption "that the facilities provided to the National Laboratories will be unique" no longer appeared to hold for research reactors, and he "expressed a hunch that accelerators were going the same way."[96] A decade earlier General Groves had warned that support of Brookhaven as a regional institution might provoke a proliferation of groups each seeking its own regional lab. The AEC's support of midwestern scientists and other university groups threatened the postwar settlement, and the GAC's priorities inspired other universities to band together and not only stake a claim to higher-energy accelerators but also seek entry to the national lab system.

Pennsylvania's collaboration with Princeton stemmed from a proposal from Penn in 1955 to establish a new facility for nuclear education and science, including reactors and accelerators, in the Middle Atlantic states—that is, the Philadelphia area; this despite Penn's (and Princeton's) relative proximity to Brookhaven and membership in AUI.[97] The AEC rewarded their gerrymandering with a new accelerator. A year later the University of Texas, Rice, and Texas A&M and the Robert A. Welch Foundation proposed, with the support of Texas congressmen, a major new lab for the Southwest, preferably around Houston. The Texans thought big: $30 to $40 million from the AEC might buy a high-flux reactor, high-speed computer, big cyclotron, medical center, and "helicopter service to and from nearby campuses." AEC staff recognized that the plan would "clearly establish another National Laboratory" and rejected it.[98]

The university groups shared the assumptions that the facilities of big science were necessary for academic functions and that the federal government should pay for them. Midwestern scientists expressed their entitlement to "their share of Federal support for the costly research facilities that are essential to promote scientific research and education in the Midwest."[99] All the groups also preferred new facilities on their campuses to ones at the existing labs, a reflection of the domination of lab facilities by in-house staff at the expense of visitors; hence neither the Penn proposal nor the Texas one mentioned the universities' proximity to an existing lab—in the latter case Los Alamos, which was pursuing an accelerator of its own and considering how it might become more of a "national" lab. In the meantime, twenty other southwestern universities were planning to incorporate, which they did as Associated Rocky Mountain Universities in May 1959 after two years of groundwork; their charter sought facilities for universities at member institutions.[100]

The trend reached its zenith in a proposal from a single university just down the road from Berkeley. Stanford scientists in 1954 had indicated their interest in a linear electron accelerator, or linac, to reach 10 BeV or more. The Cambridge accelerator edged them out, but they returned in 1957 with a request for $78 million, plus another $14 million in annual operating costs, to build a two-mile-long linac at Stanford to reach at least 15 BeV. The expense forced examination of AEC policy on the national lab system and its relation to universities. Warren Johnson of the GAC wondered whether a university campus should have such a machine but also accepted new members in the lab system: "It should be built at a National Laboratory either now in existence or at a new National Laboratory, if it is desirable to establish one." After the proposal survived interagency squabbles to reach Congress, the Joint Committee feared the precedent in the creation of a new national lab; it approved the plan only with assurances that Berkeley had no room for the long linac and Livermore had unstable geology (although the Stanford site sat on the San Andreas fault).[101]

Lab scientists viewed the Stanford proposal less as a worrisome precedent than as a drain on the budget—more of a practical threat than a threat to principle. Wolfgang Panofsky, a leader of the Stanford project, was a Rad Lab alumnus and a close friend of McMillan, the new director at Berkeley. High-energy physics in the national labs seemed secure: Argonne and Brookhaven were building their new machines and Berkeley was planning for the next generation. More important, although it would be available (in theory) to outside users, the Stanford accelerator would not require a multipurpose lab.[102] Finally, Berkeley again provided a precedent: just as the Rad Lab looked like a national lab that also trained students, from the opposite perspective it appeared as an organized research unit of a university campus that also served as an AEC lab. But the trend represented by Stanford, which affected reactors as well as accelerators, threatened to erode one of the justifications for the labs, even as the labs tried to shore up their foundations with educational functions borrowed from the universities. Hence AEC staff in 1958 suggested a reorientation of the Brookhaven program, perhaps to applied work, since universities were acquiring large facilities themselves.[103] The example of Berkeley as a hybrid national lab and campus research unit, however, might also have served as a warning of the difficulties of balancing the mission of the AEC with the demands of pedagogy; or, of defining the border between the labs and academia.

Work for Others

The Berkeley Rad Lab provided a precedent for incursion across another boundary. Although most of its support after the war came from the AEC, because of its history and hybrid status it had always accepted support from other agencies: for instance, work for the navy on the biological hazards of radiation and for the Public Health Service on atherosclerosis. Some of the other labs coveted the arrangement and sought what the AEC called "work for others." As the labs diversified beyond their original missions, work for others increasingly tempted the labs, which sought funding wherever they could; the AEC, which wondered why it was paying for work outside its mission; and other parts of the government, which saw in the labs a source of expertise.

The AEC and the labs had extensive interactions with the military from the outset (or earlier, as in the Manhattan Project). The question of military versus civilian control of atomic energy did not end with the passage of the Atomic Energy Act of 1946. Much of the programmatic work of the labs for the first several years went to military programs, not only in weapons but also in reactors for the navy and air force. The AEC still had to pay for this work out of its budget, owing to its legal monopoly on atomic energy; it relied on requirements provided by the military to justify these costs in the AEC budget to the Bureau of the Budget and Congress. For example, to sustain its support of the aircraft reactor work at Oak Ridge, the AEC needed a statement from the military Joint Chiefs of Staff conveying their need for the reactor.[104] The process did not encourage restraint on the part of the military, which could request all manner of dream machines without having to pay for them. At least so it seemed to the Bureau of the Budget, which in 1956 perceived "a tendency for technical feasibility to be converted almost automatically into military requirements." It found examples in the proliferation of nuclear weapon designs and in the projects for aircraft and rocket propulsion. To impart restraint, the director of the bureau suggested that the Department of Defense fund portions of military reactor programs. The AEC resisted: yielding budgetary control might set it on the slippery slope toward control of programs by another agency. The military did not want responsibility either, preferring its free ride under the current process.[105]

There were a few projects under the AEC which other government agencies funded, albeit through transfer of funds to the AEC: for instance, weapons effects tests, for which the military paid for AEC participation through the Armed Forces Special Weapons Project, or the joint task forces created for tests in the South Pacific. Other agencies also received support from the AEC, such as weapons development work at army and navy labs and the joint program with the Office of Naval Research for accelerators and nuclear physics research at universities.[106] Laboratory budgets remained under the AEC, although these transfers would open the door to more direct support from others.

In 1952, as the AEC research budget for 1953 was taking hits from congressional appropriations committees, Brookhaven's administrators considered taking on projects for other agencies "to provide 'more strings to the bow.'"[107] Lloyd Berkner, the president of AUI, was already stringing one up, in the form of a proposal to the Office of Naval Research to study the ionosphere and radio wave propagation at Brookhaven. Berkner and Haworth pointed to the precedent of Project East River, which AUI performed under an army contract, though not at Brookhaven. E. L. Van Horn, the local AEC representative, instead suggested a setup similar to the ones already in practice, where the navy would transfer funds for the study to the AEC. Van Horn, noting that direct funding from other agencies for work at Brookhaven would set a precedent, sent the proposal for "more thorough consideration at the Washington level."[108]

Shields Warren, laboratory coordinator for Brookhaven at the AEC, presented several arguments against the plan. Turning to other agencies, which were not necessarily in better budgetary shape than the AEC, might not provide a steady flow of funds and would only make it harder for the AEC to defend its own appropriations. Work for others would lose the advantage for the labs of the partisanship of program managers, the AEC, and the Joint Committee. If, Warren warned, Brookhaven "becomes essentially an all-purpose laboratory for a number of government agencies and private industry, I can see dilution of its aims and loss of continuity and homogeneity. Furthermore, I can see increased financial instability rather than increased sureness of annual support inasmuch as the Laboratory would cease to be a primary interest of one agency and would rather become a peripheral interest of several agencies."[109]

AUI's trustees conceded halfheartedly that work at Brookhaven should be under the AEC.[110] But other labs were also looking outward to support declining programs. As the aircraft reactor lost altitude in the late 1950s, Oak Ridge had idle facilities for research on shielding; enterprising scientists put them to work on irradiating army vehicles and on shielding studies for NASA, the Ballistic Missile Office, and the Defense Atomic Support Agency.[111] The entry of the labs into diverse fields encouraged work for others, as did the emergence of new agencies such as NASA, which set up a joint office with the AEC modeled on the Naval Reactors Branch, and the military's Advanced Research Projects Agency (ARPA). The latter would fund an electron accelerator at Livermore and optical masers at Oak Ridge as antimissile devices.[112]

While diversification encouraged consideration of sponsors beyond the AEC, the possibility of multiple sponsors fostered diversification. In its discussion of the weapons test moratorium and the need to retain key personnel at Los Alamos and Livermore, the GAC suggested that they take up work for NASA and ARPA.[113] Pluralism promotes autonomy, which explains the pursuit of multiple sponsors by lab scientists, despite a possible loss of unified representation. But work for others could also promote the view of the labs as "job shops"; again, Weinberg, the originator of the term, did not restrain Oak Ridge scientists from seeking outside support and indeed would lead the effort to win direct support from outside agencies. Finally, the same agencies courted by the labs could threaten their programs. The old concern about overlap intensified as the labs diversified beyond atomic energy and new agencies emerged with their own programs and laboratories. The question of work for others thus raised another fundamental issue about the purpose of the labs: Should they restrict their attention to problems in atomic energy, or were they indeed national labs available for work on any matter of national interest?

The Future Role

In early 1959 the AEC asked the labs to review their programs and forecast their development for the next five years. The Joint Committee soon requested that the AEC expand the predictions to cover a ten-year period and provide its own summary of the status and missions of its labs. The committee published the results in 1960 in a report titled *The Future Role of the Atomic Energy Commission Laboratories.*[114] The report examined the blurry boundaries on the three sides of the national lab system, along with the question of the growth and diversification of the labs, but it provided no easy resolution.

The Future Role consisted of programmatic statements submitted by all of the AEC's labs, with the input of the GAC and the commission, and responses to the resultant volumes solicited by the Joint Committee from interested parties in academia and industry. The multiprogram labs characteristically proposed more growth and diversification. Argonne and Berkeley expected to double in size over the decade, Brookhaven to increase staff by 50 percent, and Livermore to grow 5 percent each year. Los

Alamos, still limited by housing on the mesa, defined slow growth as 10 to 15 percent over the decade; and Oak Ridge, which expressed "no desire to expand significantly except in some areas of biology and chemistry," also found "no important objection to expanding at an expected rate of 20 percent in 5 years."[115]

The GAC and AEC staff were already disturbed, before these predictions, by "the general tendency in the AEC laboratories to suggest the optimum size is 20% greater than the present size." At some point, the committee believed, labs could become too unwieldy to administer, and individual scientists and programs could get lost in the shuffle.[116] The GAC suggested that the labs emphasize "quality rather than size" and slough off unproductive programs and scientists, the better to promote good ones. Lab directors supported their expansive tendencies with the argument, enunciated by McMillan of Berkeley, that "a certain amount of growth is essential to maintain the health and vigor of any laboratory." After AEC commissioner John Floberg questioned this theory, McMillan allowed that perhaps Berkeley might not need to double in size, but anything less than a 40 percent increase would "seriously restrict its progress," thus doubling the figure that had disturbed the GAC.[117]

In one sense, systemicity might have eased problems of size: the presence of several labs allowed the AEC to support many more scientists in smaller labs than it could have in a single large one. By contrast, the British struggled to keep their atomic energy lab, Harwell, at a manageable size; the GAC noted that Harwell had 7,200 personnel and was trying to get rid of staff.[118] Maintaining several multiprogram labs, however, entailed the support at each site of scientists in various fields, which tended to diversify; restricting each lab to a single program, as the AEC had sought to do at the outset with weapons and reactors, and as was later suggested for fusion, might have limited the pressures for growth. The growth of the several sites also threatened to increase duplication, as commissioner Floberg pointed out in a meeting with the lab directors. As an example, Floberg noted "that at every AEC installation he visited there seemed to be 'rat houses' and that each place had a hundred thousand or more 'rats' or other animals. . . . He wanted to know why all of this duplication was necessary and if this wasn't evidence that the same kind of work was going on in numerous places without regard to duplication. Fortunately," noted the AEC staff member recording the discussion, "several of the laboratory directors attacked this thesis vigorously," arguing that no "undesirable duplication of research" existed. Although they did not define the adjective, Floberg "backed down on this point."[119]

While Floberg worried about duplicative research, the rest of the AEC questioned the presence of any basic research at the labs. The commission considered its response to the Joint Committee in the light of the AEC mission, which the commissioners took to be the development of weapons, reactors, and radioisotopes; research did not make the list. The priorities suggested to commissioner John Graham that "more of the basic research work should be done by the traditional sources of learning, the universities, than in our laboratories."[120] The GAC protested; it now supported basic research in the labs on principle and urged the AEC to include it as a core mission. The advisers still worried about competition with academic research, as universities acquired large facilities and as growing in-house lab staffs duplicated research in university departments. Nevertheless, they identified several types of basic research appropriate for the labs: research related to applied projects, or requiring costly facilities and large teams, or involving expensive, hazardous, or classified programs; and, accepting the scientists-on-tap theory of lab scientists, "research needed to make a laboratory an attractive place to a reasonable number of first-rate scientists." The AEC's final draft included basic research at the labs.[121]

The GAC's justifications of research included a new one: research requiring large teams. The advisory committee accepted the argument, offered by James McRae, that the "interdisciplinary arrangements" of the labs allowed them to do research that universities could not.[122] McRae, a former director of Sandia Laboratory, was echoing the invocations of interdisciplinarity by his colleagues in the multipurpose labs, who recognized a characteristic that could distinguish and defend their programs from academic research. The academic arrangement, where faculty underwent peer review within their department to attain or maintain tenure, did not promote transgression of disciplinary boundaries and tolerated it only reluctantly. The laboratories learned to encourage it. Their appeals to interdisciplinary research perhaps helped to popularize the term itself: although the use of "discipline" to describe a branch of knowledge dates back to the Middle Ages, "interdisciplinary" seems to have first appeared in the 1930s, primarily in the social sciences, and reached wider circulation in the late 1950s, when lab scientists found it apropos.[123]

Thus Norman Hilberry cited "the combined talents of many disciplines that are brought together in sufficient numbers and quality only in national laboratories." Gerald Tape pointed to the Brookhaven medical center: "In addition to persons trained in the field of medicine, the Brookhaven medical center also utilizes specialists in reactor technology, radiation testing and numerous other related fields. . . . No single university presently had the wide range of talents necessary to operate such a medical center." Bradbury, supported by Weinberg, sought "adequate varieties of intellectual disciplines . . . so that Government laboratories always have very much more than a 'critical mass' of people." He continued, "The only large-scale programs which have been really well done have been done by the Government in Government laboratories operated by industrial or academic contractors. . . . Universities, by themselves, have done nice but uncoordinated research programs."[124] Weinberg would make interdisciplinarity a prime criterion for government funding of science in a famous manifesto of 1963.[125]

Multidisciplinarity as a justification for the labs filtered upward through AEC staff, who recognized the "advantage and benefit . . . of the 'team' approach, with particular recognition of the value of the 'cross fertilization' process which makes available to the life scientist the unique advantages of discussing his problems with the chemists, physicists, engineers, metallurgists, etc., who are also concerned with the problems of nuclear energy at the same installation."[126] It made its way thence into the AEC's report, which perceived that "each multiprogram laboratory is organized to concentrate, at one center, varied scientific disciplines together with the requisite facilities. These disciplines support, augment, and stimulate one another."[127] In 1961 Haworth, newly appointed to the AEC after his tenure as director of Brookhaven, brought the message to the AEC himself: "One of the most unique contributions which a 'National' Laboratory can make lies in its ability to perform interdisciplinary research."[128]

Multidisciplinarity, and the interdisciplinary research it promoted, thus distinguished the labs from universities in basic research. The report suggested no comparable feature to define the border between the labs and industry in the case of applied research. Although the AEC conceived the relationship as "complementary," Oak Ridge continued "to insist on its prerogative to build and to operate experimental devices of any size if such are needed to carry out the Laboratory's technical responsibilities," including prototype reactors. Industrial representatives, in their responses to the report, continued to deny the prerogative and instead sought greater support of industrial development.[129]

Incursions across the third border, that of other federal agencies, helped to relieve the pressure on the border with industry. Diversification into fields besides atomic energy took effort away from reactor projects, which were the main bone of contention with industry. The AEC recognized an expanded mission in its summary of the roles of the multiprogram labs: "The strong capabilities of the laboratories are not the exclusive resources of the atomic energy field; they are held in trust for the Nation as a whole. Urgent work for other Federal agencies on matters of national concern will be accommodated in the laboratories when their skills are needed." The Joint Committee, perhaps sensing some erosion of its political turf, solicited responses to the possibility that the labs take up non-nuclear research. The GAC replied in the affirmative: at the moment there was no shortage of work in nuclear energy, but if in the future pressing problems arose in non-nuclear fields, the labs should be free to tackle them. "The Commission's laboratories, especially the multiprogram laboratories, are important national assets and they should be doing what, at the time, best serves the National welfare and security."[130]

Continued Contestations

The Future Role provoked much discussion but little action. Examiners in the Bureau of the Budget lauded its "moderate and sensible tone" but noted that it presented no alternatives to the status quo. On the contrary, it accepted the existence and functions of the labs and only encouraged current trends toward growth and diversification, especially the volume containing the labs' own statements.[131] The report espoused complementarity but did not resolve the contested boundaries with universities, industry, and other agencies. Meanwhile, the organizational Laboratory Problem also lingered, and the agitation of the lab directors was renewed in 1961. That year the GAC, at the request of new AEC chair Seaborg, considered the several questions surrounding the national labs in a series of wide-ranging discussions that spilled over into 1962.

The general push for science education reforms after Sputnik had further encouraged the labs toward academic functions. In November 1960 a President's Science Advisory Committee (PSAC) panel chaired by Seaborg issued a report on science education that endorsed the participation of laboratory scientists in graduate education at nearby universities.[132] The AEC established a Division of Nuclear Education and Training and considered how the labs could contribute; Russell S. Poor, appointed to direct the division, did not want to see the labs become universities or even graduate schools. The GAC concurred and thought postdoctoral training more appropriate, akin to the function of hospitals for training doctors emerging from medical school.[133]

The labs tested the limits of AEC policy. Weinberg noted that the labs employed around 10 percent of scientific Ph.D.s in the United States. Why not use them to turn out more scientists? Oak Ridge, after first contemplating a National Institute of Science and Technology under the lab and then an independent Oak Ridge Graduate School of Science and Technology, arranged a joint program with the University of Tennessee to train graduate students at the lab and have lab staff teach at the university.[134] The Associated Rocky Mountain Universities, initially formed to seek facilities for universities, by 1962 had reached a similar agreement with Los Alamos.[135] Argonne proposed to the University of Chicago that the lab serve as a graduate school under the university, with about eighty Argonne scientists teaching several hundred students at the lab. The lab and the university, however, failed to consult other midwestern academics still leery of the lab after the accelerator episodes of the 1950s, and the plan fell victim to their opposition.[136]

At Livermore, scientists offered extension courses at the lab toward a master's degree, and in 1960 Teller sought to establish a college of applied science at Livermore under the Berkeley campus of the University of California. The creation of a closer campus at Davis made the proposal more feasible, and in 1963 the lab revived the plan for a Department of Applied Science, directed by Teller and administered by UC Davis.[137] The department would use lab facilities on a non-interference basis, including the swimming pool reactor, accelerators, computers, and Sherwood devices, and the university would reimburse the AEC for the facilities and the teaching time of lab staff. Graduate students would pursue unclassified research, although some doctoral candidates might seek clearance for access to restricted facilities. AEC staff worried that foreigners might use the program to gain access to the lab and that overhead costs might slip through the cracks; but they also noted that the program would train future lab employees and that students might stimulate lab programs. The AEC approved the plan after the university agreed to limit foreign student enrollment and have all students submit to FBI file checks (as uncleared AEC employees had to do).[138]

The expansion of graduate training at the labs in the early 1960s intensified the resistance of universities. The representatives of northeastern universities in AUI continued to object to any formal educational role for Brookhaven.[139] University scientists on the GAC, such as John H. Williams of Minnesota, protested that "too large a percentage of Federal funds was being channeled to the National Laboratories, at the expense of universities."[140] Williams remained a leader of the academic physicists in the Midwest who were agitating for the next generation of high-energy accelerator beyond Brookhaven's Alternating Gradient Synchrotron; MURA and Berkeley, the leading contenders for the machine, were joined by a new coalition of universities from southern California.[141] As labs assumed educational functions and universities pursued large expensive facilities, the definition of the border between the labs and academia blurred.

The national labs also continued to contest their role in applied development with industry. Large reactor projects remained the main point of contention, with industrial representatives urging that the labs relinquish reactor designs at the pilot plant stage.[142] Industry still had to demonstrate its willingness and ability to take over projects from the labs; the will, at least, certainly emerged in the proliferation of proposals in the early 1960s for power reactors from electrical utilities, not all of which performed to expectations when constructed. For projects besides reactors, industry still proved a reluctant partner. For instance, the AEC sought to promote the development of radioisotope and radiation sources for commercial uses and solicited proposals from several industrial firms; when none wanted to make the investment, the AEC turned to Brookhaven, which accepted the project despite its professed reluctance to take on what it considered technological development instead of research.[143]

Once again, expansion into non-nuclear fields offered some relief to pressure on the border with industry. Oak Ridge in particular, whose statement in *The Future Role* did little to appease industry, sought to till fertile fields outside the realm of the AEC. The lab had suffered from the cancellation of its homogeneous reactor in 1959; in the fall of 1960 Weinberg convened a series of Advanced Technology Seminars for senior staff at Oak Ridge, to consider new tasks and sponsors. Weinberg laid out some criteria for possible programs: they should be "big and expensive" to keep "many members of the ORNL usefully occupied for a long time"; "in the national interest, interpreting national interest broadly"; and "of a rather long-range character."[144] Oak Ridge received more incentive to consider alternative programs after President Kennedy canceled the Aircraft Nuclear Propulsion project in March 1961. Following the cancellation, PSAC thought that Oak Ridge might be "marking time without contributing much to urgent needs." Seaborg identified a few—desalination, environmental pollution, oceanography, geology, molecular biology, and long-range electric power transmission—and suggested that national labs "might assist in the solution of some of them, even though this goes beyond their present atomic energy missions."[145]

That April the GAC considered the particular problems faced by Oak Ridge and the general question, posed by Kenneth Pitzer: "When a major Laboratory no longer has an applied mission, is it supposed to receive a new applied or a general research mission, or should the Laboratory be closed?"[146] The GAC recalled its support of non-nuclear research in *The Future Role* and, encouraged by Libby to adopt "the viewpoint that the labs are an asset to the entire nation, not only to the AEC," examined the possibility of work for others. The advisers perceived several of the same problems pointed out by Warren a decade earlier: loss of administrative control, budgetary insecurity, loss of programmatic focus. Perhaps, the committee considered, a federal "Department of National Laboratories" could control funding and assign diverse problems to the labs. The committee reached no conclusions, but it did not dare to answer Pitzer's question with a recommendation to close any labs.[147]

The GAC recognized that lab scientists might not possess the expertise to solve every problem that confronted the nation. The labs might have answered that diversification would help their chances; in the meantime, they sought to apply what they already knew. All of the labs would pursue work for other federal agencies later in the 1960s, from the National Institutes of Health to the Department of Transportation, with the approval of the AEC and Congress. Diversification of sponsors foiled attempts to solve the Laboratory Problem; at the same time the AEC was addressing the complaints of the lab directors that the labs were treated as "job shops," the labs were expanding the variety of clients. But if it came down to suffering the indignity of being a job shop versus losing a job altogether, the decision was clear. Lab directors and scientists sought ways to sustain their institutions in the face of declining programs.

III

CONSEQUENCES

6

ADAPTIVE STRATEGIES

The diversification of sponsors was made possible by diversification of programs, another consequence of systemicity and the adaptation of the labs to their changing environment. The labs also displayed the adaptive strategy of specialization, which together with diversification would redefine the concept of a national lab.

SPECIALIZATION

From the addition of high-energy accelerators to lab programs in the early Cold War to the international competition in power reactors, accelerators, and controlled fusion in the 1950s, the national laboratories pursued parallel, but not overlapping, efforts. It is an important distinction and a characteristic of the national lab system. It applied not just to crash programs like the Manhattan Project itself or fusion in the 1950s, but to lab programs in general. With the watchful eyes of budget examiners on the lookout for duplication, lab scientists and AEC program managers differentiated their programs, on the well-founded assumption that the Budget Bureau or Congress would do it for them by elimination if they did not. Hence, when Livermore entered the fusion program, it deliberately chose a different approach from experiments already under way at Los Alamos and Princeton.[1] One finds similar specialization in large programs such as weapons, power and research reactors, and nuclear rockets, which the labs achieved by mutual agreement or had imposed from above. The rarity of redundancy extended from the specifics of the scientific and technological programs to general institutional philosophies, which depended on the competitive position of each lab in relation to the others.

Weapons

The addition of the lab at Livermore brought competition to the weapons program. From its creation Livermore faced concerns that it would duplicate the work of Los Alamos. Lawrence assured the AEC in September 1952 that "excellent cooperation between Los Alamos and Livermore precludes the possibility of overlapping efforts." Cooperation for the first few years meant that Livermore steered clear of turf already staked out by Los Alamos. For its first program in the fall of 1952, Livermore chose a fission device, relevant to thermonuclear weapon design, that Los Alamos had earlier considered and abandoned. The following year Livermore embarked on a program to provide small-diameter weapons to the army for atomic artillery, which Los Alamos had spurned in order to concentrate on high-yield designs for the air force.[2]

Specialization extended beyond specific programs to the general philosophy of the laboratory. As Los Alamos successfully developed thermonuclear weapons, the original justification for a second weapons lab, Livermore adopted as its mission the pursuit, as the AEC vaguely referred to it, of "new ideas." Von Neumann perceived in 1954 that Livermore's "objectives are being defined essentially as to do something more risky than Los Alamos."[3] Livermore's risky approach resulted in fizzles in the first several designs tested in 1953 and 1954; not until the Teapot test series in spring 1955 would a Livermore device perform as designed.[4] The test failures only encouraged Livermore to up the ante. At a GAC meeting in July 1954, Teller described the lab's fizzles and then shocked the advisers with a proposal daring in the extreme: a 10,000-megaton design, over six hundred times more powerful than the Castle-Bravo device that had erased the island of Elugelab at Bikini atoll three months earlier. Rabi advised the GAC that the proposal "was an advertising stunt, and not to be taken too seriously."[5] Los Alamos, then entering its teenage years, played the role of the wiser older sibling, with a more conservative outlook. Livermore continued to offer optimistic designs, as it did to the navy in 1956; Los Alamos continued to dismiss the designs as "gleam-in-the-eye" advertisements. In 1963 AEC staff members were still noting Livermore's "typically enthusiastic approach."[6]

Livermore's enthusiasm may have derived in part from its connection to Berkeley; some of Lawrence's boosterism and penchant for big, risky experiments may have rubbed off on the weapons annex at Livermore.

As with Berkeley's cut-and-try accelerators, Livermore preferred to pursue and test designs unproved by theory, whereas Los Alamos reduced the need for tests through its stronger theoretical tradition. Livermore's experimental emphasis would lead it to promote the use of computer calculations based on empirical models.[7] But Livermore's adventurousness also stemmed from its relation to Los Alamos and its role as the upstart lab, which could use stunning successes to justify its continued existence.

By 1955 the new lab was better established, with successful tests and a staff of 1,500, a third of them scientists and engineers. For the remainder of the decade, neither lab lacked for work. Weapons budgets ballooned to provide nuclear devices for a variety of delivery vehicles for all three military services, in order to satisfy Eisenhower's New Look defense and its reliance on nuclear weapons. In prosperous times the weapons labs could agree to divide assignments for new designs between themselves. In 1956 Norris Bradbury at Los Alamos suggested to Herbert York at Livermore that the lab directors decide "that one laboratory or the other will conduct necessary future tests and development of each of the specific systems." The AEC and GAC supported a division of labor, and senior representatives of the two labs met in August 1956 to negotiate. Los Alamos got the biggest plum, the warhead for the Atlas missile; Livermore won as consolation the design for the Titan missile that backed up the Atlas program. Livermore would also continue to work on atomic artillery, and the two labs divided responsibility for warheads for smaller rockets for the navy and army. The following year York proposed another meeting with Bradbury to decide "who does what," and the AEC agreed to make the meetings a regular event. The next one took place in December 1958, "on neutral ground" in Los Angeles; staff from the two labs "split up the pie" of twenty-eight weapons projects requested by the Department of Defense.[8]

The two labs did not always reach amicable agreements, however. After Livermore found a customer in the army for atomic artillery, the head of fission weapon research at Los Alamos urged Bradbury not to neglect the potential for a competing program there. Likewise, until they agreed to award the Atlas warhead to Los Alamos, the two labs pursued parallel designs. By competing the labs satisfied the desires of the congressional Joint Committee, which had urged the creation of Livermore with the rhetoric of competition and whose chairman, Clinton Anderson of New Mexico, sought to ensure, as he did in May 1955, that the labs satisfied the committee's intent. Anderson's allegiance to the ideal of competition proved stronger in this case than the practical concerns of his constituents at Los Alamos, who had the most to lose to the upstart. He may have perceived that when the labs competed instead of cooperating, they allowed program managers and policy makers to apply their own priorities. For example, where Livermore did choose to take on Los Alamos in its early program, in small fusion weapons, its test fizzles led the AEC to cancel its program for that yield and weight and instead select Los Alamos to design the weapon for the new B-47 aircraft. The military services learned the lesson when the labs colluded: the air force voiced its displeasure when Los Alamos, its favorite, relinquished the Titan design to Livermore; the army and navy also had less confidence in Livermore and preferred that Los Alamos do their work, but could not control assignments when the labs split the pie themselves.[9]

Even when the labs agreed to divide the weapon design pie, institutional competition remained. Bradbury railed against the redundancy of Livermore from the outset: "The original concept for the Livermore Laboratory was that it should explore ideas or systems not receiving attention at Los Alamos. It was also believed in some quarters that brilliant new ideas would flow from the establishment of competition. . . . The brilliant new ideas have not appeared." He argued that scientific and technological constraints limited the number of design options, and that the weapons design process was "identical with the 'competitions' set up for aircraft manufacturers in the design of a new plane. . . . All the designs submitted are basically the same because aerodynamics is a science and not an art, and no manufacturer will propose to produce an airplane which will fly twice as far, twice as fast, for half the weight his competitor will propose." Thus scientists in a single lab, such as Los Alamos, could arrive at the optimal design without the benefit of competition. Bradbury's comments served the interests of his lab, but Rabi, now chair of the GAC, also perceived in 1956 that "the two laboratories are converging in design and weight." Although science constrained the options of the labs, however, it also entailed disincentives to duplicate results achieved elsewhere; and the politics that required competition for aircraft contracts also acted to sustain institutional competition between the weapons labs, while ensuring that they did not duplicate programs. Science and politics together enforced differentiation. Bradbury's actions belied his words. Around the time he sent his aspersions of Livermore to the Division of Military Application, in the fall of 1954, he was arranging to coordinate with Livermore their programs and presentations to convince the AEC that the two labs took different approaches to the design of small fusion weapons.[10]

Reactors

The reentry—or, rather, the continued presence—of Oak Ridge in the reactor program after the collapse of centralization restored competition, and hence specialization, to reactor development. When Oak Ridge proposed a new reactor program in the late 1940s, it took care to appeal to its special capabilities and distinguish its design from the ones under way at Argonne. The AEC had directed Oak Ridge to focus on chemical technology when it removed reactors from its mission. Weinberg interpreted the directive to support his long interest in homogeneous reactors, which Argonne neglected in favor of boiling-water and fast-breeder reactors. Oak Ridge thus embarked through the 1950s on a series of designs of fluid-fuel reactors, while Argonne pursued solid-fuel designs, especially the plutonium breeder using fast neutrons. The emphasis at Oak Ridge on fluid-fuel designs extended from power reactors to the Aircraft Nuclear Propulsion project, whose scientists seized on molten-salt fuels as a way to provide the higher temperatures needed by jet engines, which solid fuels could not withstand. As Weinberg told the Joint Committee in 1953, "I myself am a homogeneous reactor man. . . . But I by no means say that Dr. Zinn, who, by the way, is a fast reactor man—that his way won't turn out to be just as good."[11]

As with weapons, specialization went beyond specific designs to affect the basic philosophy of each lab. In the debate over the central reactor lab in 1947, Oppenheimer had predicted that centralization would "make Argonne the Los Alamos of reactors."[12] The remark proved accurate in a way opposite to that intended. Argonne, like Los Alamos, assumed the role of the older, conservative lab in a competitive program, and Oak Ridge, like Livermore, played the brash upstart. Argonne and Los Alamos by the early 1950s had well-established programs and relationships with the benefactors and beneficiaries of their programs—the Division of Reactor Development for the former, the Division of Military Application and the military services for the latter, and the congressional Joint Committee for both. To overcome their initial disadvantage Oak Ridge and Livermore differentiated their programs from those of their well-entrenched rivals and adopted a high-risk, high-payoff approach. Hence Weinberg contrasted the reactor projects pursued by the two labs and claimed that "while the Argonne program was 'on a safe track' and would almost certainly be successful, the Oak Ridge program was less conservative, had a larger element of risk, but if successful would lead to greater returns."[13] Aircraft Nuclear Propulsion in this respect characterized the Oak Ridge mission, at least as the lab defined it in 1958: "to take on large, complex projects, even though they have a high risk of failure."[14] Again, program differentiation reinforced the institutional individuality of each lab within the system.

Fusion

The partisans of Livermore and Oak Ridge invoked the benefits of competition to sustain multiple approaches to weapons and reactors. The same rhetoric appeared in the Sherwood program, where the AEC supported four main groups—at Princeton, Oak Ridge, Los Alamos, and Livermore—in the international race for controlled fusion. But as with weapons and reactors, competition prevailed in the fusion program only after the AEC toyed with the possibility of centralization.

In September 1953, after Lewis Strauss had assumed the chair of the AEC and resolved to promote the nascent fusion program, AEC staff prepared a report on the pros and cons of a centralized lab for fusion. On the one hand, the arguments for centralization again appealed to the example of Los Alamos: a single central lab would focus attention on the work, ensure quick and easy communication while better protecting classified information, and avoid duplication of facilities, services, and scientific effort. On the other hand, the AEC would have to overcome the inertia of existing projects to relocate them to a central lab, for which there was no obvious site. The AEC had firsthand experience with institutional inertia and was reluctant to create a new lab: "It would be difficult to terminate a laboratory if someone put an end to the project by proving its goal impossible to reach." But the main argument against centralization was competition: "Until the field narrows down and many of the schemes have been eliminated, decentralization appears to provide the best basis for healthy competition and rivalry which will insure that each scheme is developed according to its merits under the leadership of its strongest advocate" and "will be reviewed, criticized, and revised by scientists with different abilities and points of view."[15] The AEC preserved competition and the status quo and decided only to arrange for better coordination of the extant programs, which it effected by appointing a Sherwood steering committee composed of representatives of the various laboratory projects.[16]

The labs had differentiated their programs at the outset on their own initiative. The group under Lyman Spitzer, Jr., at Princeton, the first entrant into the field in 1951, had settled on a torus modified to a figure-eight configuration, dubbed the stellerator; current-carrying coils around the circumference of the doubled torus provided an axial magnetic field. News of Spitzer's project soon reached Los Alamos and James Tuck. Tuck had returned to Los Alamos from Britain to work on the hydrogen bomb; his work in Britain had explored another configuration, the "pinch," which he revived at Los Alamos in 1952. Unlike the stellerator, the pinch effect confined the plasma with circular magnetic fields generated by the passage of a current down the axis of the torus. Also, the pinch was a fast-pulsed device, as opposed to the steady-state stellerator. Tuck thought the pinch more promising than the stellerator, and it satisfied the main concern of his group, to avoid duplicating the work already under way at Princeton. The same concern motivated York to choose yet another approach when Livermore established a program. Like Princeton but unlike Los Alamos, York decided to use externally produced, axial magnetic fields; unlike both Princeton and Los Alamos, York chose a straight cylinder. To cap the ends and confine the spiraling plasma particles, he proposed radiofrequency waves (supplemented, after these proved too weak, with stronger magnetic fields at the ends). Oak Ridge, when it entered the large-scale effort for the second Geneva conference, focused its program on the use of molecular instead of atomic ions in the plasma source, albeit using Livermore's magnetic mirror geometry.[17]

Specialization allowed the AEC to continue to celebrate the benefits of competition: "The rivalry between groups stimulates rapid progress."[18] It also allowed the labs to make the most of their particular abilities, while falling back on the expertise of colleagues elsewhere when necessary. Los Alamos could measure the millisecond signals from its pulsed pinch devices using fast diagnostics developed for weapons tests. Livermore capitalized on its connections to Berkeley's accelerators and the radiofrequency fields they used; York, a Berkeley graduate who had worked on the 184-inch accelerator and the Materials Testing Accelerator, recruited Richard F. Post ("radiofrequency Post") from McMillan's synchrotron group at Berkeley to the Livermore fusion project. Oak Ridge based its entry into the field on its long experience with ion sources, which had earlier helped to justify its development of heavy-ion accelerators; Oak Ridge scientists got their start in the fusion program preparing new ion sources for Livermore and then developed a full-fledged project of their own based on their new ideas.[19]

The four main fusion laboratories, which absorbed almost all the fusion budget of the AEC, would continue on their separate paths into the 1960s. Continued pressure from within and without the AEC maintained differentiation: in 1958 the GAC warned that "the programs at the different sites have tended to converge, with the result of duplication of effort," and the Bureau of the Budget questioned the need for multiple approaches.[20] Contracting budgets in the early to mid-1960s brought threats to close down one or more of the fusion projects. Lab scientists sought new directions. For example, Oak Ridge, which from the outset had duplicated Livermore's mirror geometry, in the mid-1960s proposed also to inject neutral atoms instead of the molecular ions that characterized its initial efforts; the steering committee, mindful of criticism of fusion spending from the Bureau of the Budget and Congress, suggested termination of the Oak Ridge program. The lab responded by embracing a new Soviet development, the tokamak, which the other labs had neglected and which thenceforth formed the centerpiece of a thriving program.[21]

Rover and Pluto

The specialization achieved in Sherwood came at the initiative of lab scientists, who did not want to duplicate the work of their colleagues. The ample budgets for fusion thereafter allowed continued support of parallel projects. The labs were not always so lucky; budgetary pressures could sometimes force the labs into separate paths from above, as projects Rover and Pluto, for nuclear rocket propulsion, demonstrate.

The possible use of nuclear energy for rocket propulsion emerged after the war, mostly in optimistic flights of fancy. It resurfaced independently at Los Alamos and Livermore around 1954 and received official sanction from the Mills committee of the air force and then from the AEC, which requested the two weapons labs to initiate small programs. Why Los Alamos and Livermore? The work required reactor design and would fall under the Division of Reactor Development, which might have suggested Argonne or Oak Ridge. Bradbury and York considered it a weapons program, perhaps because of early designs after the war that used explosives for propulsion but more likely because of the close connection with the air force. Despite the precedents in Aircraft Nuclear Propulsion at Oak Ridge and Naval Reactors at Argonne, AEC staff noted that the weapons labs were already studying the problem and decided to let them continue.[22]

They could still turn to one of the reactor labs for expertise. Los Alamos took its inspiration from a design presented by Robert Bussard, a reactor physicist from Oak Ridge who often visited Los Alamos. Bussard had proposed a so-called heat exchanger, a reactor operating at high temperatures (above 2,000°F) that transferred its heat to a propellant such as hydrogen that flowed through the reactor core and then was expanded and expelled through the rocket nozzle. Bussard elaborated this idea with Los Alamos scientists into a staged system using a chemical booster to launch the rocket and a nuclear-powered, hydrogen-propelled final stage once the rocket was up to speed.[23]

Los Alamos presented its plans to the Mills committee in March 1955. At the same meeting York and other Livermore scientists sketched preliminary designs for an all-nuclear rocket motor. Darol Froman of Los Alamos was thus "very surprised," at another meeting a few weeks later, to hear York claim that Livermore had decided that a chemically boosted, heat-exchanger nuclear rocket—that is, the same system Los Alamos was pursuing—would be best. Froman noted that Los Alamos had already described such a system and had sent Livermore a copy of the detailed technical report on it: "I had been under the impression that we had made this point clear." Livermore now had the gall to squat in the territory already staked out by Los Alamos. But the two designs were not identical, as York pointed out: Los Alamos used an all-graphite core loaded with uranium 235, surrounded by a beryllium reflector; Livermore mixed the beryllium with graphite and uranium in a heterogeneous core.[24]

The dual programs raised the possibility of centralization. The AEC worried that nuclear rocket research would distract the weapons labs from their main mission. Commissioner Willard Libby suggested that the AEC "establish a separate laboratory for this work when large capital expenditures became necessary"; W. Kenneth Davis, the director of Reactor Development, pointed out that the small initial scale of the programs made centralization a decision for the future.[25] Rabi twice questioned before the GAC whether parallel efforts were the most efficient path; Colonel Alfred Starbird, director of the Division of Military Application, defended competition as a consequence of "the initiative of the laboratories" and noted that the two labs were also competing for the Atlas warhead.[26] In both cases members of the AEC and GAC suggested centralization and AEC program managers defended parallel programs. The AEC agreed to continue, for the time being, the dual approaches to nuclear rocket propulsion and called the program "Project Rover."

The two labs needed no explicit directives to differentiate their programs. Los Alamos stuck with its original plan but adopted ammonia as a propellant. Livermore proposed to use first nitrogen, then hydrogen gas, in a test not of a small-scale model of the whole system but in a full-scale model of only the central "blowpipe" section. When the question of centralization arose again at the GAC meeting in July 1956, GAC members could point to the different approaches and agree, with James Fisk, that any overlap was "no more than is healthy and justifiable"; a narrow approach might neglect promising options. Rabi agreed, but insisted that eventually the two designs must converge. Warren Johnson added that "convergence should occur before too much money is expended in Nevada," where the labs would build and test their designs. York and Raemer Schreiber, who headed the Los Alamos effort, both claimed that informal coordination between the labs prevented duplication. York made an analogy to the weapons program: the labs engaged in competition with informal exchange at the research stage and then one lab won the responsibility for specific designs. York admitted that the analogy implied "convergence" —that is, the selection of only one lab for development—but insisted that it should not happen for another three years at the earliest.[27]

The committee members reached a consensus that "we are not concerned at the present time with parallel and possibly duplicated effort." But they were not satisfied with informal exchange and suggested "that the laboratories consider the possibility of cooperating strongly, rather than competing," by the establishment of a joint committee from both labs to coordinate the work.[28] The AEC did not set up any formal committee, perhaps because it trusted the Division of Reactor Development, which administered the program, to provide its usual active oversight. York and Bradbury wrote a joint letter to the division promising cooperation and pointing out the different, "complementary" approaches taken by the two labs. Their programs "covered the range from small reactors with low carbon to uranium ratios and complex moderation and control schemes, to the large reactors with high carbon to uranium ratios and simpler moderation and control problems. The Los Alamos Scientific Laboratory has started its studies at the low end of the size spectrum and worked upward while the Livermore Laboratory has started at the high end and worked downward."[29]

The labs differentiated not only their specific designs but also their general approach to development, which again reveals the philosophy of each lab: the ambitious empiricism of Livermore versus the careful reliance on theory of Los Alamos. From the start, Los Alamos scientists had presented their plans as "fairly straightforward," in contrast to the "exotic" systems Livermore was exploring. Even after Livermore settled on the heat exchanger embraced by Los Alamos, it continued to explore "advanced designs" such as the so-called slow-bomb system, which would harness the impulse from a small nuclear explosion or explosions. The slow-bomb idea fizzled—launching a rocket with a nuclear bomb produced, not surprisingly, excessive shocks and accelerations and would require too much weight in the form of containment and fissile fuel.[30] Nevertheless, subsequent plans, if less extreme, displayed the same tendencies. Livermore's blowpipe tests at Nevada of a full-scale portion of the reactor relied "on early nuclear heating tests rather than on an extensive series of local component tests done with electrical heating prior to attempting any nuclear powered tests"; the tests "aimed at developing a relatively large, high-power test engine." Los Alamos stuck with a smaller, more conservative design and would wait a year to test it, preferring "extensive and detailed analysis and testing of separate design features and components," with the component tests to use electrical instead of nuclear heat.[31]

The enthusiasm of the military, and the budgets it provoked, allowed the continued support of parallel approaches. In early 1955 the National Security Council made the intercontinental ballistic missile the highest-priority R&D program in the United States. The AEC noted the directive, and several members of the GAC concluded that the AEC therefore should have two top priorities: lightweight, high-yield thermonuclear weapons to arm the ICBM, and nuclear rocket propulsion to power it.[32] York and Froman had pointed out the advantages of nuclear power for the requirements of the Atlas program: nuclear power provided higher exhaust velocities and was limited in range only by the amount of propellant carried, and thus could cut the weight of the rocket at least in half for intercontinental ranges or allow for larger payloads; it also avoided the problem, faced by chemical combustion, of igniting the second stage above the atmosphere. Both noted only in passing the hazards of radioactivity; as Froman put it,

"Incidentally, a nuclear rocket with a chemical booster presents no contamination or radiation problem at the firing site."[33]

Their arguments convinced the Department of Defense, which in April 1956 requested that the AEC ground-test a nuclear rocket motor by early 1959 and test one in flight by 1962. In response, the AEC boosted lab budgets into orbit. Operating costs ran to $2.2 million at Los Alamos and $1.5 million at Livermore for fiscal year 1956; the AEC sought to double the Rover budget for fiscal 1957 and also rushed a $25 million construction item into the request. The crash program supported dual, but not duplicative, programs and sustained the rhetoric of competition: the AEC stressed to the Joint Committee that the expansion supported a "non-overlapping, bilateral laboratory approach."[34]

The early enthusiasm soon waned, as rapid development of chemical ICBM motors made an expensive, long-range program for nuclear propulsion less attractive. By the fall of 1956, budgeteers had noticed the budget trajectories and sought to correct them. In October the AEC submitted its requests for fiscal 1958: $30 million for operations and $20 million for construction to support the two-lab approach. The Bureau of the Budget balked: it sliced the operations request to $12 million, axed the construction item altogether, and chopped $10 million from the construction budget already appropriated for 1957; it also asked the military to reconsider the priority of Rover. The Department of Defense appointed the inevitable committee, which concluded that the military could justify only a modest effort.[35]

Without military support the AEC could not appeal the budget decisions. The directors of Reactor Development and Military Application decided to pull the plug on Livermore's Rover work and leave Los Alamos to pursue nuclear rockets alone. A modest program did not necessarily mean a frugal one: the operating budget at Los Alamos rose to $8 million for fiscal 1958, and the lab enjoyed a spacious ceiling of $15 million on its construction budget, much of which would be spent at the Nevada Test Site. The director of Military Application estimated that the AEC would save at most 30 percent from consolidation of Rover by the end of fiscal 1959.[36]

Some of the potential savings went to Livermore, which did not lose out entirely: it kept $2 million from Rover to spend on another project, whose presence at the lab probably determined the award of Rover to Los Alamos. Since January 1956 Livermore had been performing materials research for a program to develop a nuclear-powered ramjet. Los Alamos had considered the possibility of nuclear ramjets in its initial studies of early 1955, and that October the secretary of defense had asked the AEC to investigate them. On the one hand, unlike ballistic missiles operating above the atmosphere, air-breathing ramjets could capitalize on the almost limitless energy available in a nuclear reactor without having to carry propellants; on the other hand, the presence of air allowed the easy ignition of chemical combustion. The AEC assigned responsibility for what it called the Pluto project to the Curtiss-Wright aircraft corporation with the materials research at Livermore and at North American Aviation. This initial setup consciously fostered competition on the project between Curtiss-Wright and Livermore.[37]

Once again, budget examiners dampened early enthusiasm: the AEC proposed to raise the Pluto operating budget from $4 million for fiscal 1957 to $10 million for fiscal 1958, with another $9 million for construction; the Bureau of the Budget approved only $3 million for operations, but allowed that the AEC might shift up to $2 million from other programs. The AEC took the hint: in March 1957, a few months after Livermore's Rover program got the axe, the Division of Reactor Development recommended that Livermore serve as the sole Pluto contractor. The Department of Defense supported the plan, and the AEC canceled the Curtiss-Wright contract. Why Livermore? Its capability in materials research addressed one of the main problems, in high-temperature ceramics, and Livermore could apply experience and facilities from Rover to Pluto.[38] It also provided a neat solution to the Rover problem. The AEC would not have relished terminating a substantial program at one of its large labs, one created just five years before under considerable congressional and military pressure. Curtiss-Wright, an aircraft manufacturer new to the nuclear business, carried no such attachment to the AEC.

The end result by late 1957 was a division of labor between the two labs, Los Alamos taking over all Rover work toward nuclear rocket motors and Livermore concentrating on nuclear ramjets for Pluto. The early differentiation between the two labs in the Rover program did not prevent the termination of Livermore's Rover work; Curtiss-Wright suffered the same fate for Pluto. In the absence of budget support for a crash program, the AEC could impose a division of labor from above, to produce specialization beyond that achieved by the labs themselves. In the case of Rover and Pluto, the coincidental programs gave the AEC an easy way to enforce specialization without having to halt the institutional inertia behind large laboratory programs. The AEC could shift the momentum behind Livermore's Rover project to Pluto and the burden of termination from one of its own labs to an industrial contractor. The presence at the labs of alternative programs gave the AEC the luxury of this option.

Research Reactors

The research reactor program displayed differentiation from above and below. Through the mid-1950s the Materials Testing Reactor, completed in Idaho in 1952, remained the state of the art; the research reactors approved for Argonne (the CP-5 reactor) and Oak Ridge (Oak Ridge Research Reactor, or ORR) in the early 1950s did not match its neutron flux. In early 1956 the Division of Reactor Development, increasingly aware that materials problems were a bottleneck to power and propulsion reactor programs, decided to support more advanced research reactors and solicited proposals. Lab scientists accepted the invitation with alacrity. Within months, or less, Argonne, Brookhaven, and Oak Ridge responded with proposals for high-flux research reactors.[39]

The designs of the proposed reactors reflected the needs of different groups and priorities at each lab. Chemists clamored for high internal fluxes to produce heavy isotopes by neutron capture. Reactor engineers and metallurgists also sought higher internal fluxes to test materials and components, which hydraulic or pneumatic "rabbits" carried to the reactor core. Physicists wanted higher-intensity external beams for neutron cross-section and diffraction studies.

Brookhaven chose to satisfy physicists. The lab from its inception had planned to build a second, higher-flux research reactor. In August 1956 Brookhaven, encouraged by the AEC's interest, formed a study group chaired by Donald Hughes, head of the neutron physics group, and including others with an interest in solid-state physics and neutron diffraction. The panelists settled on a reactor to suit their interests: a small reactor core surrounded by a moderator of heavy water to concentrate thermal neutrons outside the core. Beam tubes tangential to the moderator further increased slow-neutron flux, since radial tubes would have conveyed more fast neutrons from the reactor core. The design, dubbed the High Flux Beam Reactor, would provide 5 to 7 $\times$ 10^{14} n/cm^2/sec external flux of thermal neutrons, about twenty times the internal flux of the lab's existing graphite reactor, at a cost of $10 million.[40]

The design of the Oak Ridge reactor stemmed from the lab's emphasis on chemistry and isotope production. Like Brookhaven, Oak Ridge was already contemplating its next research reactor when the AEC solicited proposals. In the fall of 1956 Oak Ridge appointed a panel to study a new reactor. The lab's chemists, who wanted to make transuranium elements, were the only group to express special interest. Oak Ridge thus proposed a "flux trap" reactor cooled and moderated by light water, with a cylindrical ring-shaped core to concentrate neutrons in the center of the reactor. The name of the proposal, High-Flux Isotope Reactor, reflected its primary purpose.[41]

Argonne took a still different tack. Meetings at the lab in early 1957 found interest in a new reactor from physicists, chemists, and metallurgists. The reactor designers elected to appease them all: instead of "a reactor designed for a single user and a single experiment," they chose versatility and provided for both internal irradiations and external beams. Like Oak Ridge they used an annular core, but cooled and moderated by heavy water to produce a thermal neutron flux of 5×10^{15} n/cm^2/sec, four times that of the CP-5 reactor. The design, "a large quantity of muscle contained in a small body," inspired the name: Mighty Mouse, which the lab obtained approval from CBS Television to use, although it declined to permit CBS to use the reactor to publicize the cartoon.[42] Fortunately so for the network, as the proposal would fail to live up to its name.

The labs arrived at their various designs after a year or so of work and sent them up the line to the AEC. Brookhaven's plan received first approval from the AEC, perhaps because Oak Ridge and Argonne already had newer research reactors in place and also, as one Argonne scientist noted, "the BNL reactor seems to have the most unique features and if one is to be chosen then that would be it."[43] In the meantime Argonne and Oak Ridge jockeyed for position, with Argonne at first on the inside track, again apparently because Oak Ridge had the newer reactor.[44] AEC staff, however, quickly reined in Mighty Mouse, whose ambitious, versatile design ran its price tag to $65 million. The proposal lacked a champion in Washington: the Division of Research could not afford development of an advanced reactor design, while the Division of Reactor Development refused to finance what it perceived as a research facility. And neither division had received any strong statement in support of Mighty Mouse from potential users, as opposed to its designers. The AEC killed Mighty Mouse in July 1958.[45]

Chastened but unbowed, Argonne scaled back its ambition but not its general approach. It returned that fall with an "austerity model" that retained the Mighty Mouse design and purposes but eliminated some of the more elaborate experimental capabilities, such as angled beam holes and cryogenics, and lab and office space for visitors. It was "assumed that the staff of experimenters [were] largely ANL staff members." Thus abandoning a principle of the national labs, Argonne knocked the cost down to $26 million.[46] The new proposal did not fail to invoke the Soviet bogeyman: "It seems very likely that such reactors will be built elsewhere, probably in the USSR. . . . If they are not built here, Argonne, or even the U.S., may fall behind in a field of science in which we have hitherto led the world."[47]

Oak Ridge in the meantime continued to back its own isotope-production reactor. Representatives of the two labs met in Washington with AEC staff in November 1958 to try to reach a consensus. The outcome of the meeting depended on the observer. John Williams of the Research Division reported that the parties agreed to specialization: "Emphasis should be given to reactors designed to be most suitable for specific research areas rather than to a reactor which would attempt to satisfy all needs." Thus, Williams deduced, the AEC could support one flux-trap reactor for chemists and another with beam holes for physicists, but should not try to combine them. According to Williams, the participants "unanimously agreed" that Oak Ridge would build the isotope producer for chemists, given its tradition in isotope production and research. Argonne would develop a proposal for a physics reactor for future consideration.[48]

At Argonne, however, Leonard Link perceived only a "tentative decision" that Oak Ridge would build the isotope producer.[49] Since the AEC seemed to assign the isotope reactor first priority, Argonne staff argued that their lab could provide as good a home, or better, than Oak Ridge, whose experience in isotope production did not extend beyond plutonium in the periodic table.[50] Link pondered whether Argonne should challenge Oak Ridge and enter into "a vigorous competitive situation to obtain AEC support" for an isotope producer, either with or without beam holes.[51] But Argonne staff preferred versatility and questioned Williams's premise of specialization. Winston Manning, the head of the Chemistry Division at Argonne, saw "no discontinuity" between a reactor for chemists and one for physicists: "It would seem unfortunate to construct a high flux reactor without *any* beam holes if a few could be included with only a fractional increase in cost."[52]

Argonne nevertheless protested when Oak Ridge proceeded to add beam holes to its design. Manning and others at Argonne anticipated the possibility and its implications. "A second reactor will be harder to justify than the first, particularly if there are a few beam hole facilities in the first one."[53] Argonne thus sought to "do everything possible to maintain differentiation of the two reactors. If this can be done it should essentially remove the 'competitive' situation and maintain the identity of two separate systems, each designed to meet the requirements of the two organizations."[54] Oak Ridge, however, succeeded in running beam tubes through the flux trap and in persuading Williams to allow their addition.[55]

Despite the apparent violation of the settlement, Williams tried to honor the commitment to a reactor for Argonne. The isotope reactor for Oak Ridge entered the AEC budget for 1961 and Argonne's proposal was scheduled to follow in 1962.[56] Williams had asked for "evidence of enthusiasm" for a high-flux reactor "from some physicists outside of Argonne," so the lab canvassed scientists at midwestern universities. The "testimonials we collected for our sales campaign" added up to tepid support.[57] The AEC omitted the proposal from its budget for 1962 and 1963. Argonne administrators lobbied the congressional Joint Committee, which included two members from Illinois, and helpfully forwarded information on a new Soviet reactor; the committee added the reactor to the budget for 1964.[58] But technical problems and cost overruns plagued the project, and four years later the Joint Committee sided with the Bureau of the Budget and rescinded its authorization. The reactors at Brookhaven and Oak Ridge had gone critical in 1965. Argonne had only a hole in the ground to show for its twelve years of work.[59]

The saga of Mighty Mouse and its successors at Argonne highlights two general trends in the evolution of the national labs: user-friendliness and specialization. Argonne's proposals failed on both counts. Brookhaven's original research reactor had provided neutron beams for many visitors from industrial and military labs, and its new reactor promised to improve the service. The High-Flux Isotope Reactor at Oak Ridge had the interest and support of an outside group, albeit from within the lab system: transplutonium chemists under Glenn Seaborg at Berkeley. By contrast, from 1958 forward Argonne acknowledged, indeed trumpeted, that its reactor would serve Argonne staff. Argonne's proposals again demonstrated the lab's neglect of outside users. AEC staff asked whether the Mighty Mouse reactor "could solve the MURA problem??!!!"[60] Mighty Mouse designers claimed to have answered AEC's expressed desire for "a machine for national rather than 'Argonne only' usage"; after its cancellation they felt "bitter about having been taken in by this 'NATIONAL' pitch. We blithely went all out on facilities and money."[61] But an internal Argonne memo noted that Argonne staff might have used 50 to 75 percent of Mighty Mouse facilities; and the initial proposal admitted, "While it is not claimed that Argonne will be the sole user of this reactor, the Laboratory would be the major user."[62] Mighty Mouse would hardly have solved the MURA problem or qualified as a "national" facility. Argonne's subsequent proposals would not even try to do so.

From the outset of the lab system, the "national" lab concept had meant regional labs: the provision of facilities for visiting scientists from nearby institutions. Argonne's dispute with MURA stemmed from this definition. But specialization implied a shift in the meaning of the term "national." Previous research reactors—CP-5 at Argonne, the Brookhaven Graphite Research Reactor, the Oak Ridge Research Reactor—were general-purpose, suitable in design, if not in practice, for use by a wide variety of constituents. In the late 1950s the new generation of research reactors specialized to suit particular needs. The names of the reactors suggest the shift: from the early "Research Reactors" at Brookhaven and Oak Ridge to the High Flux Beam Reactor and High-Flux Isotope Reactor, one reactor for physicists and one for chemists. Argonne's successive proposals for what it finally called the Argonne Advanced Research Reactor tried to please everyone and failed. Argonne did not follow a trend toward single-purpose facilities unique in the United States, the result of specialization. If the labs did not duplicate facilities of the other sites in the system, each particular facility would serve not just a regional constituency but a national one.

DIVERSIFICATION

In June 1952, in the midst of the national emergency and the rapid expansion of the nuclear weapons complex, AEC chairman Gordon Dean drafted a memo forecasting the future. The AEC's production facilities would surely produce a sufficient stockpile of nuclear weapons by the early 1960s. "What does this mean to the National Laboratories now supported by the Commission with public funds? . . . Suppose we were in a position to state flatly today to the directors of these laboratories that the United States will have achieved by 1964 such a stockpile of weapons that it is going out of the atomic weapons business." How would they respond? The government might continue to support research at the labs, with the exception of Los Alamos: "Is it not almost inevitable that the weapons research laboratory would shrink considerably at such a date?"[63] Dean was not alone in considering such questions. Lab scientists always kept a weather eye on the future, which would help them to survive and thrive in changing contexts.

They did so by diversifying. The nuclear rocket and fusion programs at the weapons labs, a power reactor program at Brookhaven, fusion at Oak Ridge—all point to this additional consequence of competition. The pursuit of multiple programs in the national labs allowed them to diversify, and the competitive ethic encouraged it. As these examples demonstrate, the labs diversified in parallel too; the AEC had defined its mission to include the support of certain fields that all of the labs felt free to enter. Hence Brookhaven and Los Alamos took up power reactor research, and Los Alamos, Oak Ridge, and Argonne entered the accelerator game. But the labs did not limit themselves to approved programs. As changing external contexts threatened their original missions—the expanding stockpile, fallout, and the test ban for the weapons program; the development of a nuclear power industry for reactors—the labs sought new ones. They found opportunities in the same contexts that constrained them.

Diversification in Parallel

The Castle-Bravo test in 1954 marked the end of the crash program to build the H-bomb and the start of an uncertain new period for the weapons labs. The H-bomb program had taxed Los Alamos to the limit and led to the creation of Livermore to assume some of the burden. The circumstances did not promote the acquisition of additional programs. When Strauss sought to expand the effort on Project Sherwood in 1953, he learned that "neither Dr. Bradbury nor Dr. York wanted the Controlled Thermonuclear Reactions program to grow because of interference with the weapons program."[64]

Amid the physical and figurative fallout from Castle-Bravo, however, the weapons labs began to recognize the handwriting on the wall that Dean had noticed two years earlier. One objection raised by the AEC against talk of a nuclear test ban in late 1954 was the effect on the weapons labs.[65] The following year Bradbury of Los Alamos addressed the future of his lab. In five to ten years, he predicted, "everyone will eventually have all the weapons in all the variety wanted." No radical advances comparable to the H-bomb appeared possible, and weapons research was devolving to tweaking and tinkering. "Because of this prospect, and because the Laboratory must have an intelligible, realistic, and exciting future for its staff if they are to remain and work hard on weapons *now*, we have increased our interest and effort in the reactor field, the nuclear propulsion fields, the Sherwood field, and in basic research."[66]

Whereas the weapons lab directors and their program manager had earlier opposed their diversion to Sherwood, they now agreed that new programs could help them. The staff paper approving the Rover program noted that, although it might compete with other reactor research for staff, "in the end, the nuclear rocket propulsion program may prove beneficial to the weapons program by attracting personnel to the laboratories."[67] The need to keep scientists on tap at the weapons labs worried special advisory panels considering a test ban in 1956 and 1958, and occupied the GAC at a meeting held over the weekend the moratorium went into effect in October 1958.[68] The AEC asked the GAC, "What general fields of research and development can the weapons laboratories pursue during a test moratorium which will keep their staffs at peak morale and effectiveness for expedited weapons development in case the test suspension is suddenly terminated?" The GAC suggested expansion of Rover, Pluto, and Sherwood, work on small mobile reactors, and more basic research on materials problems, which might help these reactor projects and also the aircraft reactor program at Oak Ridge. Since the AEC was already supporting work in all of these areas, the GAC could conclude "that the laboratories would not be at a loss for work in the event of a test moratorium."[69]

The test moratorium thus spurred the expansion of Sherwood, already boosted by the Geneva conference of 1958, and of Rover and Pluto. An additional justification for the last two had appeared in the sky a year earlier. Amid the general public panic in response to Sputnik, the AEC called for a crash program for aircraft nuclear propulsion in general and an increase in the Rover program in particular.[70] Thanks to Sputnik, the anticipated cost savings from the division of labor on Rover and Pluto failed to materialize. Far from it: operating budgets for both programs tripled from 1957 to 1960, and doubled again in the three years after that. The Joint Committee, in particular Senator Clinton Anderson of New Mexico, championed the programs as a way to get the AEC, and hence the Joint Committee, into the space business. As Bradbury put it, "If something is raining dollar bills on you, you're not going to say, 'No, I don't want them.' . . . So you swallow hard and get out your bucket."[71]

Rover also indirectly aided the entry of Los Alamos into power reactor research. Since the expansion of the production complex in the early 1950s, the AEC had been the largest consumer of electric power in the United States (some 12 percent of all generated power in 1957).[72] The labs consumed their share: accelerators, pumps for vacuums and to circulate reactor fuels and coolants, and more mundane power requirements taxed local power grids. In 1956 the Division of Reactor Development projected the power needs of Los Alamos to reach 30 megawatts by 1960, with the additional burden of electrical heating tests of Rover components and proposed accelerators. The division agreed to sponsor an experimental power reactor program at the lab to help provide power.[73] Los Alamos had built a series of research reactors and criticality experiments going back to the war and more recently had undertaken a couple of small experiments on power reactors. The new directive resulted in the Los Alamos Molten Plutonium Reactor Experiment (LAMPRE), which went critical in April 1961. LAMPRE, a fast breeder reactor using molten fuel, borrowed elements from the Argonne and Oak Ridge programs but differed in its use of liquid plutonium; Bradbury could claim that Los Alamos "knew more about plutonium than anybody else in the business."[74]

Project Pluto also gave Livermore entry into power reactor development. A prospectus on the future program of the lab from 1959 noted its Pluto experience and proposed building a small-scale prototype for civilian power. Livermore was careful to distinguish its proposal from existing reactor programs at Argonne, Brookhaven, and Oak Ridge, "with which as a general rule we do not desire to compete. Consequently we want to limit our contribution to the special field of ultra high-temperature, gas-cooled reactors." The lab's efforts led to a small program on power reactors for a few years in the early 1960s.[75]

The moratorium and the international race for fusion helped fusion and reactor development to make increasing inroads into the programs of the so-called weapons labs. In fiscal year 1957 both Los Alamos and Livermore gave one-quarter of their programs to non-weapons work (table 4); by 1961, when one factored in physical research, Los Alamos gave only one-third of its effort to weapons and Livermore little more than half.[76]

The diversification of the weapons labs alarmed the GAC, some of whose members wondered in 1960 what Los Alamos was doing in the power reactor business. Bradbury had to lobby the AEC over the head of the GAC in order to win approval of the successor to LAMPRE. The following year the GAC worried that Livermore and especially Los Alamos

Table 4. Operating budgets for Los Alamos and Livermore, in millions, for FY1956–1963

	1956	1957	1958	1959	1960	1961	1962	1963
Los Alamos								
Reactors for space	2.8	5.9	7.9	12.2	13.4	20.3	27.3	29.0
Power reactors	1.5	2.2	3.0	3.2	5.5	5.9	5.7	7.3
Sherwood	1.1	1.8	2.3	3.0	3.3	2.0	1.9	2.0
Biomedicine	0.8	0.9	0.9	1.0	0.9	1.0	1.0	1.0
Total lab operations	46.4	43.7	52.0	57.9	58.4	65.9	79.3	82.1
Livermore								
Reactors for space	1.5	3.3	2.9	6.9	13.9	18.6	19.2	21.5
Power reactors	—	—	—	—	—	0.1	0.2	0.2
Sherwood	3.2	4.1	5.6	5.3	5.5	6.1	6.5	7.0
Total lab operations	22.0	29.7	38.3	48.8	66.0	75.5	100.5	103.7

Source: "Multiprogram Laboratories, Summary of Operating Costs," May 1966 (GM, 5625/15); Sherwood figures through 1960 from *The Future Role.*

had lost their focus during the moratorium and were unprepared in case the moratorium should end; it recommended "that the AEC again tell both Los Alamos and Livermore that their primary mission is weapons research."[77] The sudden resumption of testing by the Soviets in September 1961 did catch the United States flat-footed; but the labs could divert scientists from their diversified programs to weapons work. Livermore quickly transferred 125 scientists from Pluto, 50 from Sherwood, and 50 from Project Plowshare to weapons design, in addition to hiring 175 new staff. It thus achieved a 40 percent increase in its weapons effort in the space of a few months, much of it from its scientists on tap; in-house staff provided most of the immediate response, then returned to their former programs as new hires came on the job. After the initial flurry of testing subsided and negotiations resumed toward the Limited Test Ban Treaty, both labs again sought to expand their non-weapons work and even make it their main priority. Several Los Alamos scientists informed the GAC that the lab was "losing its enthusiasm for weapons research and would be happier if its primary role were that of a multipurpose research laboratory with close ties to southwestern universities, on the pattern of Argonne or Brookhaven."[78]

The increasing diversity of programs at Los Alamos and Livermore diluted their designation as weapons labs. One might instead categorize them with the reactor labs. By 1960 Livermore was approaching and Los Alamos had passed Argonne in their levels of reactor work; in 1962 Los Alamos surpassed Oak Ridge to become the largest reactor program of the AEC (fig. 3). Even better, one might eliminate distinctions altogether. The diverse programs of both labs had brought them closer to the fold of the titular national labs. And Los Alamos and Livermore were not the only labs to diversify in this period.

The entry of Los Alamos and Livermore into power reactor development belied the uncertain long-term future of reactor work in the national labs. Atoms for Peace and the revision of the Atomic Energy Act in 1954 aimed to promote the quick development of nuclear power and its integration with electrical utilities. Like Los Alamos and Livermore, the reactor labs were victims of their own success: reactors were moving from research to development and construction. In 1954 the AEC approved the first commercial nuclear power plant at Shippingport, Pennsylvania, based on the pressurized-water reactor designed initially by Argonne for naval propulsion, and new industrial firms were joining Westinghouse and General Electric in the reactor business, some from the aviation industry.[79]

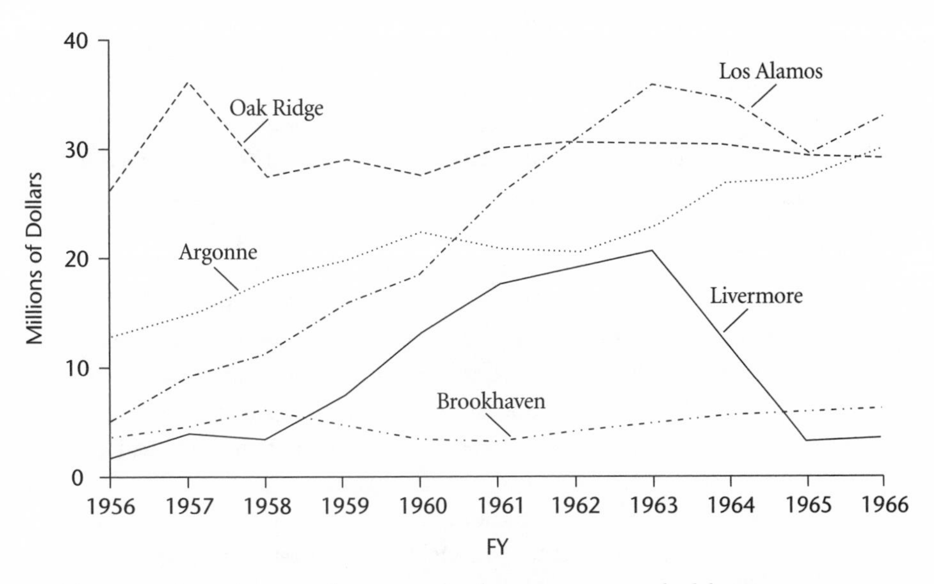

Figure 3. Operating costs for reactor development program by laboratory, FY1956–1966. Source: Material for lab directors' meeting, May 1966 (GM, 5625/15).

Weinberg noted the trend and asked what would happen to the national labs when the main reactor problems had been solved, which he predicted would occur within twenty years. He saw two choices: "They can very gracefully concern themselves only with the most basic aspects of nuclear energy; or . . . they can seek out large-scale problems quite apart from nuclear energy but which involve, in common with nuclear energy, the most fundamental questions of our biological and geological survival." Weinberg preferred the latter. "It is reasonable to expect that the basic studies will gradually diversify, and that the laboratories, whose frame of reference originally was nuclear energy, may, after a generation, be concerned with many other issues. For example, work in photosynthesis or in cancer."[80]

Weinberg did not note that Oak Ridge and the other labs already had programs under way in both photosynthesis and cancer research. Like the weapons labs, Oak Ridge and Argonne found it easiest to enter fields already assured of AEC support. Hence both labs squeezed into the fusion field, Argonne at a level of effort of one or two hundred thousand dollars a year in the late 1950s, small relative to the crash program at Oak Ridge for Geneva in 1958.[81] Argonne also put out a bucket to catch some of the dollars raining on Rover—another example of the centrifugal force of crash programs. In 1962 the lab proposed a gas-cooled, ceramic-metal fueled rocket reactor as an alternative to the Los Alamos design, which the AEC then supported with about $3 million a year until 1966.[82]

Argonne and Oak Ridge also trespassed on the domain of Berkeley and Brookhaven. The discovery and development of strong focusing encouraged the other labs to contemplate high-energy accelerators. Argonne had developed its initial proposal of early 1953 into a fixed-field, alternating-gradient machine whose second stage aimed at 25 BeV with one hundred times the intensity of the Brookhaven AGS. Argonne reluctantly settled for the more conventional 12-BeV accelerator to beat the Soviets to high energies, which would still give it twice the energy available in Berkeley from the Bevatron. As with the weapons labs, diversification threatened primary missions: in recommending the accelerator for Argonne, the AEC had also stressed that reactors remained the lab's first priority.[83]

Oak Ridge likewise staked a claim in high-energy physics. In 1953 Weinberg appointed a committee to consider the "question of what ORNL might be doing in physics in ten years," in particular whether that should include high-energy physics. The committee's affirmative response launched a design program under the lab's Electronuclear Research Division.[84] By early 1956 T. A. Welton had developed a cyclotron design based on new ideas about sector focusing, a principle first proposed in 1938 by L. H. Thomas as a means to overcome relativistic limitations on the energies of accelerated particles. Thomas's principle employed a magnetic field that increased with radius in order to bend the faster, heavier particles in larger orbits, but with successive pie-shaped pieces of the cyclotron alternating between strong and weak fields to keep particles in planar orbits. Berkeley had confirmed the principle in a small cyclotron model built under the MTA project, and midwestern scientists working under MURA, Argonne's competitor, were combining it with strong focusing.[85] Welton had collaborated with the MURA group and designed a cyclotron using spiral instead of pie-shaped sectors, which Oak Ridge pitched to the AEC in March 1956.[86]

Sector-focused cyclotrons offered an alternative route around relativity to that provided by synchrocyclotrons and proton synchrotrons, which achieved the same end by varying the frequency of the electric field. The fixed frequency of sector focusing allowed continuous operation and hence higher intensity than pulsed machines like linacs or synchrotrons. Both the initial proposal from Argonne for a 25-BeV machine and Welton's design for Oak Ridge used fixed electric fields to achieve high intensities, and thus differed from the designs of Berkeley's and Brookhaven's accelerators and also the experimental aims: high-intensity beams gave high statistical accuracy and could capitalize on electronic logic counters, whereas lower-intensity machines relied on photographic images from cloud or bubble chambers or nuclear emulsions to identify rare "golden events." Cold War politics had soon overtaken Argonne's differentiated design and forced the lab to build a machine similar to Brookhaven's AGS; but a year later, after the initial reaction to the revelation of Dubna had subsided and a panel for the National Science Foundation had identified intensity as a valuable goal in addition to high energy, the AEC amended the proposal to allow Argonne to build a high-intensity, weak-focusing machine.[87]

The Oak Ridge design also was pulled in several directions. Welton's cyclotron aimed at 1–2 BeV, which did not guarantee that it could produce K mesons at their threshold energy of 1.8 BeV. Weinberg and Welton pushed ahead and argued that even below 1 BeV the accelerator would still produce pi mesons, which Hans Bethe had recently suggested might be more important than K mesons at binding energies—and "ORNL's basic mission is in binding energy nuclear physics." They also stressed that Oak Ridge should maintain an "active part in the most advanced fields of physics" and that an accelerator would not divert effort from reactors and other applied programs. The GAC encouraged their plans, but the Argonne and MURA proposals continued to occupy the AEC's attention on accelerators.[88]

In the meantime, another group criticized the lab's modest, mission-oriented approach. Unlike Argonne, Oak Ridge enlisted the support of regional universities, but it too found that university scientists might not share the priorities of the lab. The lab formed a joint accelerator committee with members of the Oak Ridge Institute of Nuclear Studies, the consortium of southern universities formed to capitalize on lab facilities. The academic scientists on the committee rejected Welton's proposed cyclotron and insisted that the lab aim for "at least" 8 BeV at high intensity, in order to study the rare antinucleons recently produced in Berkeley's Bevatron. The lab agreed to investigate the possibility and sent its scientists to consult accelerator designers at Brookhaven, Berkeley, and MURA. But Oak Ridge scientists also sought to satisfy the criterion, held by "the management of ORNL and presumably the AEC, . . . that we should come up with a proposal which has unique features."[89]

Oak Ridge sent its revised proposal to the AEC in May 1957. It incorporated a spiral-ridged, sector-focused cyclotron at 900 MeV injecting into a strong-focusing synchrotron to reach 8 BeV, at intensities "by far the highest in the world."[90] The design tried to satisfy everyone, and failed. Despite Weinberg and Welton's earlier admission that Oak Ridge "is not necessarily equipped to pursue 20 BeV phenomena," the lab found itself, at the insistence of southern academic scientists, pushing toward 12 BeV.[91] Berkeley physicists defended their territory and their "doctrine" of high energies instead of intensities; Luis Alvarez, who was building a mammoth bubble chamber to capture rare high-energy events, circulated a programmatic declaration that "high intensity is not important."[92] The AEC balked at the $20–25 million price tag. It was already spending comparable amounts on accelerators at Brookhaven and Argonne, with the MURA group waiting in the wings and smaller projects under way at universities, and had also just received the Stanford linac proposal for an estimated cost of $78 million.

Perhaps to make the proposal more palatable to budgeteers, Oak Ridge focused on the cyclotron injector and implied that the higher-energy second stage could wait for the future; in other words, the lab reverted to its original plan, which was closer to its mission and sidestepped difficult technical problems of beam injection. The regional scientists then reminded the lab of the "indivisibility" of the two stages, and the GAC judged that the proposal "continues to reflect the enthusiasm of a small group of machine builders rather than the needs of the scientific community."[93] With only lukewarm support from the GAC, the AEC rejected the whole plan.[94]

Meanwhile, not to be left out, Los Alamos and Livermore also pursued high-energy physics programs. Livermore deferred to its parent lab at Berkeley and did not propose new accelerators for itself, but it maintained a small high-energy program using the remnants of the MTA.[95] At Los Alamos, in October 1955 two physicists, H. V. Argo and F. L. Ribe, drew up plans for a spiral-ridge, sector-focused cyclotron to reach 2 BeV. Argo and Ribe thought to use frequency modulation at large radii to keep the beam focused, and Darragh Nagle then suggested the use of several superimposed radio frequency fields either to increase the intensity of a fixed-field machine or improve the duty factor of pulsed machines.[96] Aside from these variations, however, the plan resembled the machine designed by Welton at Oak Ridge. Los Alamos scientists admitted the derivative character of their design but argued that novelty was not the point: "For us the work has meaning only if it leads to the erection of a high energy research facility."[97] Also like Oak Ridge, the Los Alamos group was soon aiming higher and designing a 12-BeV proton synchrotron, which it estimated would cost $15 million.[98] Duplicative designs would not have found favor at the AEC; although Bradbury beseeched Strauss not to lose sight of Los Alamos in the accelerator derby, the Los Alamos plans did not yet make it that far.[99]

By the summer of 1956 an optimistic Nagle was surveying possible locations for an accelerator at the lab.[100] But not everyone at Los Alamos wanted to jump on the accelerator bandwagon. Some physicists thought the lab was too late, "that 'all the cream will be skimmed off,' that the whole subject of new particles is and will remain recondite, that interest in it is a passing fad." An accelerator would siphon scientists from Sherwood and other projects. The chairman of the Physics Division, J. M. B. Kellogg, preferred an "all-out Sherwood effort"; "what are we doing fooling around with accelerator planning when we've got Sherwood?" An accelerator program would also exacerbate the "chronic" housing shortage. Accelerator proponents took care to recognize "the present overriding priority of the weapons program, and . . . the high priority of certain other programs."

They admitted that an accelerator would compete "for men, money, and housing with other secondary projects."[101] But, like their colleagues at Oak Ridge, they did not want to miss the bandwagon when it pulled out of town. High-energy physics would be "the most productive and the most glamorous activity" in the next five to ten years. And well supported: "Certainly great amounts of governmental funds will be spent in building accelerators in the coming years. Some of this money should be directed to conserve the exceptional national asset which now exists at Los Alamos."[102]

Appeals to the scientists-on-tap model intensified the next year after Eisenhower announced, on 22 August 1958, that the United States would stop nuclear testing on 31 October. Less than a week after the announcement, Los Alamos physicists revived plans for an accelerator: "The present turn of events would indicate that the topic of multibev machines is again germane."[103] They noted that nuclear physics research had attracted scientists to the lab, who had then helped the programmatic effort. "Many of these people could probably not have been hired directly for their present activities." The language of fashion persisted: "Low-energy nuclear physics is no longer the glamorous field which it used to be." What was? "At present the most glamorous field of physics is high-energy and elementary-particle physics."[104] Without an accelerator "many staff members interested in these fields [of nuclear and sub-nuclear physics] will probably become discouraged and leave." To enhance the chance of acceptance, the lab might enlist the support of regional universities, as Oak Ridge had done: "Would it be possible that making part of the Los Alamos Laboratory a 'National Laboratory' might solve some of the problems, budget or otherwise?"[105]

The arguments for an accelerator at Los Alamos could be distilled to an essence: "because people here want one."[106] The AEC would need stronger reasons. AEC staff and the GAC did not consider accelerators among the alternative programs that might sustain the weapons labs during the moratorium.[107] But the AEC based its rejection of accelerators at Los Alamos and Oak Ridge not on any official policy to limit the accelerator program; the multiplication of machines elsewhere, such as Argonne, belied any such policy, but at the same time soaked up funds that might have supported further multiplication.

Hence, the cold water thrown by the AEC on high-energy efforts at Oak Ridge and Los Alamos failed to extinguish enthusiasm. Laboratory scientists again demonstrated their refusal to accept bad news, and proposals for new machines at both labs proliferated faster than the AEC could reject them. In 1959, only months after the AEC shot down their proposal, Oak Ridge scientists considered with regional colleagues the possibility of a scaled-down version of Brookhaven's AGS, to reach 12 BeV, before reverting to their plan for an 850-MeV cyclotron for pion physics.[108] Los Alamos returned with its own plans for a pion-generating linac, to reach 800 MeV with 1 milliamp intensity.[109]

Both labs took care to justify future proposals based on their traditional strengths: Oak Ridge pitched its cyclotron as "the natural extension of our interest in nuclear structure rather than a major excursion into the field of high-energy physics," and the Los Alamos linac aimed at research in "nuclear physics—as distinct from high energy physics."[110] These justifications rested on a redefinition of "high energy," which by the mid-1960s did not mean what it had ten years earlier.[111] They rested also on the continued diffusion of the categories of the national labs, which brought former weapon and reactor labs into the accelerator business as their original missions dwindled.

Berkeley and Brookhaven, the first-class passengers on the accelerator bandwagon, saw the others as contenders for their seats. Hence they looked to other established missions into which they might diversify. Fusion offered one possibility. Berkeley maintained a small program in support of Livermore's work, which it doubled between 1957 and 1959 to a $1 million effort.[112] At Brookhaven, Haworth in 1958 preferred that the lab pursue small-scale theoretical work on fusion but not large-scale engineering projects; AUI's trustees agreed but urged that the lab's long-range plan include fusion work.[113]

Fusion would not gain much of a foothold at Brookhaven, but not because the lab refused to engineer large projects. A reactor program thrived there despite the misgivings of Haworth and other administrators. After completion of the research reactor, Brookhaven's metallurgists and chemical engineers began to explore liquid-metal fuels and breeder blankets for power reactors; they soon developed a design using a molten alloy of uranium 233 and bismuth that flowed through a perforated graphite cylinder, with a surrounding blanket of thorium to breed more uranium 233 through neutron capture. Haworth noted that the program did not duplicate the reactor work of the other labs. The AEC's Research Division supported the early chemical and metallurgical work, but after the lab's engineers had settled on their design, the Division of Reactor Development took over.[114] The Liquid Metal Fueled Reactor (LMFR) at Brookhaven would consume $1.1 million a year through 1955, or around 10 percent of the total lab budget. In 1955 the lab and AEC reactor staff accelerated the effort; at its peak in 1958 the project consumed $4.2 million, or one-fourth of the total lab program. Although the AEC canceled the project the next year, Brookhaven continued with reactor research and by the mid-1960s had recovered to its earlier peak and added a small program on rocket propulsion.[115]

The LMFR project at Brookhaven illustrates the ability of lab scientists to follow a fruitful line of research into areas beyond the lab's program, despite the resistance of lab administrators. In this case the support of basic research at a national lab fostered diversification into a programmatic activity supported by the AEC. As the pursuit of accelerators by Argonne, Oak Ridge, and Los Alamos demonstrates, diversification could flow in the opposite direction, from programmatic development toward basic research. Both of these directions, toward reactors and accelerators, still fell within the mission of the AEC. But the labs would also push into new fields and test the boundaries of their accepted missions.

Diversification by Divergence

The national labs had justified basic research for both of their main purposes: to keep on tap scientists who might work on applications of atomic energy, and to support the facilities provided by the labs to visiting scientists. Both supported diversification. As to the first, Haworth argued, "it is impossible, and indeed improper, rigidly to channelize the thinking of first class scientists." Hence, beyond the labs' "somewhat restricted focus on 'atomic energy,' born of necessity, their interests will develop toward much broader horizons." As to the second, he suggested, "In view of the cooperative nature of the Brookhaven program, it seems appropriate also to maintain a broad and diversified base" in order to serve "the varied interests of potential visitors."[116] Low-flying basic research helped lab scientists maneuver beyond the AEC's airspace—for instance, into astrophysics and molecular biology. Even large programs accepted by the AEC could start small: discretionary funds allowed the initial establishment of fusion research at Los Alamos and Oak Ridge, and Brookhaven's reactor program started out as basic metallurgical and chemical research. One catches only fleeting glimpses of stealthy small science in the historical record, but they provide important evidence for diversification.

Small science still had to pass at some point under the eyes of AEC program managers, and the easiest route to new fields thus led from accepted missions. Reactor waste disposal and fallout justified ecological research; fallout and reactor emissions sustained meteorology; underground weapons tests spurred seismology. As Bradbury explained his strategy at Los Alamos, "[You] put your public attention primarily on this [weapons work], and you then say, 'I can't keep on doing this unless I have a big active basic research program and all the fields which are relevant to nuclear weaponry.' . . . And, boy, I can sure stretch relevant."[117] A few examples will demonstrate how lab scientists stretched the boundaries of accepted programs to encompass new fields.

COMPUTERS One of the tools that the labs used to till new fields was the electronic computer. The weapons program of the AEC had fostered the early development of computers. During the war, Los Alamos scientists had strained the capacities of mechanical desk calculators and had turned to an electronic computer at the University of Pennsylvania, the ENIAC (Electronic Numerical Integrator and Computer). The ENIAC was initially intended to calculate ballistic tables but instead was first used for hydrodynamic calculations for Los Alamos in late 1945. After the war, the computing needs of Los Alamos led the AEC to support the development of the second generation of electronic computers, one under John von Neumann at the Institute for Advanced Study in Princeton and a similar machine built by Nicholas Metropolis at Los Alamos, dubbed the MANIAC, both of which were first tested with calculations for the hydrogen bomb in 1952.[118]

The importance of computers to the weapons program led the AEC to subsidize the development of a computing industry. From his post on the General Advisory Committee and then on the AEC, von Neumann declared that "high-speed computers are just as vital to the AEC development programs as . . . high-speed particle accelerators." The AEC provided a double intellectual and financial subsidy to the computing industry: lab scientists developed the technology, transferred it to industry for production, and then bought it back. In 1955 Livermore commissioned a computer from Remington Rand, the Livermore Automatic Research Computer (LARC), which Livermore programmers helped to develop. Likewise, Los Alamos scientists helped design the so-called STRETCH computer that the lab bought from IBM. The future product lines of both IBM and Remington Rand would reflect the innovations developed in collaboration with lab scientists.[119]

Systemicity fostered the spread of computers and the diversification of their uses. Livermore at its inception lacked computers and sent its staff to Los Alamos to use an IBM 701 machine there; the midnight-to-eight A.M. shift assigned to the visitors from Livermore undoubtedly encouraged them to obtain their own machine.[120] Oak Ridge had set up a computing program in the late 1940s under Alston Householder and Cuthbert Hurd, again under the impetus of the weapons program—in this case, calculation of the iterations in the gaseous diffusion process of uranium separation.[121] Argonne benefited from the presence on its staff of J. C. Chu, who had helped build the ENIAC, and Donald "Moll" Flanders, who had organized the mechanical computing effort at Los Alamos during the war. Chu and Flanders designed a copy of von Neumann's machine for Argonne, dubbed the AVIDAC, and then collaborated with the Oak Ridge group on a more powerful version for Oak Ridge, the ORACLE. When the new computers came on-line in 1953, scientists at Oak Ridge and Argonne applied them to reactor calculations but soon found other interesting problems. Oak Ridge used ORACLE to plot particle orbits in its design of a high-energy accelerator, and Argonne physicists tracked elementary particle interactions with AVIDAC; ORACLE was used for work on fusion and health physics, AVIDAC on radiobiology and solid-state physics.[122]

The pursuit of computers by the national labs stemmed from programmatic needs. But increasing recognition of their scientific usefulness also suggested their inclusion as research facilities. Early suggestions by Philip Morse that Brookhaven pursue a computer provoked no "very intense interest" at the lab or the AEC.[123] Then in 1952 the AEC decided to buy a UNIVAC from Remington Rand and locate it somewhere in the Northeast, primarily to serve John Wheeler's work on the hydrogen bomb at Princeton. Wheeler thought a computer too large a service function for a university, and Brookhaven's trustees thought the lab might provide one as a "unique facility," like the reactor, for scientists at regional universities. The AEC noted the computers at the other national labs and urged Brookhaven to take on the UNIVAC. Haworth, however, yet again reported "no great enthusiasm for a computer" from Brookhaven staff, especially since most of the first two years of operating time would be reserved for weapons problems. The AEC gave up and sent the UNIVAC to New York University.[124]

Brookhaven would pay for its apathy. Literally: its physicists soon found the UNIVAC useful for, among other things, calculating the design of the Alternating Gradient Synchrotron, and the AEC surprised the lab with a bill for $45,000 for computer time in 1954. That year Haworth reported "decidedly more interest in computing work at the Laboratory." After first considering the purchase of a commercial machine, Brookhaven decided to capitalize, like Argonne and Oak Ridge, on the computer expertise within the lab system, and in 1956 it embarked on the construction of a copy of the MANIAC II at Los Alamos.[125] The interest in computing came from the accelerator program and the growing reactor work, and also from the solid-state physicists, who sought to calculate the many-body problems of radiation damage and neutron scattering in crystal lattices.[126]

The presence of powerful computers at the labs, justified by programmatic requirements, thus encouraged lab scientists to play with them. And administrators too: at Oak Ridge, the ORACLE analyzed lab salaries, personnel categories, and other arcane aspects of lab administration.[127] In the weapons program, the nuclear test moratorium increased the need for computer simulations to replace experimental tests, but it also stimulated the weapons labs to find uses for computers beyond their original purpose. George Gamow and Metropolis had already applied MANIAC to calculate the generation of amino acids from DNA elements; Los Alamos and Livermore scientists then turned their computers to accelerator design, elementary particle physics, materials science, celestial mechanics, and astrophysics.[128]

METEOROLOGY One of the opportunities Livermore perceived for its computers during the test moratorium was meteorology. Computers and codes developed for multidimensional fluid dynamics of weapons design could also model complex weather patterns. The labs had already made inroads into meteorology, again via the weapons and reactor programs. Los Alamos and Livermore scientists studied weather patterns to predict fallout from nuclear tests and picked up their efforts during the fallout controversy after Castle-Bravo.[129] Argonne, Oak Ridge, and Brookhaven had early established meteorological stations to help monitor emissions from their reactor stacks.[130] In 1952 the AEC revised its radiation tolerance standards upward and thus reduced the need for monitoring, but Haworth was reluctant to shut down Brookhaven's station. "Meteorology, and specifically micro-meteorology, is far from a closed book." The lab had facilities and personnel in place; the flat geography of the lab site would allow generalization of local results; and the AEC needed meteorology for weapons tests and civil defense. AUI's trustees doubted the last argument—meteorology was "not particularly close to research programs of the Commission"—but supported the program anyway, and suggested that weather facilities might be available, at cost, to outside users. The AEC agreed to support a staff of eight to ten scientists, who would give two-thirds of their time to basic meteorological research and the remainder to routine monitoring.[131] Argonne also maintained its group but at the end of the decade worried that it had "neglected" meteorology; to correct the oversight it sought to establish its own micrometeorological research facility, with a small wind tunnel followed later by a full-scale version. "Costs will run very high but [are] believed warranted." The AEC rejected the idea.[132]

The labs were not content to predict the weather. They also sought to change it. In the technological enthusiasm that swept the United States after Sputnik, scientists answered the old saw "Everybody talks about the weather, but nobody does anything about it." The possibility that nuclear explosions affected the weather had arisen in the early 1950s, especially after the AEC started nuclear testing at the Nevada Test Site. In 1953 a congressman from Alabama, George Andrews, asked during the AEC's budget hearings, "Could there be any connection between these tests you had out in New Mexico [sic] and these tornadoes that have been ripping through the southwestern section of the country and the excessive rainfall we have had in the southwestern and southeastern sections?" The AEC saw no evidence for it and backed up its denial with a report from the Weather Bureau that found no relation between nuclear tests and the weather.[133]

In the late 1950s, however, lab scientists embraced the opposite view. Teller touted the possibility of "weather modification" in his programmatic statement on the future of Livermore. Livermore scientists had in mind the atmospheric detonation of thermonuclear weapons to change local weather patterns: bombs might melt polar ice to produce warmer temperatures, modify atmospheric pressure systems, dissipate hurricane centers, or convect ocean water to the atmosphere for rain clouds (the convected moisture "might be used to clean the lower air of pollutants," not referring, apparently, to fallout).[134] In 1961 Argonne scientists advanced their own plan to use airborne nuclear reactors, in a "hovering-type aircraft or balloon," to apply heat to rain clouds and thus affect precipitation.[135] The schemes of Argonne and Livermore were a particular example of a more general hubris that proposed dams across the Bering Strait or the Strait of Gibraltar, in addition to thermonuclear weapons, as a means to climate control.[136]

PLOWSHARE The technological enthusiasm of Livermore received official sanction in Project Plowshare for the peaceful uses of nuclear explosives. The project also represented the impulse to find redeeming uses of nuclear energy in general and, more difficult, nuclear weapons in particular. It took its name from the biblical prophecy, marked by Strauss in 1953 at his swearing-in, that "they shall beat their swords into plowshares." Teller, preaching to the choir in *Popular Mechanics* in 1960, conveyed the evangelical fervor of the enterprise: "We're going to work miracles."[137]

Plowshare had its efficient cause in a proposal of October 1956, by Herbert York at Livermore, for a conference to discuss peacetime uses of nuclear bombs. York noted that scientists at Livermore, Los Alamos, and Sandia were already considering several possibilities: electrical power production, plutonium production, and excavation and mining, including oil extraction from shale beds. The AEC, despite worries that a peaceful program could divert effort from weapons work, approved the classified conference for February 1957, at which the conferees concluded that the subject deserved further study. The Division of Military Application agreed to support a small program, under $100,000 a year. Spurred by interesting results from the Rainier nuclear test, an underground shot, Livermore scientists were soon agitating for a rapid expansion to $3 million by 1959. Their plans piqued the interest of commissioners Strauss and Willard Libby; the AEC agreed to the expansion and called it Project Plowshare.[138]

Plowshare had a material cause in MICE and BATS. The AEC had supported these two projects under the Division of Research since at least the middle of 1955; they stemmed from ideas of Theodore Taylor, a weapons designer at Los Alamos, to use nuclear weapons tests to produce more tritium and plutonium, for instance, by neutron capture in a surrounding layer of thorium oxide. Project MICE studied underground tests, BATS aboveground. Argonne and Oak Ridge, which had long experience in chemical engineering and the production of weapons material, represented the national labs in the programs; Argonne sponsored a conference on MICE chemistry in September 1955, but Oak Ridge came to assume a larger role. By 1958 the AEC was spending $200,000 a year on MICE, about half of which went to the U.S. Geological Survey for site studies (suitable media included fresh water, such as very deep lakes or the Greenland ice cap, and salt domes) and less than a quarter to Oak Ridge. The AEC planned to test the idea in the fall of 1958.[139]

York cited MICE and BATS in his original proposal, and the two projects provided a basis for discussion at the conference in Livermore. MICE continued in parallel with Plowshare but at different laboratories and under different AEC staff divisions. By 1958 the two projects were converging: the AEC was planning a second MICE shot in a salt dome in the spring of 1959, which would also test the potential of power production from water boiled to steam by the blast. A 1-megaton bomb might run a 300-megawatt plant for a couple of months, if the steam pipes could withstand the shock of the blast; the calculated cost of 5 or 6 mills per kilowatt-hour would be higher than the 3 mills of conventional plants but below the 8 mills of the first nuclear plant at Shippingport. The power production aspect brought Plowshare into the picture, and what began as a MICE project became a dual-purpose test.[140] The moratorium, however, postponed the proposed test and threatened the survival of MICE, BATS, and Plowshare. The AEC and Teller turned it around: the potential of Plowshare should preclude a moratorium.[141] The enthusiasm of Livermore scientists for the undemonstrated potential of Plowshare could not overcome the geopolitical momentum behind the moratorium, but the moratorium also encouraged research, if not field tests, on Plowshare as a way to occupy idled weapons scientists.[142]

York's initial proposal had cited several possible applications of nuclear explosives. The higher costs of electrical power production as compared to conventional power plants and the difficulties of protecting a power plant from blast discouraged the first. The second, production of weapons material, did not exactly qualify as a peaceful application. That left excavation, mining, and oil extraction as the most promising avenues to explore. The increasing dependence of postwar America on oil had spurred attempts to develop energy independence through domestic sources (a similar impetus to the push for nuclear power), including efforts in the late 1940s to extract oil from shale. In particular, the Western Slope of the Rocky Mountains in Colorado and Utah had sizable reserves of oil locked in shale formations, which might be freed by pulverizing and heating the rocks—for instance, by a nuclear explosion. The availability of cheap imported oil by the 1950s made gasoline production through nuclear explosions too expensive and inefficient; it would only again receive serious consideration in the energy crisis of the 1970s, by which time Plowshare efforts focused instead on stimulation of natural gas wells.[143] Thus, although the Division of Military Application expected, in an apparently unintentional metaphor, "mushrooming" requests from the petroleum industry, they failed to materialize.[144]

Plowshare scientists instead would heed the advice of the GAC and focus on "ditch-digging."[145] Livermore scientists by mid-1958 were planning Project Chariot for 1960, which aimed to use six bombs to blast a harbor into Alaska's arctic coast. Project Chariot did not meet with enthusiasm outside of Livermore. The proposed harbor site would be icebound nine months of the year (assuming that no climate controllers tried to dam the Bering Strait); Teller toured Alaska to drum up support but "found no one who could justify at this time the harbor on an economic or military basis." The delay from the moratorium allowed local opposition to organize, and the AEC canceled Project Chariot in 1962.[146]

Ditch-digging fared better, for a time, in warmer climes. In 1947 the governor of the Panama Canal had recommended construction of a new sea-level canal to replace the lock system of the original, which was taxed by increasing ship sizes and traffic. The recommendation lingered until 1955, when the Bureau of the Budget asked Strauss's opinion on the effect of nuclear weapons on the proposed canal, "in view of the major developments in nuclear weapons since the earlier report." Strauss replied that a bomb crater would block any canal, but that a sea-level canal might be less vulnerable to nuclear weapons than the lock system.[147] The Suez crisis of 1956 highlighted the advantages of an alternate canal and also set Livermore scientists to thinking about how to dig one; their thoughts turned to nuclear explosives. The possibility of nuclear excavation for a canal would become a major focus of the Plowshare program, even though plans for a canal took a back seat to the space program after Sputnik; they revived in the late 1960s before finally fading in 1970.[148]

All of these Plowshare programs fostered the diversification of the labs. Excavation and oil extraction stimulated the study at Livermore of geology and seismology, which also were necessary for monitoring the test moratorium.[149] Plowshare motivated the establishment at Livermore of a Biomedical Division in 1963, primarily to examine the biological and environmental effects of fallout.[150] Plowshare also encouraged lab scientists to think more broadly about energy problems beyond nuclear or thermonuclear power. Finally, it helped sustain Livermore through the moratorium: by 1960 Livermore was devoting about one hundred scientists and 5 percent of its budget to Plowshare.

Plowshare centered on Livermore but was not confined to it. Frederick Reines at Los Alamos had considered the possibility of peaceful uses in 1950, after the Soviet Union had claimed that the nuclear detonation Joe 1, detected by the United States, had been only for the purpose of construction projects "which necessitate large-scale explosive work with the application of the most up-to-date technical methods." Reines concluded that the Soviets had less innocent intentions and that the uses of nuclear bombs for peaceful engineering "appear at best to be extremely limited in scope."[151] By the late 1950s Reines had changed his mind and was studying electrical power production, although he thought that a moratorium would and should preclude the development of peaceful uses. One of his colleagues at Los Alamos, Stanislav Ulam, used the forum of Plowshare to revive the idea of nuclear bombs as a means of rocket propulsion.[152] Argonne, noting its earlier involvement in the MICE project, sought to participate in the thriving Plowshare work.[153] Plowshare was also not restricted to engineering. Donald Hughes at Brookhaven planned to use the Plowshare test in the salt dome, known as Gnome, as a source of neutrons for high-resolution spectroscopy. After Hughes's death in 1960, other Brookhaven scientists consulted on experiments for Gnome on fission and neutron capture cross-sections and gamma distributions.[154]

SPACE Perhaps the most popular direction for the labs to look for new opportunities was upward. Even before Sputnik the labs had several reasons to consider outer space and the upper atmosphere a suitable place to explore. Cosmic rays continued to offer a means to attain energies unattainable in lab accelerators or to attain them for less money. Projects Rover and Pluto brought Los Alamos and Livermore into rocket propulsion. The plasma physics problems of Sherwood related to research in stellar physics. After Sputnik the incentives would only increase. The expertise and organization already present in the national labs encouraged the AEC and the congressional Joint Committee to suggest that the AEC take responsibility for an American space program. The possibility of a space program within the AEC raised anew the question of centralization. The GAC once again favored centralization: if the AEC got the space program, the advisers recommended that it assign one lab, preferably Livermore, as the central facility for space research. The designs of the AEC and Joint Committee on outer space came to naught, after political maneuvering in Congress and the executive branch led in the end to NASA.[155] Neither the creation of NASA, however, nor the talk of a centralized lab could avert the gaze of any of the labs from space.

The test moratorium gave Los Alamos and Livermore cause to consider space research. When the GAC in the fall of 1958 suggested alternate programs to sustain the two labs, it included space research among the possibilities.[156] At the same time, weapons tests gave them reason to pursue it. The final rounds of tests in 1958 included several high-altitude shots to study the electromagnetic effects of nuclear weapons and the possibility of ballistic missile defense. The high-altitude tests included three fired as part of Operation Argus, the brainstorm of Nicholas Christofilos at Livermore. Sputnik had galvanized Christofilos to find a defense against Soviet missiles, and the adventurous atmosphere of Livermore nourished his imagination. He came up with the idea to use nuclear explosions to inject high-energy electrons into the magnetosphere; the electrons, trapped by the earth's magnetic field, might provide a radiation shield against ballistic missiles. By August 1958 the United States had a satellite in place to measure the effect, and the experiments went off three hundred miles above the South Atlantic. Operation Argus failed to provide a means of missile defense—the earth's magnetic field was too weak to support enough electrons—but it did introduce Livermore to atmospheric physics.[157] The need to monitor compliance during the test moratorium provided more experience and allowed Teller to assert, "In connection with the observation, detection and hiding of nuclear explosions we have acquired some expertness in high atmosphere physics, [and] in electrical phenomena of the atmosphere."[158]

High-altitude nuclear tests also led Los Alamos to consider some far-out phenomena. A list of subjects submitted for study with weapons tests included geomagnetism, lunar and interplanetary magnetic fields, the effect of solar winds on particle trajectories, pulsing of the solar corona and prominences or the aurora of Venus from a nuclear explosion, and the use of a bomb to punch a hole in the atmosphere of Venus to allow observation of its surface.[159] The absence of nuclear tests also spurred space science, which offered, in the words of an anonymous staff member, a "future for Los Alamos" during the test moratorium.[160] In order "that the laboratory should stay at the frontier," one Los Alamos physicist, John Brolley, Jr., proposed that the lab form a new A Division, for Astronautics, to include Rover but also study of matter-antimatter propulsion, gravity and relativity, astronomy and astrophysics, and cosmic radiation. Other lab physicists suggested that space flight could justify basic research into plasma physics, energy and heat transfer, and radiation damage in solids.[161]

The Sherwood program had already brought Los Alamos scientists into problems in plasma physics and magnetohydrodynamics and thus provided another route to astrophysics, which offered, along with high-energy physics, an alternative to low-energy nuclear physics as the focus of physics research at Los Alamos.[162] Los Alamos did not set up a new division for astronautics, but in early 1959 the physics division created a group for high-altitude physics out of the old Cockroft-Walton accelerator program, whose work had been dwindling for a couple years. Another Space Sciences Group developed instruments for satellite and rocket measurements of possible weapons tests in space and also for cosmic radiation measurements, and later in 1959 the Weapons Test (J) Division set up its own group for high-altitude phenomenology.[163]

The space race gave a new lease to programs at the other labs. At Oak Ridge, research on reactors and radioisotopes as power sources for satellites (assigned the acronym SNAP, for Space Nuclear Auxiliary Power) helped to offset the declining fortunes of Aircraft Nuclear Propulsion.[164] At Brookhaven, a few months after the AEC canceled the liquid-metal reactor project, the lab's nuclear engineers sought support from NASA under SNAP and reoriented their program to consider heat transfer at zero gravity. O. E. Dwyer, a Brookhaven engineer, noted that several other organizations, including Oak Ridge and Argonne, were entering the field and urged Brookhaven "to get into the swim": "Space technology is on the rapid ascendancy, and it is a good idea for the [Nuclear Engineering] Department to become involved in its development. We already have the high-temperature alkali-metal corrosion program in Metallurgy. A boiling heat transfer program would widen the bridgehead." The enterprise of Brookhaven's nuclear engineers instead persuaded the AEC's Aircraft Reactors branch to sustain the program, with more money than they could spend. They thus suffered in the long run when President Kennedy canceled the aircraft reactor project the following year; in the short run, the prospects of space had revived and reoriented their defunct program.[165]

Space also stimulated biomedical research and, through biomedicine, accelerators. Space flight would subject astronauts to high but intermittent fluxes of high-energy protons and lower but continuous fluxes of heavy nuclei. Biomedical researchers at the labs took advantage of the presence of accelerators and used them to evaluate the hazards of space flight. Biologists at Brookhaven under Howard J. Curtis used the lab's 60-inch cyclotron to simulate damage in tissue from cosmic radiation.[166] Part of Berkeley's biomedical group under Cornelius Tobias had continued wartime work on aviation medicine and investigated radiation hazards of high-altitude flight. The long experience of Berkeley with the biological effects of high-energy particles jibed with needs of the space program, and by 1962, with the support of NASA, the lab's HILAC and 88-inch and 184-inch cyclotrons were spending about thirty hours a week combined for biomedical studies. Competition for beam time on these machines led Tobias's group at Berkeley in 1961 to propose a biomedical accelerator, to provide 1-BeV protons and heavy ions with 125 MeV per nucleon; the proposal promised to give "special consideration" to NASA problems.[167]

Berkeley's plans inspired Oak Ridge, whose biologists were using a cyclotron at Harvard for cosmic radiation studies while waiting for the lab's 76-inch isochronous cyclotron to come on-line in 1962. The Berkeley proposal suggested that Oak Ridge appeal to space biology to justify its own long-delayed high-intensity cyclotron, but the lab's biologists and physicists agreed "that it would be unwise to climb on any biological accelerator bandwagon unless this is the only way in which the Laboratory could obtain interest and support in its 850-MeV accelerator."[168] Two more years of rejection persuaded the accelerator builders to climb aboard, and in 1963 the lab's cyclotron proposal cited space biology as one, though not the only, justification.[169]

MOLECULAR BIOLOGY The biological research for the space program suggests a final example of a field the labs entered. In the late 1950s, following the discovery of DNA, a discipline of molecular biology began to make institutional inroads, under various guises. Lab scientists were always quick to spot a trend, as their obeisance to the glamour of high-energy physics indicates, and physicists and biologists recognized an opportunity in molecular biology. The labs had gained an early foothold in biological research, including a strong genetics component, and were well stocked with the methods, machines, and members of the physics discipline perceived to undergird the new field.[170] Berkeley's biomedical group was already moving away from nuclear medicine toward molecular biology (or what they called "biophysics").[171] In November 1958 John Manley suggested "biophysics" in addition to astrophysics and solid-state physics as a replacement for waning work in nuclear physics in the Los Alamos Physics Division.[172] The following November Bradbury asserted that, since

> discovery of the fundamental role of DNA and RNA in living matter, molecular and theoretical biology are now at the same state as was nuclear

> physics at the time of the discovery of the neutron. During the next several decades, great advances in the biological and medical sciences may be expected through the fundamental and theoretical approach. In the event of a de-emphasis of nuclear weapons or a long-standing test moratorium, the laboratory's emphasis on theoretical and cellular biology may well be increased materially in the next 10 years.[173]

The same month the Brookhaven Biology Department urged that "greatly increased emphasis should be placed on molecular biology," including free radical chemistry, protein and nucleic acid biosynthesis, protein structure, and enzyme action mechanisms. The biochemistry group in the Oak Ridge Biology Division likewise planned "considerable expansion . . . on the interrelationships and genetic control of nucleic acid and protein biosynthesis, biochemical studies of gene action, and studies of cellular differentiation at the molecular level."[174] Molecular biology appealed to interdisciplinarity, such as crystallographic techniques shared with the solid-state physicist. At Brookhaven, a young biologist named Daniel Koshland, Jr., proposed an interdisciplinary institute combining physics and biology. A couple of years later, Brookhaven's trustees questioned the presence of molecular biology at the lab, since universities were equipped to pursue it. They accepted it because of its use of neutron diffraction techniques with the reactor and its interaction with the physics and chemistry divisions of the lab.[175]

Implications of Diversification

Diversification complicated the organizational Laboratory Problem. It affected Los Alamos and Livermore in particular, as it dragged them, protesting, from the shelter of the Division of Military Application. The entry of Los Alamos and Livermore into reactor work via Rover and Pluto raised the question of cognizance. From the start, Bradbury and York urged that Military Application have responsibility for Rover. Both viewed the project as weapon related, and the division by this time had primary responsibility for both labs. Further, any initial tests would take place at the Nevada Test Site, which the division controlled. But the division's staff paper instead perceived the problem as closer to reactors than weapons and cited the precedent of Aircraft Nuclear Propulsion, which fell under the Division of Reactor Development. The report also noted that Military Application had primary but not exclusive responsibility for Los Alamos and Livermore and that other divisions, such as Biology and Medicine, directed certain programs at the two labs. Hence Reactor Development in fiscal 1957 assumed budgetary and technical responsibility for the rocket programs.[176] Similarly, early fusion research was funded by the Division of Research for Oak Ridge and Princeton, Military Application at Los Alamos, and Reactor Development at Livermore, before the AEC consolidated the program in the Research Division. In both cases the diffusion of programs through the system further tangled lines of responsibility, and both exemplify the difficulty in deciding where new programs might fit within the AEC.

None of the labs abandoned their original priorities. Lab scientists were consistently careful to emphasize that new research would not detract from programmatic work. Despite the test moratorium and the accumulating American arsenal, weapons scientists pursued new ideas, such as the so-called clean weapons advocated by Teller and other Livermore designers, which ensured that weapons budgets would not fall too far (see appendix). Reactor research also continued to enjoy strong support; the entry of Los Alamos and Brookhaven into power reactor work shows that the decline forecast by Weinberg was only a long-term possibility. To preserve their base the labs took on new work by adding scientists instead of diverting them from existing programs. All of the labs ended the 1950s with many more scientists than they had had at the start of the decade; Los Alamos and Oak Ridge almost doubled their technical staffs, and Livermore had leapfrogged Argonne, Berkeley, and Brookhaven (table 5).

External pressures again promoted diversification. Fallout fears, the weapons test moratorium, and the development of a nuclear power industry squeezed the missions of the labs in new directions. The space race also challenged the labs: NASA threatened to cut into funding and provoked skeptics in the labs to deride the prospects of space exploration. In the initial reaction to Sputnik, Rabi warned that space efforts might sap resources from basic research. Weinberg, speaking before the American Rocket Society at a joint space-nuclear conference organized at Oak Ridge, argued that manned space flight was a dangerous waste of time and money. Instead he urged support for nuclear power—and molecular biology.[177] If Weinberg was right, the space program was wasting some of its money on Oak Ridge: $3 million in 1962, which supported 160 scientists on research relating to space flight.[178] Teller's trajectory also proceeded from obstacle to opportunity: after Teller crusaded against a test ban, he came to extol the expertise

Table 5. Scientific/engineering personnel and total employees at each lab, 1951–1959

Lab	1951	1952	1953	1954	1955	1956	1957	1958	1959
ANL: sci/eng	513	521	519	488	531	577	698	800	844
ANL: total	2,827	2,983	2,709	2,316	2,351	2,488	2,914	3,299	3,579
BNL: sci/eng	285	319	331	342	366	368	379	408	440
BNL: total	1,377	1,415	1,423	1,387	1,458	1,503	1,585	1,676	1,835
LASL: sci/eng	464	529	602	679	725	778	857	934	979
LASL: total	3,195	3,739	3,812	3,910	3,727	3,846	3,995	4,061	4,073
ORNL: sci/eng	763	884	1,011	1,028	1,143	1,223	1,350	1,398	1,457
ORNL: total	3,257	3,836	3,985	3,971	4,178	4,414	4,763	4,728	5,023
UCRL-B: sci/eng	546	556	519	493	495	536	588	624	693
UCRL-B: total	1,581	1,580	1,668	1,541	1,524	1,709	1,765	1,907	2,194
UCRL-L: sci/eng	—	—	275	428	535	654	778	885	931
UCRL-L: total	—	—	625	1,039	1,504	2,072	2,661	3,026	3,642
Total sci/eng	2,571	2,809	3,257	3,458	3,795	4,136	4,650	5,049	5,344
Total employees	12,237	13,553	13,597	14,164	14,742	16,032	17,683	18,697	20,346

Source: Compiled from figures for individual labs in *The Future Role.*

Note: All figures as of June 30 of each year. Figures for Argonne include staff at National Reactor Testing Station in Idaho (total Argonne staff there never numbered more than 100).

of Livermore in seismology, atmospheric science, and other fields needed to enforce it.

The competitive structure of the lab system encouraged scientists to exploit opportunities wherever they led. Talk of glamorous subjects and bandwagons indicates the perception of lab scientists that they might lose out if they did not keep up. Beyond the individual or program level, the need to ensure the continued viability of each lab as an institution also drove diversification. This is most apparent in the case of Los Alamos and Livermore, whose scientists consciously cast about, with the help of the GAC and AEC staff, for alternative programs to sustain them during the test moratorium; but it applies to all of the labs, for which diversification offered better hope of institutional survival in changing external environments. Despite, or because of, their constant concern about their future, they were quite successful. In a meeting with several directors of European labs in 1960, the Americans heard their colleagues worry about the survival of their institutions in the face of declining programs, especially in nuclear power. "The American laboratory directors, however, did not indicate any lack of faith in a worthwhile workload for the near future."[179]

7

EXEMPLARY ADDITIONS

Two fields exemplify the ability of lab scientists to identify and exploit interesting opportunities: biomedicine, and solid-state and materials science. Their addition and growth as programs roughly coincide with the first two periods in the history of the labs: biomedicine for 1947 to 1954, solid-state science for 1954 to 1962. Both fields derived support from programmatic goals, but scientists tested the limits of laboratory missions and convinced their overseers that diverse programs deserved support. Biomedical and solid-state programs benefited from the smallness of their science and the latitude it allowed them to diversify. They also serve as examples of the fertility of interdisciplinary research, as demonstrated in the cross-pollination of molecular biology at the labs with the techniques of the solid-state scientist.

BIOMEDICINE, 1947–1954

Alongside accelerators, one of the first fields pasted on to lab missions was biomedicine. All of the national labs eventually developed biomedical programs, even though this did not necessarily fulfill their main purposes. Biomedicine did not promise any applications of atomic energy for national security, with the exception of radiological warfare. Nor did it seem to require big, expensive equipment such as reactors and high-energy accelerators. Thus when in 1945 Walter Bartky's subcommittee of the Scientific Panel proposed the provision of large facilities at regional labs, biomedical scientists presented the only objections. But radiological warfare and the hazards of the production and use of nuclear weapons justified secret programs on the biological effects of radiation; and lab scientists quickly perceived opportunities for novel biomedical research using the reactors and accelerators already acquired by physical scientists. Unexpected political support for life science in the labs solidified its presence.

From Manhattan to Bikini

As with the labs themselves, to find the roots of biomedical research in the labs one must look back to the Manhattan Project, and in the case of Berkeley farther back to the 1930s. Ernest Lawrence had recognized the potential of biomedical applications of his cyclotron in the radiation beams it emitted, the radioisotopes it produced, and the philanthropic and government funds it stimulated. His Radiation Laboratory spun off two satellites for biomedical research: the Crocker Laboratory with its 60-inch "medical cyclotron" under Joseph Hamilton, and the Donner Laboratory under John Lawrence, Ernest's brother, to capitalize on the products of the Crocker cyclotron. World War II diverted the Crocker and Donner labs into military research with the rest of the Rad Lab. While Donner scientists concentrated on aviation medicine, the Crocker cyclotron was put to work for the Met Lab and Los Alamos making neptunium, plutonium, and lighter elements produced in the fission process.[1]

In the meantime, however, scientists at the Met Lab had perceived the biomedical implications of atomic energy. The study of x rays and radioactivity in the first decades of the century had demonstrated the health hazards of radiation, and Met Lab staff recognized that their chain-reacting pile might pose an even greater threat. Arthur Compton set up a Health Division within the Met Lab in July 1942 and chose Robert Stone to head it. Stone, a radiologist at the University of California Medical School in San Francisco, had collaborated on therapeutic tests of high-voltage x rays and neutrons at the Berkeley Rad Lab and thus was quite familiar with the medical effects of radiation. The Health Division consisted of sections for biological and medical research and health physics; a fourth, short-lived Military Section studied radiological warfare. The Health Division began studying the special hazards of radiation and toxic chemicals present in the project, developed detectors and dosimeters to measure them, and designed procedures for screening the safety of workers.[2]

The diffusion of the bomb project outward from its central location in Chicago carried along the biomedical work. The work of the Health Division suffered at first from lack of the materials it intended to study and of experts familiar with their effects. It turned to Hamilton in Berkeley, who had kept some of the elements produced by the Crocker cyclotron for studies of the metabolism of fissionable elements and fission products. The construction of a larger pile and isotope separation facilities at Clinton spurred the creation there of another Health Division; as with the rest of the Clinton staff, many of the new group came from Chicago.[3] At Los Alamos, Oppenheimer recognized the need for health precautions and recruited Louis Hempelmann, Jr., a young doctor who had studied radiobiology in Berkeley, to head a health group.[4] To coordinate the proliferating programs, the MED set up a project-wide Medical Section in the summer of 1943 under Stafford Warren, a radiologist at Rochester.

Little of this work qualified as basic research. The Met Lab's Health Division, including its biological research section, focused on "work directly connected with immediate problems of the Project." In May 1943 Stone proposed a "marked expansion" of biological research at the Met Lab, which the MED refused to fund, as it also did for other long-term programs.[5] Hamilton in Berkeley commented on his group's preoccupation with the protection of workers' health.[6] In Los Alamos, Oppenheimer resisted attempts by Hempelmann to undertake basic research: "We are not equipped for biological experiments." Concern among lab chemists over their exposure to plutonium persuaded Oppenheimer to support a small research program with additional scientific staff in August 1944, but he continued to seek ways to delegate the program to other sites and limit distractions from the "urgent problems" of the lab; he placed the health group under the Administrative Division of the lab with other service groups.[7] From the headquarters of the MED, Warren, "a dictator not always benign," kept noses to the grindstone—that is, on short-term problems of radiation safety and monitoring. "The efforts of the District must not be diverted in any way to post-war problems."[8]

The end of the war did not bring an end to health programs. The Jeffries report of November 1944 on the prospects of nucleonics, to which Stone contributed, addressed applications to biology and medicine.[9] By January 1945 Stone was taking "active steps . . . to see that no hiatus will occur between Army and other sponsorship."[10] The end of the war found Stone still pressing for a postwar research program against Warren's resistance.[11] But Stone's colleagues were deserting his position: Howard Curtis at Clinton and Raymond Zirkle of the Met Lab urged the termination of their programs and the return of biological research to the universities, with financing from either the government or private corporations. Paul Henshaw of Clinton suggested that the U.S. Public Health Service might take over the work.[12] Other life scientists spoke with their feet and joined the postwar exodus from the labs. Among those leaving were Stone, who returned to San Francisco, and Warren, who was named dean of the new medical school at UCLA. Both Stone and Warren would capitalize on their MED ties and obtain AEC support of their research. Stone took care to note that his proposed program "definitely does not involve fundamental research."[13]

If Stone appeared to capitulate, however, other scientists took up the cause of biomedical research. Another subcommittee of the Scientific Panel in June 1945, this one on the MED's research program, included biology in its postwar plans for reciprocal reasons: nucleonics needed biology to understand the effects of radiation for health and safety, and biology could use new tools provided by nucleonics.[14] Hamilton in Berkeley did not plan any expansion into basic research, but he perceived the opportunities at Chicago and Clinton as "an entirely different situation" in which expansion of research facilities might be warranted.[15] Henshaw at Clinton had also come around. He had "not until recently visualized a postwar program in biology at this laboratory that I would care to participate in." But after the end of the war, he could "think of no place better suited for certain fundamental research in biology."[16]

Cancer research stood at the top of Henshaw's list of possible research topics. The field offered particular opportunities: study of the mechanism of carcinogenesis through external radiation exposure, use of radioisotopic tracers to explore internal pathways and mechanisms of carcinogens, and development of radiotherapy. The MED tried to contain the enthusiasm for basic research in general and cancer research in particular. Berkeley had started cancer research before the war, but now an MED administrator reminded Ernest Lawrence that the district "cannot support such work," which he described as "academic biological research."[17] In July 1946, however, the MED announced that it would allow up to 15 percent of each medical program to go to basic research. After some confusion, the MED tightened its definition: "The words 'basic research' and 'fundamental academic research' are to be interpreted to mean only research in the treatment of malignant tissues."[18] Groves also restricted the 15 percent clause to research contracts with a strong biological or medical research component, such as Berkeley or Chicago; it did not apply to sites "directly connected with operational problems," such as Clinton Labs and, presumably, Los Alamos.[19]

The 15 percent rule received support from the Medical Advisory Committee, appointed by the district to consider the postwar program and chaired by Warren. At its meeting in September 1946, the committee endorsed the rule and broadened the definition beyond cancer to embrace "remotely related problems which may not culminate in immediately applicable information." Basic research would exploit the evident opportunities and also stem the flow of scientists away from the project, since "a considerable portion of the required research is not of the type spontaneously chosen by investigators of the desired caliber." This early appearance of the scientists-on-tap model, combined with the desire of scientists to explore promising avenues of research, underlay the committee's support of basic research.[20] The labs added a third purpose that would become standard: Argonne, Clinton, and Los Alamos all included training of new personnel in the techniques of radiobiology and radiation safety in their projected research programs for 1947.[21]

The MED adopted the broader definition of basic research, beyond cancer, although the clause continued to confuse administrators.[22] It limited basic research to a small though not insignificant fraction of growing biomedical budgets: for fiscal year 1946–47 Argonne spent $1.2 million on biomedicine; Berkeley, $235,000; Clinton, $180,000; and Los Alamos, $100,000.[23] As the AEC took over the project, biomedicine had gained a toehold at the labs, but long-term basic research was in a precarious position. Life scientists would have to convince the AEC that biomedical research belonged among the missions of the labs.

Biomedicine and the AEC

Biomedicine lacked sponsors in the fledgling AEC. The General Advisory Committee was not general but instead limited to the physical sciences and engineering, and the lone scientist on the AEC was a physicist. The Division of Research, led by another physicist, included no representatives for biomedical research. The commission recognized this shortcoming and in January 1947 reconvened Warren's committee, now called the Interim Medical Committee, with wider membership. The committee urged expansion of biomedical research in the long term; for the present it recommended continuation of current programs. It also suggested the establishment of an advisory committee for "Health-Safety" comparable to the GAC and a Division of Health Safety to go along with the other statutory staff divisions of the AEC.[24] After a subsequent meeting the committee persuaded the AEC to raise the limit for basic research to 20 percent of each program, in order to attract and retain top-flight scientists. A Medical Board of Review of eminent academic life scientists repeated these suggestions in June.[25]

The support of biomedicine by independent and in-house life scientists still had to overcome ambivalence within the AEC. As he had for accelerators, James Fisk, the director of research, found the "relevance of biological research to [the] over-all program not at all clear. We can, however, do almost anything in that line under the law."[26] The Atomic Energy Act of 1946 allowed research on "utilization of fissionable and radioactive materials for medical or health purposes" and "the protection of health during research and production."[27] The AEC found reasons to support biomedical research under these provisions. To Congress, it cited its responsibility to "extend fundamental knowledge of the interaction of nuclear radiation and living matter," on the grounds that basic research could help improve health protection and reduce costs from excessive safety margins.[28] This rationale appealed also to lab contractors: "We feel keenly," said William Harrell of the University of Chicago, "about the health hazards involved in this work."[29] Both the Interim Medical Committee and the Medical Board of Review had urged that, in addition to research in support of programmatic responsibilities, the commission support basic research to attract top scientists and to train new scientists in health physics and radiobiology.

But while the Interim Medical Committee, consisting of representatives from the wartime sites, sought to continue research at the labs, the academic scientists on the Board of Review echoed the early position of the GAC on physical research. Although the labs could help train students and provide equipment to universities, the committee preferred academic investigators over "the initiation of broad new programs of research in laboratories which are not specially qualified to deal with such basic investigations. Scientific manpower will in this way be conserved, scientific accomplishments will be more significant, and teachers will be conserved to the universities."[30] Like the GAC, the board reasoned that the program would profit more in the long run from training new scientists in universities than from diverting scarce talent to the labs.

Instead of incorporating biomedicine within its existing framework, the

AEC assembled a separate apparatus for it. It appointed an Advisory Committee for Biology and Medicine, which first met in September 1947, and began to establish what would become the Division of Biology and Medicine, which Shields Warren agreed to lead until the AEC could find a permanent director.[31] The committee would have an "unusually active advisory capacity"; although it technically advised the commission and left operating decisions to the Division of Biology and Medicine, the division's staff generally reviewed programs and some specific decisions with the committee.[32] Like the early GAC, the advisory committee had the confidence of the AEC, but unlike the GAC, the biomedical committee maintained its influence: in 1955 an AEC staff study noted that the AEC "practically always" accepted its recommendations.[33] Also like the GAC, despite its objective advisory role the committee often acted as a partisan of the labs, in particular in defense against budget cuts. It would, in addition, serve as a force for further diversification. In December 1948 and again the next year it urged expansion of the biomedical program beyond radiobiology, since "there is a considerable deficit in our understanding of normal structure and function in living materials, and . . . without such knowledge the problem relating to radiation effects cannot be resolved."[34]

As the AEC got its house in order, it deferred the question of basic research at the labs and agreed only to continue current projects through 1947.[35] Lab scientists did not sit idle. All of the labs were drawing up ambitious programs in biomedicine, which taxed the limited mandate of interim committees. Argonne and Berkeley had the best-established programs, which they sought to continue and expand. Argonne's Biology Division under Austin Brues proposed for 1947 the second-largest research program at the lab, behind only chemistry; it had more scientific staff and twice the budget of the experimental nuclear physics group, which, unlike the biomedical program, had been depopulated to staff Los Alamos during the war. Argonne also supported divisions for health physics and medicine, both of which undertook some research in support of their service functions.[36] Berkeley's biomedical programs under John Lawrence and Hamilton doubled their effort in the first half of 1947 and proposed to increase by another third for the following year; by mid-1948 Rad Lab staff included ten biologists, six physiologists, six medical doctors, and six biochemists.[37]

Clinton Lab maintained its health physics group after the war under Karl Z. Morgan, which, in addition to its service function for radiation monitoring, devoted about a third of its effort to research.[38] To advance biological research, the lab hired Alexander Hollaender from the U.S. Public Health Service in November 1946 to head a Biology Division.[39] Hollaender—"a remarkable scientific impresario," "a business man [who] knew how to back people, how to get money"—quickly set about building an empire in Oak Ridge.[40] He proposed a minimum staff of 47, including lab technicians, to expand to a permanent staff of 75. The MED Medical Committee rejected the plan as too ambitious; two months later Hollaender returned with much the same proposal.[41] By early 1948 his division had a total staff of 66, of whom 57 had at least a bachelor's degree and 26 had Ph.D.s or M.D.s. A new group in bio-organic chemistry, brought over from the Chemistry Division, and the mouse genetics program helped fuel the expansion and joined existing groups in cytogenetics, microbiology, physiology and pharmacology, and biochemistry. Hollaender, unsatisfied, planned a further expansion to 112 total staff, with about 80 scientists.[42] The division took two years to reach the target. In March 1950 Hollaender estimated that over half the effort of the 115 staff members was going to basic research, although he did not want to publicize the fact—not for fear of tipping off the AEC, but because a 50 percent programmatic load would seem a lot to his scientists. "The great advantage of our setup here is that our people do not necessarily have a feeling that they are doing programmatic research where they actually are doing a very great amount of it."[43]

At Los Alamos, Hempelmann continued to push for research at the expense of programmatic activity. An advisory board to the AEC on industrial health and safety reported on his program in the fall of 1947: "The original health group has had much greater interest in potential research problems than in the establishment of good routines of laboratory safety."[44] Hempelmann's proposed program for fiscal year 1947–48 provided for a permanent research staff of thirty-five, plus six to twelve trainees.[45] Thomas Shipman, who took over the health group after Hempelmann's departure in 1948, perpetuated his approach. "The question is sometimes asked," said Shipman, "'Why is medical research carried out in a weapons development laboratory far remote from any other similar institution?' The people at Los Alamos would phrase the question differently, and would like to know if any one can think of a good reason why biomedical research should not be carried out there." Los Alamos had laboratory facilities and experts in diverse disciplines; "the workers in these fields, and the fabulous equipment which they use, much of which can be duplicated nowhere else, can and should be used in medical research."[46]

To take advantage of these opportunities, the lab's health group included a section on radiobiology with twenty-two scientific staff (out of a total technical staff for the health group of sixty in November 1947). In 1948 the radiobiology section estimated it spent about half its time "on work directly related to problems of the Los Alamos Laboratory," plus some service work for other divisions. That left 27 percent for "work of a fundamental nature and indirectly concerning the Laboratory" and 14 percent for "work unrelated to the problems at hand." Most of this basic research fell under organic chemistry and biochemistry, including new programs in the synthesis of compounds labeled with carbon 14 and study of their metabolism.[47] The AEC noted the size and diversity of the program and sought, according to commissioner Robert Bacher, "to ensure that only the most necessary activities in the biological and medical field were carried on at a laboratory whose primary objective was weapons research and development." The AEC's biomedical advisers pondered the "desirable extent" of the Los Alamos program, especially in light of the isolation of the lab and the housing problem, but they recommended its continuation primarily on the scientists-on-tap theory. The AEC accepted their advice.[48]

Brookhaven provides a particular example of the resistance to biomedical research and the enterprise of life scientists required to overcome it. Unlike the other sites, Brookhaven had no foundations from the Manhattan Project; its initial purpose, to provide large facilities for northeastern academic scientists, did not require a biomedical research program; and the planning committee of the lab consisted of a "preponderance of physicists," a condition continued on the board of trustees of AUI.[49] Nevertheless, perhaps because Brookhaven's founders had looked to existing labs for models, the initial lab program included biology and medical departments, which it tried to assemble from scratch.[50] The first head of the Biology Department, Leslie Nims, possessed ambition similar to that of his colleagues at other sites but less political acumen. He drew up plans for an initial staff of 100, of whom 40 would be scientists and most of the remainder lab technicians; he expected this skeleton crew to grow by 1952 to a staff of 250, with an annual budget of $1.4 million.[51] The plan matched the great expectations of the rest of Brookhaven and met the same fate. The AEC's Advisory Committee on Biology and Medicine wondered why a lab dedicated to visitors needed so many in-house biologists and questioned "the maturity of judgment displayed in Dr. Nims' program"; the AEC agreed and froze the department's permanent staff at 24 scientists.[52]

Brookhaven's Medical Department was faring little better against opposition closer to home. F. William Sundermann, who agreed to serve as interim head during the establishment of the department, like Nims made big plans, including a two-hundred-bed research hospital at a cost of almost $5 million, plus another $2.5 million for equipment.[53] Lab director Philip Morse doubted the potential of reactor neutron beams or high-energy particles for clinical research and hence the need for a clinic.[54] The trustees of AUI perceived no demand for a medical research program and cut the budget for it, leaving only enough to meet service needs; it was the only program they would cut in the lab's proposed $29 million budget.[55] Although some doctors from regional institutions welcomed the presence of a program at Brookhaven, others, notably those in New York City, thought that the isolation of the lab would deter both investigators and patients. C. P. Rhoads of Memorial Hospital in New York deigned to instruct the newcomers: "Clinical Investigation should be carried on principally where there exist the best Clinics and the best Investigators, since both require years to develop. They are found principally in the metropolitan teaching centers." Rhoads proposed instead a "Radiobiological Institute" in the city, perhaps with support from the AEC and staffed by local medical schools and teaching hospitals.[56] Similar sentiment at Argonne and Oak Ridge led to the establishment of medical clinics independent of the labs, under the University of Chicago and the Oak Ridge Institute of Nuclear Studies.[57]

Medical research at Brookhaven also faced the opposite problem to that of other sites: the perception that it was not fundamental enough. The founders of Brookhaven conceived their lab to be "engaged primarily in pure basic research," under which they included biology but not medicine: the trustees thought it "inadvisable . . . to make the theoretical biologists subordinate to applied scientists" in a medical program. "Developments at the Laboratory are more likely to be made in biology than in medicine."[58] This attitude would foster the eventual embrace of both biological and medical research at the lab. The turnaround began in early 1948. Sundermann had worn out his welcome with the AUI trustees and resigned, leaving the Medical Department in the hands of R. D. Conrad, a retired navy engineer. To salvage the program AUI prevailed on Donald Van Slyke, a recently retired eminence from the Rockefeller Institute for Medical Research, to join Brookhaven as a part-time consultant. It would turn into a full-time job.[59]

Van Slyke had the confidence of AUI from the outset: the trustees derived "tremendous hope from Dr. Van Slyke's appearance on the scene. . . . [H]e will draw an improved caliber of scientist, and that is what we need most of all." The AEC's biomedical advisers and staff also considered him "a great asset" and stemmed their criticism of Brookhaven.[60] Van Slyke also harbored big plans for a biomedical program, but the men he lured to lead it knew better than to advertise them.[61] Van Slyke first hired Lee Farr from Du Pont to head the Medical Department in the fall of 1948. Whereas Sundermann had been, in the words of Haworth, "unwilling to start from small beginnings," Farr aimed low: the Medical Department needed no special clinic, and it would limit clinical research to a few special cases in order to avoid competition with the institutions in AUI.[62] His caution won over the AUI board and AEC staff, who approved limited clinical research.[63] Van Slyke also eventually helped find a replacement for Nims in the Biology Department: Howard Curtis, a veteran of Clinton during the war and familiar to AEC staff, took over in mid-1950. From these more modest beginnings, Farr and Curtis would build a thriving program, including a new medical center with its own reactor.

Shields Warren identified some of the effects of systemicity in the establishment of biomedical research at Brookhaven. Diversification, he perceived, derived from knowledge of programs at other sites. "There is in all the national laboratories a continuing trend to multiply facilities in order that each laboratory may engage in some of the same type of activities as the other laboratories." Warren urged that Brookhaven resist this trend and "develop a distinctive type of work." His staff at the AEC "emphasized the desirability of the biology program being centered around the special facilities, especially the pile and cyclotron."[64]

The AEC acceded to the ambition of the labs and accepted biomedical programs, although lab staff still had to battle the impression "that biological work under the Atomic Energy Commission is looked on as a luxury or an invasion into such fields as cancer treatment or cell biology," as well as to convince "doubtful representatives of the Bureau of the Budget."[65] As the early history of Brookhaven demonstrates, ambition alone was insufficient; lab scientists had to back it up with ability and achievements. The AEC trumpeted their accumulating results in its semiannual report to Congress in the summer of 1949: radioisotopic tracer work at Berkeley on blood volume and total body water, at Brookhaven on radio-iron in blood cells and radio-phosphorus in sugar metabolism, at Los Alamos on B vitamins, and the biosynthesis of organic compounds labeled with carbon 14 at Argonne; genetics programs at all of the labs, including the extensive mouse program at Oak Ridge; agricultural research such as the "gamma field" at Brookhaven, whose cobalt 60 source irradiated plants in the field; biochemistry programs on radiation damage to nucleic acid, enzymes, and blood cells; and cancer research, such as the use of phosphorus 32 to treat leukemia and polycthemia at Berkeley and Argonne's use of radioarsenic on leukemia.[66]

Peacework

The AEC could allow biomedicine to take root and grow because of resources provided by the political framework. The report to Congress on the life sciences responded and appealed to public and political sympathy for peaceful uses of the atom.[67] During the investigation of the AEC by the congressional Joint Committee on Atomic Energy in 1949, Representative Melvin Price complained that the hearings had focused on weapons and suggested some presentation "on the peacetime application of atomic energy." The AEC obliged with a summary of its biomedical program, in which it estimated that one-third went to basic biological research and another third to basic medical research—a far cry from the 15 percent rule of the Manhattan Project. The Joint Committee applauded the program and blasted recent cuts in the biomedical budget by appropriations committees.[68]

The committee's attitude jibed with that of Lilienthal, who found biomedicine a "rather hopeful" subject with "a different aura to it than the subject of weapons," and also Strauss, who likewise early embraced the peaceful prospects of biomedicine and continued to support it when he chaired the commission.[69] Despite the Joint Committee's criticism, appropriations committees usually gave biomedical representatives and budgets easier treatment than those of other programs. The AEC's biomedical staff made a convert out of Representative Gore: "Instead of feeling fright that we have made this stupendous discovery, as I previously did, my attitude has changed to one of enthralldom at its possibilities and potentiality for mankind."[70] The spirit even infused the Bureau of the Budget: for fiscal year 1948 the bureau cut the initial physical research budget of the AEC roughly in half, to $5.3 million; at the same time it knocked only 8 percent off the biomedical budget, to $4.6 million.[71]

Sympathetic treatment persisted into the 1950s: for example, the biomedical program for fiscal year 1954 emerged unscathed from Eisenhower's Bureau of the Budget and both appropriations committees, the only AEC program accorded such protection.[72] Amid the national emergency, when biomedical programs shifted to national security topics—among others, neutron and tritium toxicity, radiological warfare, and fallout—with the rest of the AEC, they still served as a reminder, as Warren put it, that atomic energy was "not merely a means of killing people; . . . it is a means and a very valuable means of helping them."[73] An article on "atomic medicine" in *Time* magazine featured a picture of John Lawrence and Hamilton from Berkeley as well as some purple prose: "In the mushroom-clouded dawn of the Atomic Age, dark as it was with fresh dangers for the human race, there shone a ray of brilliant promise; the sudden abundance of radioactive elements gave medical researchers their most important new tools since the invention of the microscope." More breathless puffery appeared in a *Collier's* article, "Atomic Miracle: Science Explodes an Atom in a Woman's Brain," which featured the Brookhaven research reactor and boron neutron capture therapy.[74]

The House Appropriations Committee not only tolerated the budgets suggested by the AEC but also provided an unsought windfall that forced the AEC into a field it was avoiding and thus sealed the status of biomedical research in the labs. After the war, cancer recaptured its hold on the imaginations of Americans, whose increasing affluence encouraged them to assume their entitlement to good health and the availability of funds to ensure it.[75] In the spring of 1947, as the subcommittee on independent offices considered the first AEC budget, Representative Everett Dirksen brought up the topic of cancer. Admitting that it was "a little bit of a fetish" and "something of a crusade" for him, Dirksen noted that cancer killed one American every three minutes and caused the equivalent of seventy-two times the casualties from Pearl Harbor every year: "If we are going to spend a few hundred million dollars in the atomic-energy field to perfect an instrumentality of death, then let us take a little of that money to develop an instrumentality to preserve life." Dirksen's colleagues agreed: cancer research "is a 'bully' idea and something has to be done"; "Is not radioactivity the only hope of the scientists for finding a cure for cancer?" Dirksen suggested 5 or 10 percent of the AEC budget as an appropriate figure and thus arrived at the nice round number of $25 million, which the committee approved.[76]

The House action forced the Senate Appropriations Committee and the AEC into the unusual position of restoring fiscal austerity. Senator Clyde Reed noted that Congress had provided for the National Cancer Institute and was considering a National Science Foundation, and that the American Cancer Society had raised its own substantial endowment: "I do not know why in the world we should drag the Atomic Energy Commission into cancer research." Lilienthal replied that the AEC had not asked for the role and did not want it, although "we will, of course, do what the Congress directs us to do." He added, "We have got quite a big job without adding periphery to it." Strauss noted that the AEC would have trouble spending the $25 million.[77] The Senate knocked the number down to $5 million and the AEC set about searching for ways to spend it. At first it could find worthy projects for only $2 million, one-fifth of which would go to the labs to provide free beds for cancer patients; the bulk of the money would go to research contracts outside the labs. Soon, however, the AEC found that it could burn another million dollars on research in the labs, three-fourths of it at Argonne and the rest at Berkeley and Oak Ridge, in addition to building cancer clinics at Brookhaven, Chicago, and Oak Ridge.[78]

The labs thereafter benefited from the booty of the war on cancer. The AEC soon was facing charges of duplicating the work of other agencies, leveled by the same appropriations committees that had gotten it into the work in the first place. In response, some commissioners would suggest returning cancer research to universities and to other agencies.[79] Appeals to "humanitarian duty" from the Advisory Committee for Biology and Medicine and "constructive uses for atomic energy" by commissioner Henry Smyth stifled dissent in the AEC; in Congress, as one member complained, "it is hard to resist a plea on the floor of the House that you need to encourage research in cancer."[80] State governments also enlisted: a tour of Crocker and Donner labs in the spring of 1947 inspired Earl Warren, the governor of California, to support a bill appropriating $250,000 for cancer research in the University of California system; Donner Lab got the largest portion of it and would return for more in the following years.[81]

Lab scientists and their supporters learned to appreciate the peaceful appeal of their discipline: "Very great importance should be attached to the influence of the Medical Department at Brookhaven upon the public attitude and support of the entire atomic energy program. This is peace-work and humanitarian effort: it is for the public a welcome release from thinking about war and destruction."[82] The resources provided by political support soon made biomedical research strong enough to justify other programs—including reactors. After the centralization of reactor work at Argonne, Weinberg cast about for reasons for Oak Ridge to reenter the reactor field and build a new pile: "The question of motivation for the ORNL pile was discussed with Zinn, and informally with Fermi. On the whole it was felt that isotope production was probably the strongest justification for the machine, especially since the proposed budget cut specifically exempts biological research, and the isotope production function of the pile falls largely within the biological and medical spheres."[83]

Political support solidified the position of biomedical research in the national labs by the early 1950s. In January 1950 biomedicine accounted for about 20 percent of the program at Argonne, the nominal reactor lab, about the same amount provided for physical research.[84] At the same time in Berkeley, biomedicine amounted to almost a third of the Rad Lab program, and its funding from the AEC by the following year had jumped an order of magnitude from the levels of 1947. Berkeley benefited further from outside funds: in 1950 Donner Lab received half again its AEC budget from state and private funds.[85] The $1.9 million provided to Oak Ridge for biomedical research in fiscal 1951 extended the lab's most optimistic estimates from 1947 and amounted to 12 percent of the total lab program; that same year Brookhaven devoted about one-fourth of its research program to biomedicine.[86]

Despite large programs such as Megamouse at Oak Ridge and the construction of a medical reactor and clinic at Brookhaven, much of the biomedical research qualified as small science. Budget requests bound for the Division of Biology and Medicine from each lab consisted largely of many small projects, each requiring a few researchers and budgets of around $100,000 or less. For example, in 1952 the Oak Ridge Biology Division divided its $1.6 million budget and seventy-three scientists among fourteen projects, the largest of which, Megamouse, occupied twelve scientists and $260,000. The lab's Health Physics Division broke down into even smaller projects.[87] That year Berkeley distributed its $1.6 million biomedical budget and sixty-seven scientists among twenty-three projects, the largest of which engaged ten scientists and $182,000.[88] Brookhaven's medical program displayed a similar distribution; only a program on boron neutron capture therapy under Lee Farr, using the research reactor, might have qualified as big science, and only the construction of a medical center and reactor attracted attention in Washington beyond the desks of AEC program managers.[89]

Their low profile helped biomedical scientists continue to diversify in basic research. Consider the pioneering program at Berkeley under John Gofman and Hardin Jones on atherosclerosis, a form of arteriosclerosis. The work derived from research on radiation damage to lipoproteins in serum, which related to Berkeley's cancer research and qualified for partial support from the state's cancer funds. Through a study of the rate of production and utilization of lipoproteins, Gofman posited a role for them as artery-clogging agents in atherosclerosis. The hypothesis held much promise for the study of heart disease. With AEC support, Gofman soon spun off a program using an ultracentrifuge to separate blood samples from Rad Lab employees and other donors, including samples sent from patients at Brookhaven.[90] The expanded program, however, no longer bore much relation to the AEC mission. Warren wondered whether the National Heart Institute might better take over the work, although Gofman and John Lawrence fought to maintain AEC support of research on radioactive-labeled cholesterol and phospholipids.[91]

Despite his reservations, Warren chose to tolerate program diversity. Berkeley staff encountered "considerable controversy among Dr. Warren's staff as to the extent [to which] this problem should be supported. During the staff meeting, Dr. Warren stated quite clearly that it was his feeling that the A.E.C. could well afford to provide additional support for good research in the physical chemistry of the large molecules involved in the process which Dr. Jones and Dr. Gofman are studying."[92] As with cancer, Warren's tolerance may have been boosted by political and personal interest in heart disease in Congress. Gofman arranged to test blood samples from Brien McMahon, chair of the congressional Joint Committee, and his staff.[93] In hearings on appropriations, Representative Gore hoped that Warren "would have some progress to report on the treatment of heart ailments"; Warren obliged with Gofman's results on cholesterol.[94]

Although Gofman obtained outside support of clinical studies from the Public Health Service, the AEC continued to support his lab research—enough to make Gofman's project the largest biomedical program at Berkeley in 1952. Gofman would recall, "I was very, very generously supported, couldn't have asked for better. I had everything I wanted."[95] One may find similar examples of small science, and its diffusion into fields at the borders of the AEC mission, in biomedical programs at the other national labs. Lab scientists enjoyed latitude within existing programs to pursue promising leads; if these strayed from the AEC mission, sympathetic program managers could elect to indulge them within the limits of their discretion. Over time, the gradual diffusion into new fields would amount to broad diversification of lab programs. The later popularity of molecular biology at the labs demonstrates the possibilities.

SOLID-STATE AND MATERIALS SCIENCE, 1954–1962

Solid-state physics began to coalesce as a discipline in the 1950s out of several fields, including crystallography, magnetism, crystal defects, metal and semiconductor physics, and quantum theory of solids. Among the first institutions to welcome the new discipline were the national labs, albeit sometimes as part of a broad approach encompassing aspects of chemistry and metallurgy and referred to as materials science.[96] Solid-state and materials science found a home in the labs because of new research tools they provided, especially neutron beams from reactors for diffraction and radiation damage studies, and programmatic needs for knowledge of materials to withstand high-radiation environments. Solid-state programs appealed to the interdisciplinary inclinations of the labs, but their interdisciplinary character complicated their institutional accommodation.

Pushes and Pulls

Programmatic needs spurred calls by the GAC and AEC for materials research which intensified throughout the 1950s. The AEC and its scientists had early recognized the practical problems in the fabrication of fuel elements, moderators, and cooling systems for reactors, which had motivated the construction of the Materials Testing Reactor. By 1951 the GAC had come to accept basic research in the labs; one of the first fields it suggested was materials research, which it identified as a bottleneck in the reactor program.[97] The high-temperature, high-radiation environments of reactors for aircraft and rocket propulsion would further push the limits of available materials, and after 1955 the GAC kept up a drumbeat of support for materials research; instead of crash projects to produce hardware, the advisers averred, the AEC should focus on long-term research into basic properties of materials. The AEC agreed to accelerate its program.[98] By

1959 the AEC had increased the support for materials research under chemistry, physics, and metallurgy in the Division of Research to $13.2 million a year, half of which was divided among Argonne, Berkeley, Brookhaven, and Oak Ridge. The Division of Reactor Development dwarfed this with its own materials program of $83.5 million, much of which went to the labs and some of which, according to the Research program manager, fell in the "'gray area' gap between pure engineering and basic research."[99]

As programmatic support of materials research from Washington pulled the labs into the field, recognition by lab scientists of the potential of solid-state science pushed the creation and expansion of new groups at the labs throughout the 1950s. Oak Ridge was the first lab to accommodate solid-state research. From the outset its program included "Solid State Research" and "Physics of Solids," which encompassed neutron diffraction and radiation effects.[100] Pioneering work on neutron scattering by Clifford Shull and Ernest Wollan, which would earn Shull a Nobel Prize in physics in 1994, and increasing recognition "that solid state physics—as much as nuclear physics—is basic to reactor technology," persuaded Oak Ridge scientists to bolster the work. The lab proposed in 1950 to form a Solid State Division; after the metallurgists protested that a separate division would splinter radiation damage work, the lab instead established a Physics of Solids Institute under Metallurgy, which in 1952 would finally emerge as a separate division.[101] In 1956 the lab's visiting committees for both the Physics and Chemistry divisions recommended that they too expand work on the solid state; by the following year, solid-state programs at Oak Ridge had grown to half the size of the nuclear physics effort.[102] In 1960 the Solid State Division had a staff of seventy-two, fifty-nine of them scientists, and the lab would add a Metals and Ceramics Division the following year.[103]

The other labs followed suit. The initial plans for both the chemistry and physics departments at Brookhaven envisioned neutron diffraction research.[104] The physics department hired Donald Hughes from Argonne, who started a neutron scattering program; the chemistry department, seeking to strengthen its presence at the lab's research reactor, also sponsored neutron diffraction work by Lester Corliss and Julius Hastings. In 1950 the physics department established a solid-state group and the next year hired George Dienes to lead it; the group focused at first on radiation damage in graphite, a topic also studied by the metallurgy department.[105] The nascent program benefited from a year-long visit by J. C. Slater and some of his students in 1951–52, which helped persuade Haworth to increase the effort.[106]

Why did Brookhaven build up a solid-state program with permanent staff? The AEC had programmatic interest in radiation damage to graphite: its reactors had to cope with "Wigner disease," the physical expansion and storage of energy in the crystal lattice of graphite from neutron radiation. The Hanford reactors and Brookhaven's research reactor suffered from it, and in 1952 the sudden release of stored energy in a British production reactor highlighted the problem, which Dienes and his group aimed to solve through high-temperature runs to anneal the reactor.[107] Reactor problems, however, did not account for the growth of in-house programs on neutron diffraction; the neutron beams from the research reactor were intended for use by visiting scientists. The beams did attract many visitors, more from industrial and military labs than universities, and would become the second-largest visitor program at the lab behind high-energy physics.[108] Nevertheless, in-house staff continued to grow, and in 1958 Dienes proposed further expansion. He seemed to justify it on the principle that the lab needed representatives of all possible fields in order to chaperone visitors in their programs: "BNL staff members are spread somewhat too thinly and cannot cover properly all the major activities." The principle suggested "that the group should become slowly more versatile," for instance, by broadening the program to include semiconductors, electronic properties and crystal growth, and low-temperature physics.[109]

Dienes's successor, George Vineyard, elaborated the principle a few years later in suggesting that "the solid state effort at Brookhaven should be made even broader." "In order to have a high grade activity in a few lines, certain collateral lines must be pursued." Vineyard added that for Brookhaven "to be attractive to the brightest young men in the solid state field, it is necessary to have a program of considerable scope." He hinted also at possible programmatic uses, for instance, of cryogenics in accelerators and bubble chambers; the lab's visiting committee for physics suggested superconductivity.[110]

These arguments for the growth and diffusion of solid-state research did not prevent criticism of work that seemed to stray too far. In 1954 a local AEC officer questioned the work of Paul Levy at Brookhaven on gamma and neutron effects on optical properties of crystals. Levy pointed to possible applications to radiation detectors. A few years later Samuel Goudsmit, the chair of Brookhaven's physics department, asked, "Does work on radiation damage in non-metals make sense as an integral part of the physics research program at Brookhaven?" A member of the physics visiting committee assured him that it did. Levy's program survived to raise questions again, this time among the trustees of AUI, some of whom wondered why Brookhaven should support any solid-state research, given the effort dedicated to it in universities and industry.[111] Resistance to solid-state may have stemmed from the focus on high-energy physics at Brookhaven within the physics department, the lab administration, and the trustees. Goudsmit admitted that "I always have difficulty evaluating some of the Solid State work"; he and lab administrators toyed with various plans to reorganize the physics department to accommodate both high- and low-energy physics.[112] The appointment of Vineyard as chair of the physics department in 1961 cemented the status of solid-state research at Brookhaven and suggested its future prospects.

Argonne did not suffer the resistance of high-energy physicists, and its route to solid-state bypassed the Physics Division, to the chagrin of lab physicists. The initial lab program in 1947 included a solid-state group under the Chemistry Division, led at first by Oliver Simpson. The program received a boost at the outset when Peter Pringsheim, an émigré from Germany and an expert on luminescence, passed through Chicago; the university would not hire him since he was over sixty-five years old, so Simpson snapped him up for Argonne and put the solid-state group under his direction.[113] Pringsheim started work on color centers in alkali halides, or the optical study of structural defects in crystals, which could be caused by radiation.[114] The color center work used ultraviolet and x-ray wavelengths and hence seemed to bear little relevance to atomic energy or big science. Simpson justified it on the grounds that it contributed to fundamental knowledge of radiation damage to crystals, which could be applied to graphite.[115] In addition to the chemists, in the early 1950s S. S. Sidhu in the lab's Metallurgy Division spun off work on x-ray and neutron diffraction of uranium reactor slugs to more basic studies of crystal structure and magnetic properties.[116]

The solid-state work in chemistry grew to occupy about fifteen scientists by 1959. Simpson was not satisfied with slow growth. With the support of the lab's review committee for metallurgy and solid-state, he proposed to expand and broaden the program to fifty staff members over the next five years and to form a separate Solid State Division to accommodate them.[117] The proposal convinced both lab director Norman Hilberry, who believed that "Solid State over the past ten years has essentially become a professional discipline of its own," and the rest of the lab's Policy Advisory Board.[118] Hilberry and the board members feared that the lab might lose

Simpson and fail to recruit strong staff without expansion and a new division, and noted the support by the AEC of materials research and the possibility of tapping into forthcoming funds for it. They unanimously approved the plan. The lab formed a Division of Solid State Science that summer out of the group in chemistry; in its first year the new division more than doubled in size.[119] Like Brookhaven, Argonne appealed to the chaperone principle to support diversification: regional labs still needed a broad in-house program, as the physics review committee claimed, since "most so-called users are really apprentices, working under the close supervision of ANL staff researchers."[120]

Solid-state physics also attracted the attention of Los Alamos scientists, who were looking for alternatives to weapons work. First they had to demonstrate that the lab required a solid-state program. Boosters of a solid-state group preferred, as their biomedical colleagues had a decade earlier, to frame a negative question: "Can the laboratory afford not to have such a group?" One staff member estimated "that at present 1/3 of all the work in physics and 9/10 of all the applications of physics are in solid state." Los Alamos should undertake a program in neutron diffraction "since there are reactors here." And a solid-state program could then help the lab diversify further: "As the emphasis of the laboratory's program shifts this [solid state] group will become increasingly important."[121] But solid-state research lagged at Los Alamos; an anonymous staff member identified it as "a major area of physics effort not represented" there, and what little there was performed a service function for other groups.[122]

As they had done with accelerators, Los Alamos scientists set out to position themselves "at the forefront of current research interest." H. H. Barschall in the Physics Division perceived that "besides nuclear and high-energy physics, solid-state physics is currently probably the most active field of physics," but "Los Alamos is one of the few large physics laboratories at which there is hardly any organized effort in solid state research." John Manley seconded the observation and likewise promoted solid-state physics in addition to an accelerator as a means to sustain physics at Los Alamos.[123] Manley continued to admonish the lab to catch "the wave of the future . . . in physical research," but Los Alamos would not establish a formal solid-state group until 1978, out of a weapons neutron research group within the Physics Division.[124]

Berkeley caught the wave. As at Argonne, chemists at the Rad Lab provided the impetus behind solid-state research. In 1955 Berkeley paid little attention to the field apart from some x-ray crystallography, magnetic susceptibilities of rare earths and actinides, and thermodynamics of heavy elements in the solid state.[125] The last program, part of Leo Brewer's work on thermodynamics and high-temperature chemistry, in the late 1950s felt the pull of solid-state and materials research, aided by the service of Berkeley chemist Kenneth Pitzer on the GAC. In 1955 the AEC had approached Pitzer, Brewer, and Wendell Latimer of Berkeley for advice on an accelerated materials program; in 1959 Pitzer conveyed the continued interest of the GAC and AEC in materials research to his colleagues at Berkeley, who responded with a proposal for a materials research institute. The AEC duly approved the addition to the Rad Lab in 1960.[126] Even before the institute opened its doors, Berkeley's materials program under Brewer had grown by 1962 to occupy twenty-four principal investigators and forty-four postdoctoral researchers, over half the size of the high-energy physics program; and the eighty-six graduate students it attracted rivaled the popularity of high-energy physics.[127]

Solid State, Fluid Boundaries

Systemicity fostered the spread of solid-state programs through the sharing of facilities, personnel, and information. In 1950, while Brookhaven's research reactor inched toward completion, the lab's solid-state group used the reactor at Oak Ridge to get their program up and running.[128] Dienes spent part of the summer of 1958 at Los Alamos, and soon afterward Los Alamos staff made plans to visit the Solid State Division at Oak Ridge in order to jump-start their own program.[129] Once established, solid-state programs broadened further with the help of other sites: for example, Argonne's group sought information on work at Los Alamos on the Mössbauer effect in graphite and on superconductivity.[130]

Solid-state research capitalized on the growing recognition of interdisciplinarity as a strength of the labs. Glenn Seaborg, in his tenure as chair of the AEC, dedicated the Inorganic Materials Research Laboratory at Berkeley as an exemplar of the interdisciplinary potential of the national labs.[131] A member of Argonne's review committee for physics likewise proclaimed the promise of solid-state physics: "It seems that just these fields, which require a high degree of interdisciplinary competence, are well suited for a regional laboratory."[132] The interdisciplinarity of the field is evident in the problems it presented the labs in accommodating it within traditional disciplinary arrangements. Brookhaven's metallurgists and physicists pursued radiation damage work and were joined later by the nuclear engineers; neutron diffraction at Brookhaven occupied chemists and physicists. At Oak Ridge, members of the Solid-State, Physics, and Chemistry divisions all pursued neutron diffraction, and metallurgists, reactor engineers, and solid-state scientists studied radiation damage and surface properties of metals. As the creation of a Solid State Division at Oak Ridge irked the metallurgists, so would the same step at Argonne raise the ire of the Chemistry, Physics, and Metallurgy divisions, none of which wished to relinquish its claim to the field or the people who practiced it. Argonne consciously avoided existing disciplines when it named its new division. Hilberry noted, "Certainly we chose solid state science advisedly and not solid state physics."[133]

Both Argonne and Oak Ridge decided that the benefits of a new interdisciplinary division outweighed the costs of disciplinary disputes; as Simpson at Argonne put it, "As important as I think it is to have mixed disciplines in the approach, then I think it makes sense to have a separate division."[134] But members of traditional disciplines could argue the converse: placing new interdisciplinary specialties in separate groups decreased their fertility by isolating them from the diversity of programs in physics or chemistry departments; if solid-state scientists associated only with other solid-state types, they might not interact with, say, physical chemists, or physicists working on the Mössbauer effect. Hence members of both the chemistry and physics review committees at Argonne criticized the "fragmentation" that resulted from the creation of a Solid State Division.[135] The organizational conundrum reflected contradictory tendencies in the formation of interdisciplinary fields: as they broadened beyond the confines of traditional disciplines, they narrowed their focus within a new specialty.

The labs had appealed to interdisciplinarity to help distinguish themselves from academic research. But the growth of solid-state science produced more skirmishes along the border with academia. In 1955 the AEC's Division of Research sought to establish a long-term program and considered whether to create a new central lab for materials work, designate a Solid State Division at one of the national labs to assume responsibility, distribute the program among several labs, or turn to university departments. To consider these questions it appointed a committee of eight scientists, five from universities and two from industry plus the program manager for materials from the Division of Research. The panel had no representatives from the labs; not surprisingly, the mostly academic committee favored research institutes at universities, on the notion that training of new materials scientists was the most pressing need. But the GAC, demonstrating the reversal of its position on research in the labs, insisted that the AEC support research in the national labs instead of creating new university institutes.[136]

The GAC's resistance shelved the plan until the late 1950s. In the meantime, John H. Williams, an academic physicist who had championed midwestern universities against Argonne for an accelerator, had assumed direction of the Division of Research. In the spring of 1959 Williams chaired a committee on materials research created by the Federal Council for Science and Technology. The committee, with representatives from various federal agencies with an interest in materials research, returned with a recommendation for interdisciplinary materials institutes at universities in order to increase the supply of scientists. Williams turned the plan into policy at the Division of Research and canceled a new building to house solid-state science at Argonne.[137]

The commission itself, however, was less keen on the idea than its research director. The AEC claimed that the plan required a departure from its policy of government-owned, contractor-operated labs: any facilities it built would "become in fact university property since the possibility of recovering or salvaging such property is remote"; and, unlike off-site contract research, the AEC would pay for facilities and thus become committed to long-term support. The AEC did not perceive a precedent in its construction of accelerators at universities, such as the Stanford linac, or in the materials research laboratory associated with Iowa State University at Ames.[138] The Bureau of the Budget did, but it also did not want the AEC budget to "be the means for aiding universities in meeting their science building needs." The bureau's opposition provided an excuse, which the AEC seemed to want, to escape pressure from the Federal Council for Science and Technology, to which the President's Science Advisory Committee and the congressional Joint Committee had added their weight.[139] The bureau could have pointed also to Berkeley as a precedent for AEC facilities at a university campus. Berkeley's hybrid status helped it obtain one of the proposed institutes: its association with a campus would satisfy the goal of training within existing arrangements.

The episode illuminates the continuing adjustments required to accommodate the diversification of the national labs within the ecology of American science. It illustrates also a general consequence of diversification: diffusion by the labs into fields coveted by universities or other government agencies. In this case, overlap with other federal agencies offered a solution to the problem of academic institutes. The Department of Defense, in addition to NASA, had a strong interest in materials science. The department's Advanced Research Projects Agency had a $17 million budget for 1960 for materials research, which it elected to devote mostly to institutes at universities. The AEC would contribute a small percentage toward an institute at Illinois, but otherwise ARPA would fund academic institutes.[140] The solution satisfied everyone—because the resources available after Sputnik allowed a non-zero-sum outcome and the evasion of difficult questions.

The pursuit of solid-state science by universities suggests that it was not big science. Although solid-state research at the labs used reactor beams and could be applied to reactor design, it often consisted of small-team, small-budget programs. Consider the physical research program at Oak Ridge for fiscal year 1958. The lab budget included separate projects for neutron diffraction and low-temperature, nuclear, and solid-state physics; high-temperature and structural chemistry and chemistry of corrosion; and several projects in metallurgy and ceramics. Of the ten or so projects roughly classifiable as solid-state science, the largest, on radiation damage in solids, occupied thirteen scientists and proposed a budget of $455,000; the rest engaged up to five or six scientists with budgets around $100,000.[141] AEC program managers reviewed these small projects before folding them into divisional budgets. Although both Argonne and Oak Ridge had new buildings for solid-state programs postponed or canceled by budget examiners, these actions limited only the space available for expansion, not the breadth of their programs.

Although questions of duplication thus arose at higher levels only for proposed facilities, lab scientists still differentiated their small-science programs, whether through suggestions from program managers or through their own informal coordination and desire to avoid duplication of results obtained elsewhere. For instance, Brookhaven and Oak Ridge developed early programs in neutron scattering. Shull and Wollan at Oak Ridge had used neutrons of wavelengths on the order of the lattice spacing in crystals to produce elastic scattering; the direction of the scattered neutrons provided information about the crystal structure. When Donald Hughes's group at Brookhaven entered the field, it used slower neutrons for inelastic scattering, in which incident neutrons lose energy to the crystal lattice; the energy spectrum of scattered neutrons elucidated lattice dynamics. The different emphases of the two programs—elastic scattering for crystal structure at Oak Ridge, inelastic scattering for crystal dynamics at Brookhaven—persisted through the 1950s.[142]

IV

EPILOGUE AND CONCLUSION

8

EPILOGUE, 1962–1974

Subsequent events confirmed the wisdom of diversification for the national labs. President Johnson's pursuit of both the Vietnam War and the Great Society as well as the space race ended the expansion of lab funding and staffs. At the same time, the original missions of the labs were dwindling. The Limited Test Ban Treaty constrained the weapons program, which was already becoming more a matter of miniaturization and customization than of fundamentally new ideas. The transfer of reactor development to industry accelerated in the so-called bandwagon reactor market of the early 1960s, and military and political support for advanced reactors for rocket propulsion eroded and collapsed, wiping out large programs. In high-energy physics, a new single-purpose facility at Fermilab emerged to end the dominance of the multiprogram labs, while research reactors continued to proliferate at individual universities and industrial firms.

In response, the labs sought new programs and new sponsors: for example, civil defense, following the crises in Berlin and Cuba and the deployment of ICBMs in the early 1960s; environmental research, after the publication of Rachel Carson's *Silent Spring* in 1962; and alternative energy sources, especially during the energy crisis of the early 1970s. Diversification coincided with wider concerns about the AEC's dual role as regulator and promoter of atomic energy and with the energy crisis. The eventual result proved larger than the labs: the Atomic Energy Commission itself folded its tent at the end of 1974, and the national labs found themselves working for a new federal agency, the Energy Research and Development Administration (ERDA), which in 1977 evolved into the Department of Energy.

As the AEC derived from the wartime labs, in that the labs created the technologies and policy issues that impelled its creation, so did ERDA and the Energy Department arise from the diversified AEC. The regulatory problem could have been solved without recourse to a new agency; the creation of the Nuclear Regulatory Commission in 1974 could have left the AEC with its promotional role intact. Nuclear power was not perceived as an easy solution to the energy crisis; it was already beset by questions of safety and waste disposal, and the American response focused more on other energy sources and conservation. Nevertheless, instead of creating an entirely new agency to address the energy crisis, the federal government poured the old wine of the AEC into the new bottle of ERDA and the Department of Energy. The diversified national laboratory system had already responded to the energy crisis and thus made it possible for the new agency to take over the labs and their parent agency wholesale.

THE FRAMEWORK

Additions to the System

The lab system was never static. Brookhaven had joined the system after the war, at the same time Berkeley gained entry, and Livermore was created six years later. The AEC had otherwise restricted membership in the system, squashing the tentative talk of regional facilities for southern California, Texas, and the mid-Atlantic states, and likewise rejecting attempts by existing labs elsewhere to bootstrap themselves into the system. But the national labs were not the only ones diversifying. The AEC's production and engineering labs expanded their programs into research, on their own initiative or with political help from above, and began knocking on the door of the system.

The system did not welcome new members. The AEC rebuffed early plans at Hanford after the war to expand its program and facilities to allow basic research, determining that "it was not wise to consider setting up Hanford as another national laboratory."[1] The weapons engineering lab at Sandia, spun off from Los Alamos, at first adopted the attitude of the head of its research division: "Real basic research can be very vague and abstract and, in many cases, useless." But Sandia scientists in 1956 began to agitate for an in-house research program, justified by nuclear radiation effects on weapons components, and the following year the lab created a small group for physical research.[2] When Sandia scientists suggested, however, that they expand basic research during the weapons test moratorium, the AEC's General Advisory Committee resisted their efforts and termed them "worrisome."[3] Soon afterward a persistent Hanford and the AEC production facility at Savannah River proposed research programs; the GAC urged the AEC "to stand firmly against very large expenditures of funds and consequent establishment of new multipurpose laboratories."[4] A few years later, in 1963, when Hanford, Savannah River, and the Idaho reactor test station proposed that they take on some educational functions, AEC staff observed that "these were not multiprogram laboratories," and the GAC feared that acceptance of the Hanford plan, or any of the others, might encourage it "to become a national laboratory."[5]

Decisions at higher political levels challenged the exclusivity of the national labs. By 1960 the nuclear production complex was at peak capacity, with the new facilities ordered earlier in the decade now on-line and cranking out around 70 megatons' worth of uranium, plutonium, and tritium a year, enough for over five thousand warheads.[6] Los Alamos and Livermore meanwhile had satisfied the desires of the military for diverse nuclear devices, from tactical artillery shells to megaton ballistic missile warheads.[7] In January 1964 President Johnson announced a reduction in nuclear materials production in his State of the Union speech, to include shutting down reactors at Hanford. The AEC, prodded by local citizens, recognized that the community of Richland, Washington, depended on the Hanford works as a one-industry town; hence, rather than curtail the operations of Hanford, the AEC elected to diversify its program.[8]

The reconstituted lab, now named Pacific Northwest Laboratory, started in 1965 with a staff of 1,800 and a budget of $20 million, small compared to the existing multiprogram labs. Its program focused on the nuclear fuel cycle, which encompassed reactor engineering and waste disposal, and health and environmental hazards of nuclear energy, which brought in radiobiology, cancer research, marine biology, meteorology, and ecology.[9] The upstart tiptoed around the established labs: the director of Battelle Memorial Institute, the new contractor, sounded out Weinberg about "a very tentative informal suggestion . . . that Pacific Northwest Laboratory should consider requesting change of status to a national laboratory."[10] The Pacific Northwest Laboratory failed to earn the "national" title by the time the AEC transferred the lab system to ERDA, but it had joined the group of multiprogram laboratories. Of that group, Argonne, Brookhaven, and Oak Ridge remained the only nominally "national" labs, as Los Alamos, Berkeley, and Livermore still lacked the designation. The AEC in 1973 continued to categorize Sandia as an engineering development lab like Knolls and Bettis.[11]

The elevation of the Pacific Northwest Laboratory indicates the increasing interest and influence of local communities and political representatives in the national laboratory system. Congressional representatives were learning to appreciate the programs of big science as substantial slices of pork, especially after the Joint Committee obtained the power to authorize the AEC's appropriations in 1963.[12] The pork potential of lab programs was evident in intense lobbying for the location of a new high-energy accelerator in the mid-1960s, which brought local chambers of commerce, state governors, and representatives of affected congressional districts into the politics of science. Most vocal of all were midwestern politicians and scientists nursing grudges from the MURA-Argonne squabbles of the 1950s, which had ended with Argonne getting the Zero Gradient Synchrotron and the midwestern universities winning only assurances of access to Argonne. Ohio's governor James A. Rhodes, in the midst of the argument over the siting of Fermilab, exemplified their attitude: "We have been discriminated against continually. . . . We're not asking for something here. We're entitled to it."[13]

The politics of entitlement helped win the Midwest the National Accelerator Laboratory, later known as Fermilab. As production and engineering facilities at Hanford and Sandia sought to enter the national lab system from one direction, new research facilities in high-energy physics tried from another. The Stanford linac (SLAC) and Fermilab continued the AEC's support of single-purpose accelerator labs, but now the new facilities equaled and surpassed the capabilities of the accelerators at the multipurpose national labs. The Pacific Northwest Laboratory and Fermilab redefined the concept of national lab. Pacific Northwest Laboratory included more developmental work, and more generally indicated acceptance that the labs were local as well as national resources and that the federal government would sustain them even if their original purposes ended. The single-purpose Fermilab expanded the definition beyond multipurpose labs and instead emphasized the provision of facilities for visitors as the core meaning of "national." The addition of new types of institutions at opposite ends of the spectrum of lab programs brought the original labs into closer contact in the center.

Managerial Meddling

As the lab system grew, so did the bureaucracy that administered it. Complaints from lab scientists over the amount of administrative oversight, or nitpicking in their view, kept pace with the growth. Several specific factors contributed to bureaucratization. The increased control stemmed from the continual pressure of executive and congressional budget examiners for more detail, which the AEC could at best resist but not reduce. The Division of Reactor Development increased its meddlesome ways in order to cope with regulation of the bandwagon market for reactors. In 1964 the AEC appointed Milton Shaw to head the division; an alumnus of Admiral Hyman Rickover's nuclear navy, Shaw applied Rickover's style and centralized management of the reactor program in Washington, enforcing strict oversight and control.

Other programs that had earlier enjoyed some latitude were increasingly centralized; for example, fusion, which starting around 1966 shifted to a "project" organization with increased oversight from AEC staff and committees in Washington.[14] Argonne officials complained in 1968, in a meeting with staff of the AEC's new Division of Plans and Reports, that in general the "AEC seemed to be tightening its direction of the work and leaving less initiative and flexibility in the laboratory."[15] The program manager for Biology and Medicine warned the lab directors in 1973 that "you can expect a closer surveillance of your activities, a tougher review of your products and a degree of interaction you have not previously experienced." The scheme for responsiveness included the reinsertion of the AEC field offices in the line of responsibility between the labs and the program offices in Washington, thus reversing one of the main achievements of the reorganization of 1961.[16] These specific factors all reflected the general political trend toward accountability, enforced by the growing apparatus of the administrative state. The trend accelerated in this period; in addition to fiscal, contractual, and procurement controls, the labs began coping with new environmental, occupational safety, and equal opportunity regulations.

Three other characteristics of the context—the technology of the electronic computer, the managerial techniques of systems engineering, and general technocratic tendencies—converged in the administration of the national labs. Among the procedures adopted by the AEC in the early 1960s was the PERT planning process. PERT, or Program Evaluation and Review Technique, was developed in 1958 to manage the development of the Polaris missile system for the navy.[17] It involved detailed planning of each step of a project, with consideration of feedback effects from various aspects of the project on costs and schedules. Testimonials to the efficiency achieved by PERT and similar systems, and the evident success of the Polaris project, persuaded the federal government to encourage their extension to other agencies, including the AEC in 1962.[18]

The new management techniques involved complicated network calculations and thus relied on the supply of digital electronic computers then emerging from commercial firms. Demand for the techniques came from increasingly complex technologies. A RAND report of 1959 highlighted the cost overruns and schedule slippages of weapons development programs; by the early 1960s, budget examiners learned to multiply cost estimates for major military and space programs by two or three.[19] Large AEC programs likewise displayed large overruns: for instance, the AGS at Brookhaven ended up costing $31 million, half again the original estimates of $20 million.

The new techniques resonated with a general trend toward scientific management, which first flourished in the techniques of Frederick Taylor and other American social engineers in the early twentieth century, but reached its zenith in the late 1950s and 1960s.[20] The quantitative techniques of operations research, developed during World War II, reached industry in the 1950s in the emergence of "management science," which applied mathematical methods and the processing power of the electronic computer to business management. The technocratic approach of the Department of Defense under Robert McNamara and his "Whiz Kids" in the 1960s exemplified the ideals of rational management.[21] The appearance of PERT expressed the same trend. Its application to the national labs indicates that science had become big and complex enough to require industrial management techniques.[22] Los Alamos used PERT for its Meson Physics Facility, and Argonne used both PERT and the Critical Path Method on the final construction stages of the Zero Gradient Synchrotron (ZGS). Los Alamos completed the linac on time and within a few percentage points of its original cost estimates; but the techniques could not prevent Argonne's ZGS from costing twice its original estimate.[23]

The Lab Directors' Club Redux

The Lab Directors' Club continued to meet through the 1960s into the 1970s. The meetings generated "a feeling of mutual sympathy" and "useless but possibly amusing discussion" about the AEC's "new methods of technical and administrative 'control,'" but few positive accomplishments.[24] By the early 1970s the reorganization of 1961 had receded in the rear-view mirror, and lab directors could again complain that "relations with the AEC are growing more formal and more distant. . . . Communication between the labs and top levels of the AEC appears to have weakened." Old problems invited old solutions: Weinberg recycled his proposal to vest official responsibility in the lab directors and thus "convert the Directors' club from a marching and chowder society into a real force."[25]

A generational shift coincided with mounting management issues in the early 1970s. The formidable figures from the Manhattan Project were fading from the scene, and the labs began to lack their earlier leadership. Norris Bradbury retired in 1970 after twenty-five years as interim director of Los Alamos; Alvin Weinberg was ousted at Oak Ridge at the end of 1972, his views on reactor development putting him increasingly at odds with reactor program staff and congressmen on the Joint Committee; Maurice Goldhaber resigned in 1972 after eleven years as director of Brookhaven; and Edwin McMillan left Berkeley in 1973 after fifteen years as Lawrence's successor. The subsequent succession of directors indicated programmatic shifts: Brookhaven chose not a nuclear or high-energy physicist but a solid-state physicist, George Vineyard, to succeed Goldhaber, and Berkeley in 1980 would select a materials scientist, David Shirley, to succeed Andrew Sessler, an accelerator physicist, as director.

A changing of the guard took place in Washington too. In 1971 Glenn Seaborg stepped down after ten years as chair of the commission, with James Schlesinger, formerly an economist at RAND, assuming the chair. New, younger faces began showing up on the GAC, which also exhibited disciplinary diversification. Through the early 1960s GAC members came exclusively from the physical and engineering sciences. In 1964 the first biologist, John Bugher, joined the committee; in the 1970s civil engineers, botanists, economists, political scientists, and financiers appeared.[26] The 1960s may represent the final high-water mark for both the wartime leaders, as they assumed positions on the AEC and President's Science Advisory Committee, and the autonomy of the laboratory system that they defended.

THE ENVIRONMENT

Social Stresses

The rapid expansion predicted by the labs in their ten-year projections of 1960 did not occur. Livermore took a 9 percent budget cut in 1965, Los Alamos absorbed a smaller cut the same year, and over the next several years Berkeley and Brookhaven worked within level or slightly contracting budgets. Constant funding at current dollars did not sustain the labs. In the late 1960s inflation in the United States rose above 5 percent for the first time since 1951.[27] Oak Ridge saw its budget erode in value by up to 25 percent between 1965 and 1970.[28] As a result, the labs had to lay off staff, some of them for the first time, a process made easier by the absence of tenure. Staff at Oak Ridge declined from 5,500 in 1968 to under 3,800 in 1973, a 30 percent drop; Argonne suffered a similar cut over the same period.[29] The new fiscal environment produced a sense of crisis in the national labs, whose scientists wondered about the long-term federal commitment to the labs and looked over their shoulders for the next wave of layoffs.

Budgetary pressures resulted from changes in the external environment. The war in Vietnam and Johnson's Great Society diverted resources from nuclear energy research. Of the resources remaining for science and technology, the crash program to put a man on the moon soaked up a large share. With the advent of the space race, the stock of nuclear science fell as that of rocket science rose; space, not nuclear physics, was the new glamour field, with budgets to match. NASA did not grow entirely at the expense of the AEC, and the national labs continued to do work for the space agency. But, in a sign of the times, the AEC invited NASA staff to Oak Ridge, Hanford, and Savannah River to see whether the booming space program might pick up the slack for dwindling activity in atomic energy.[30]

The labs also contended with a general antiscientific trend in American society, fueled by the counterculture of the 1960s, protests against Vietnam, and humanist critiques of technocratic government. Reaction against the military uses and environmental hazards of technology focused in particular on nuclear weapons and reactors.[31] The national labs were thus a natural target for protest, but the geographical remoteness of most of them, limited access owing to security restrictions, and dependence of local economies on the labs provided a measure of protection from public demonstrations. Weinberg, after a visit to Boston in 1969, noted: "We in Oak Ridge, living as we do in a sheltered and pleasant scientific lotus-land, just don't know what our colleagues in the beleaguered universities are up against. What a shock it is to go to the hub of the intellectual universe for what one expects to be a rather routine scientific meeting, and to run smack into a full-scale confrontation between the scientific establishment and the angry young people."[32] But the more accessible sites saw their share of protests—especially Livermore, a target for demonstrations against the University of California's management of nuclear weapons research. The protests helped persuade the university to separate Livermore from its parent lab at Berkeley in 1971.[33]

The isolation of the labs could not shelter them from the activism of their own staff. Brookhaven's director Goldhaber allowed a protest to proceed as part of a national demonstration against the Vietnam War on 15 October 1969. The protest arrayed a large group of antiwar activists against a vocal crowd of hecklers and spawned subsequent activism, albeit subdued, at the lab. Liberal activism at the lab generated friction with the surrounding community, which was conservative politically; one local resident wrote to President Nixon about the "pinks, finks, and other democrat types infecting Republican Long Island. . . . Correct this situation please."[34] The protests thus also divided the lab internally along the lines of transplanted scientists versus local support staff. Other current issues reached the labs, including race and gender relations. Argonne, for instance, instituted a vigorous affirmative action program, but still failed to meet new AEC standards for recruitment of minority staff and equitable pay for women.[35]

The American political system ensured that wide public movements would affect the labs, indirectly if not directly. Brookhaven's director George Vineyard perceived that distrust of experts in general, and nuclear scientists in particular, had led to increased oversight by Washington: "As public concern with technological issues has increased and as this concern has been reflected in Congress and the Federal agencies, tighter management from above is being imposed."[36]

Programmatic Pressures

In the 1960s the main initial justifications for the national labs—development of nuclear technology, and provision of big, expensive equipment to outside users—grew increasingly tenuous. From weapons and power reactors to research reactors and high-energy accelerators, the labs faced pressures on their primary programs that would increase the incentive to seek secondary missions.

WEAPONS The weapons program entered a new phase in the 1960s. The Limited Test Ban Treaty of 1963 constrained the development of new designs, although the labs could still test weapons underground. And there were fewer new designs to test: after the revolution in nuclear weapons brought about by the hydrogen bomb, attested Darol Froman of Los Alamos, weapons design became a matter of "slow and rather insignificant improvement with the addition of frills—as in the automobile industry." Livermore designers, typically, were more optimistic about the possibilities for "clean" bombs and atomic hand grenades, but the military—and Los Alamos staff—did not share their enthusiasm.[37] The Kennedy administration was reorienting national security policy to the strategy of flexible response, which reduced dependence on nuclear weapons and hence decreased demand on the national labs.[38] Spending on atomic energy defense activities peaked around 1960, in terms of both current and constant dollars, and declined until the mid-1970s. Only one new warhead design entered the stockpile in the late 1960s, in contrast to the previous pace of a few each year.[39]

The plans to reconstitute the AEC as ERDA included discussion of shifting the weapons labs to the Department of Defense.[40] But the possibility never passed beyond discussion, perhaps because of a realization that the so-called weapons labs gave substantial effort to non-weapons work. The decline of demand for new weapon designs encouraged weapons scientists to look for new problems and sponsors, although they did not abandon weapons research entirely; in the midst of the Cold War, American commitment to nuclear weapons meant a continuing source of work for weapons scientists, and Los Alamos and especially Livermore staff continued to pursue a few new ideas, including enhanced- and suppressed-radiation devices.[41] But while demand dwindled for new weapons, the budget for military applications did not necessarily decline. For example, the labs found several fruitful programs in lasers. In 1961 the weapons scientist Ray Kidder at Livermore thought to use the recent development of lasers to simulate thermonuclear explosions. Later in the decade John Nuckolls and Lowell Wood at Livermore started to apply lasers for the controlled production of power, as did their colleague Keith Boyer at Los Alamos. As usual, the two labs differentiated their approaches: Livermore used a neodymium-doped glass laser with a short wavelength (1,060 nm); Los Alamos used a carbon dioxide gas laser, whose long wavelength (10,000 nm) provided higher power but more plasma instability than Livermore's lasers. In the early 1970s both Livermore and Los Alamos began research into the use of lasers for the separation of isotopes for weapons and reactor material, again pursuing different approaches.[42] Livermore's laser work would eventually lead to "Star Wars" in the 1980s. To take another example, work at Los Alamos on the Vela program to enforce the test ban involved satellite and ground-based detectors of atmospheric nuclear tests.[43]

Los Alamos and Livermore, in other words, diversified within military applications, not necessarily to new sponsors. And the cancellation of other programs, especially nuclear rockets, skewed the distribution of programmatic effort back toward weapons. Hence the proportion of weapons work, measured by the source of funds, increased to 68 percent at Los Alamos and 80 percent at Livermore by 1973; but this included more than just design of nuclear weapons.[44] And it still left from one-fifth to one-third of the programs of the so-called weapons labs in research outside military applications.

REACTORS Demand for nuclear reactors did not flag and instead reached new heights. From the three reactors operating under electrical utilities in January 1962, fourteen were running in the United States and Puerto Rico by January 1965, with five others under construction and three more in the planning stage.[45] The demand came from a utility industry eager not for advanced research but for construction of existing designs, and from a political coalition that anointed the fast breeder reactor as the design of the future and sought a crash program to realize it.

With the apparently successful transfer of the light-water reactor to industry in the early 1960s, the AEC looked to the breeder as the next generation of reactor, one that could produce not just cheap power but even its own fuel, a sort of nuclear *perpetuum mobile.* The breeder worked by surrounding the fissile reactor fuel with non-fissile thorium 232 or uranium 238, which could capture neutrons from the reactor core to produce uranium 233 or plutonium 239. The AEC designated the breeder its highest-priority reactor and assigned prime responsibility for it to Argonne, as the Liquid Metal-Fuelled Breeder Reactor, or LMFBR, project. The assignment proved a mixed blessing. It helped Argonne regain its lead role in power reactor development and, as one of the few programs to enjoy increased budgets in the late 1960s, helped offset reductions in other fields and softened the blow of staff cuts. But Argonne ran afoul of Milton Shaw and AEC reactor staff for continuing to emphasize longer-range research over immediate development. Through the late 1960s Shaw constantly pressed Argonne for more disciplined organization, stricter accountability, and quicker results.[46]

The crash project for the breeder reactor renewed pressures for centralization in the power reactor program. The centripetal force of centralization, aided by the interest of the congressional Joint Committee, pushed against the centrifugal forces that had dispersed reactor work to Los Alamos and Brookhaven. Brookhaven had pioneered the use of liquid-metal fuels in the 1950s, although in a thermal instead of fast breeder, and continued reactor research into the 1960s. Shaw, however, viewed Brookhaven as a "country club" and eventually canceled its power reactor work.[47] Los Alamos fared better: in the mid-1960s the lab had reoriented its program in high-temperature, molten-plutonium reactors to support the LMFBR program, and the performance of Los Alamos scientists and engineers earned praise from AEC reactor staff, even as Shaw was raking Argonne over the coals.[48] And Argonne itself, no more keen to become a single-purpose lab for breeder reactor development in the 1960s than it had been to become the centralized reactor lab in 1948, instead sought to maintain multiprogram status.[49]

Hence, although the AEC designated Argonne the center of the LMFBR program, other labs joined in the work—including the Pacific Northwest Laboratory, to which the AEC awarded a fast-flux test reactor originally slated for Argonne. Oak Ridge in the meantime pursued a parallel path. After President Kennedy canceled the Aircraft Nuclear Propulsion project in 1961, Oak Ridge engineers turned their molten-salt technology to breeder reactors. In 1965 they began experiments with a breeder reactor fueled with molten uranium fluoride and thorium fluoride, with the breeding achieved by thermal neutrons instead of the fast neutrons of Argonne's LMFBR. After promising results Oak Ridge proposed to scale up its design, but Shaw's emphasis on Argonne's reactor left the molten-salt reactor to dribble away; it lasted on limited funds from 1969 to 1973, when the AEC put it out of its misery. Together with the cancellation of the gas-cooled reactor in 1966, it left Oak Ridge without a major reactor project. Oak Ridge reactor engineers, however, had meanwhile enlisted in the LMFBR program, which was expanding at the lab just as the molten-salt project was dwindling. Decentralization won out again: the AEC, in one of its last acts, decided to build the full-scale demonstration of the LMFBR on the Clinch River in Oak Ridge, with support from both Oak Ridge and Argonne.[50] Emerging concerns over the safety of nuclear power would delay and finally cancel the Clinch River plant; either way, the shift of nuclear reactors from research and development to production, from the laboratory to industry, combined with increasing public protests against nuclear power, signaled the shaky status of reactor development for the national labs.

BIG EQUIPMENT The other main mission of the national labs, the provision of big equipment, also faced an uncertain future. Research reactors, the original sort of equipment envisaged for the labs, continued to proliferate at individual universities and industrial firms, and universities continued to encroach on the lab system's dominance in high-energy accelerators. The 20-BeV linear accelerator at SLAC, which came on-line in 1967, rivaled the energy of the AGS at Brookhaven, and MURA and a new university group centered on Caltech were clamoring for the right to build the next generation of high-energy accelerator. In 1962 the GAC and the President's Science Advisory Committee convened a joint panel, chaired by Norman Ramsey of Harvard and hence known as the Ramsey panel, to recommend a policy for high-energy accelerators in the United States. The panel's report of April 1963 recommended immediate construction of a 200-BeV proton synchrotron, with a second machine aiming for 600–1,000 BeV down the road. The panel in effect ratified the accelerator agreement of 1948 by presuming that Berkeley would build the first machine and Brookhaven would design the second.[51] Then politics intervened, and the outcry raised by midwestern scientists and politicians resulted in the location of the 200-BeV accelerator at Fermilab.[52]

The national labs did not meekly cede the ground of high energy. The next step toward 1,000 BeV remained on the table, and Brookhaven moved to claim it. But as the high cost of Fermilab and delays in its construction pushed the second machine into the future, Brookhaven reverted to earlier suggestions for a colliding-beam proton accelerator. Such a device would capitalize on the greater energy provided by a head-on collision between particles than by impact on a stationary target, but it required precise control of the beams to produce collisions. In the early 1970s Brookhaven designed a collider called Isabelle, with superconducting magnets and the AGS as injector, whose storage rings would run protons up to 200 BeV.[53] The machine spent a decade on the drawing board before its demise in 1983 in favor of the Superconducting Super Collider, which enjoyed a similarly stunted ten-year existence.

The existing national labs retrenched in low- and medium-energy nuclear physics, where Berkeley and Brookhaven, their game of leapfrog interrupted by Fermilab, now competed on level ground with the other national labs. In the early 1960s nuclear physicists began to appreciate the potential of particles with energy below 1 BeV for study of nuclear forces and the distribution of nucleons, electric charge, and magnetic moments in the nucleus. At the very small distances between nucleons, high-energy probes with short wavelengths could provide information about local nuclear structure that lower-energy electrostatic accelerators could not. In particular, nuclear physicists focused on the production of secondary beams of pi and mu mesons from high-intensity proton accelerators, and several groups rushed plans for "meson factories" into circulation in the early 1960s.[54] A group at Oak Ridge led by Arthur Snell and Alexander Zucker redesigned the sector-focused cyclotron injector from its canceled high-energy device into the so-called Mc^2 cyclotron, which aimed to provide 100μA of protons at 800 MeV.[55] Louis Rosen at Los Alamos proposed a linear accelerator for the same energy but with a higher intensity of 1 mA.[56] Brookhaven, while it pushed on to 1,000 BeV, hedged its bets by suggesting an injector to increase the intensity of AGS, which might then serve as a meson factory.[57] MURA, grasping at straws before the final cancellation of its plan, proposed that its high-intensity, 10-BeV accelerator could provide pions of comparable intensity to the proposed meson factories.[58]

Competition, as usual, led to differentiated designs. Rosen and his colleagues had first considered a spiral-ridge cyclotron for Los Alamos before settling on the linac, which promised higher intensities and variable energy. Oak Ridge argued that the pulsed linac of Los Alamos, as opposed to the continuous output of the cyclotron, would limit research; Bradbury replied that duty factor was both overrated and overestimated by Oak Ridge, that the cutting-edge cyclotron remained an unproven design, and that, as before, the Oak Ridge proposal represented more the interests of accelerator builders than of users.[59] The labs, also as usual, did not just compete: Los Alamos scientists collaborated at the technical level with groups at Argonne, Berkeley, and Brookhaven designing 200-MeV linacs as injectors for their synchrotrons, similar to the first stage of the Los Alamos linac. The AEC's Research Division encouraged the collaboration through a formal coordinating committee.[60] Los Alamos eventually won the competition for a medium-energy meson factory. Another PSAC-GAC panel, this one under Hans Bethe's direction and focused on meson factories, preferred a machine that could produce variable energy, thus favoring the Los Alamos linac. A lobbying campaign orchestrated by Rosen eventually persuaded Congress and the Bureau of the Budget to approve the $55 million budget for the Los Alamos Meson Physics Facility (LAMPF), which came on-line in 1972.[61]

The usefulness of the machines extended beyond nuclear physics to cancer therapy using negative pions and study of the radiation effects of space flight on humans. Oak Ridge included a medical facility for cancer research in its original proposal, and Los Alamos would add a biomedical facility to LAMPF.[62] The Los Alamos proposal also cited its relevance to the weapons program, through the production of short, high-intensity bursts of neutrons, useful for production of tritium and other isotopes and studies of neutron-induced damage and weapon hardening. The meson factory would ensure "that no *new* avenue of scientific application to national defense is overlooked—or found earlier by some other country." A Weapons Neutron Research Facility, funded by the Division of Military Application, had access to the proton beam from the linac; and Rosen stressed the military as well as the medical applications of the machine to Congress.[63] The LAMPF proposal also emphasized the importance of the accelerator, in the time of the test ban, to convince lab staff "that Los Alamos has a future with or without weapon testing."[64]

But Los Alamos also presented the accelerator as a "National Facility," with half—but only half—the beam time available to visiting experimenters and with a committee of Los Alamos and outside researchers to evaluate experiments. The presence of classified weapons applications complicated this aspect of the national laboratory concept. Visitors on the oversight committee might need security clearances, and visiting experimenters might enjoy only limited access to the accelerator. The issue of classification raised some resistance among prospective users to the location of the facility at Los Alamos.[65] Although the lab managed to segregate the weapons research and minimize the effects of secrecy, the presence of biomedical and weapons scientists would also enhance the competition with nuclear physicists for the lab's half of the beam time.

The Oak Ridge cyclotron, though canceled in 1964 by AEC staff, popped up in proposals in various guises throughout the 1960s, until the lab in 1969 switched course and proposed an accelerator for heavy ions. Theoretical predictions of "islands of stability" in the farthest regions of the periodic table suggested that new super-heavy elements might have long lives, and nuclear chemists and physicists began to explore the acceleration of uranium ions on uranium targets for synthesis of super-heavy elements. The Oak Ridge APACHE device (Accelerator for Physics and Chemistry of

Heavy Elements) drew on the lab's experience with heavy ions and appealed to the interests of AEC chair Seaborg, as Weinberg recognized, but it also soon faced competition from a similar proposal from Argonne. In 1969 each lab proposed a tandem Van de Graaff injecting into a cyclotron, with the combination to cost $25 million.[66]

The AEC instead chose to fund a cheaper upgrade of the Berkeley Hilac to the SuperHilac with a new ion source, which would help evaluate the promise of the super-heavy region. Argonne and Oak Ridge returned with their proposals in ensuing years; the near-identical designs brought other factors into play. Oak Ridge stressed its experience in transuranic chemistry and isotope production, boosted by the High Flux Isotope Reactor; Argonne pointed to its own strength in transuranic chemistry and nuclear physics. Oak Ridge complained that it was the only national laboratory without a major accelerator. Argonne pointed out that Oak Ridge had the high-flux reactor, while Argonne's A^2R^2 research reactor had recently been canceled; it was Argonne's turn, in its own eyes, although Oak Ridge blamed Argonne for fumbling the reactor project and added, "Why blame ORNL for ANL's blunder?" More perceptively, Oak Ridge noted that Argonne claimed that its heavy-ion facility would be "regional," serving midwestern universities. Oak Ridge instead proclaimed that its accelerator would be a "national" facility, welcoming users from midwestern universities and indeed any region. To emphasize the point, Oak Ridge renamed its proposal the National Heavy Ion Laboratory. It won approval in 1974.[67]

Heavy ion accelerators also had their applications: Oak Ridge cited possible uses in cancer therapy, supported by recent results from the Berkeley Hilac.[68] In 1972 Richard Taschek at Los Alamos suggested the relevance of heavy ions to nuclear weapons: stable, super-heavy elements might prove fissionable and hence provide new sources of weapons material. Los Alamos thus proposed its own heavy-ion device, a helical linac based on ideas explored in the early design of LAMPF. Snell at Oak Ridge, having already lost the meson factory to Los Alamos, could not believe that Los Alamos was not content with the newly activated LAMPF: "We cannot let this pass, and we have a fight on our hands."[69] Los Alamos entered too late in the game, but some duplication would persist in the field even after the approval of the Oak Ridge accelerator: Berkeley soon thereafter merged the recently upgraded Hilac with the Bevatron in a combination called the Bevalac. Brookhaven, after the cancellation of Isabelle, would also enter heavy-ion research: its flagship accelerator of the 1990s was the Relativistic Heavy Ion Collider.

The national labs also found opportunity in detectors for high-energy physics. New generations of particle detectors entered the realm of big science in the late 1950s, as the size, complexity, and expense of the largest bubble chambers and electronic counters surpassed the ability of small teams at universities to build them. Detectors thus joined the category of big equipment provided by the national labs in the 1960s. In the mid-1960s Argonne and Brookhaven proposed similar large liquid-hydrogen bubble chambers, Argonne a twelve-foot-diameter and Brookhaven a fourteen-foot device, both in the neighborhood of $17 million. The AEC, prodded by Fred Schuldt at the Bureau of the Budget, noted the similarities between the two proposals and asked if two chambers were necessary, and if so, whether they might just build identical chambers more cheaply. The detector designers from each lab explained their need for a machine of their own and the reasons for customizing it to the accelerator at each site. They also argued that "the bubble chamber art could be advanced further by having two separate groups pursuing their problems in their own ways." Paul McDaniel, the manager of the AEC's Research Division, allowed that "a certain amount of rivalry between the Laboratories was healthy."[70]

Appeals to the benefits of competition and testimonials to informal cooperation did not sway the Bureau of the Budget, which approved only Argonne's twelve-foot chamber in 1964. Brookhaven returned with revised proposals for the next four years, to no avail. When Fermilab's builders began to shift their attention from the accelerator to detectors, they capitalized on systemicity by hiring away several members of Brookhaven's bubble chamber group, to the irritation of the group's leaders.[71]

The development of national instead of regional accelerators enhanced systemicity. Instead of duplicating machines across the lab system, scientists at a particular lab were forced to use the facilities of other sites in the system. For instance, after Brookhaven shut down the Cosmotron, the lab's medium-energy physicists began using Berkeley's Bevatron and planned experiments for the meson factory at Los Alamos; and high-energy physicists at both labs would undertake experiments at Fermilab.[72] Nominally a national lab intended to provide facilities for visiting academic and industrial researchers, Fermilab played host instead to staff from the other national labs. To complete the role reversal, Berkeley would build detectors for SLAC; scientists from a national lab were thus designing detectors for experiments at a university accelerator.

The trend undermined a traditional justification for permanent staff at the national labs, who were supposed to raise the quality of the lab as well as chaperone visitors: Why retain in-house staff if they did not work in-house? Allowing permanent staff to work at other labs did keep talented scientists on tap within the *system,* if not always at a particular laboratory. But it also forced lab scientists to compete with their constituents—in the case of accelerators, university scientists seeking beam time on facilities at the national labs. As Fermilab neared operation, university scientists expressed concern over the number of proposals submitted by Brookhaven scientists.[73] Goldhaber stated that Brookhaven staff should not compete with academic experiments but should submit proposals only for special experiments, or only if the work contributed to Brookhaven's program. He did not specify what constituted a special experiment, nor did he note that Brookhaven's program was by and large the program of the other national labs.[74]

THE RESPONSE

Mission Creep: Beyond Atomic Energy

The labs made up programmatic and fiscal shortfalls by turning to other missions and sponsors in federal agencies and institutions outside the federal government. In this effort they were encouraged by the earlier declaration of tolerance for work outside the field of atomic energy, as given in *The Future Role* in 1960, and subsequent statements by Glenn Seaborg, chair of the AEC, suggesting new directions for the labs. In 1964 the AEC issued a directive approving work for other government agencies and defining acceptable programs. The directive noted that work for others was not new but that the AEC had never established an official policy. The directive found statutory precedent in the Economy Act of 1932, and allowed work for others so long as it did not distract from AEC programs or require an increase in staff or facilities, and if it could not be done as conveniently or cheaply by private industry: "As a general rule work for other agencies will be performed when it is important in the national interest and the AEC facility has unique or close to unique capabilities for carrying it out." Spofford English, the assistant general manager for research and development, noted that work for others would reduce the impact on the labs of AEC budget cuts; but he warned that "our facilities will not become 'job shops.'" To prevent this possibility English set an upper limit on work for others at twenty man-years of effort for any one lab in a year. He also noted that Oak Ridge was exempt from the limit since the lab already exceeded it, with about 110 scientific man-years of work for others in progress.[75]

All of the labs would take advantage of work for other agencies, from cancer research for the National Institutes of Health to laser research for the Pentagon's Advanced Research Projects Agency. Two fields exemplify how the labs identified opportunities for diversification in new national priorities: environmental and energy research.

ENVIRONMENTALISM The environmental movement derived from the campaign against nuclear fallout. The publication *Nuclear Information,* started in the 1950s during the fallout debate, was transformed into *Environment,* a main vehicle of the movement. Barry Commoner, a driving force behind both magazines and a leading ecological critic of the 1960s, declared, "I learned about the environment from the United States Atomic Energy Commission in 1953." Rachel Carson, whose *Silent Spring* launched the movement, likewise noted a connection to fallout: "We are rightly appalled by the genetic effects of radiation; how then, can we be indifferent to the same effect in chemicals that we disseminate widely in our environment?"[76] But if environmentalism reacted directly against the products of the labs, that did not prevent the labs from pursuing environmental research. On the contrary: lab scientists proclaimed that they had the expertise to address the problems of the environment. Far from espousing the anti-technology sentiment of the environmental movement, lab scientists proferred technological solutions for smog, radioactive waste disposal, and other pressing environmental problems.

Alvin Weinberg was, as he described himself, "king of the technological fixers."[77] Environmental research had not come up at Weinberg's Advanced Technology Seminars at Oak Ridge in 1960, but several years later a similar series of staff meetings identified it as a promising field. The lab had some help in recognizing the public interest in environmentalism. In June 1966 Representative Chet Holifield of the Joint Committee suggested that the national labs might tackle environmental research, and the AEC's general manager followed up later that year with a letter soliciting proposals. Holifield perhaps had politics as well as pollution in mind; adding responsibility for the environment to the mission of the AEC would expand the mandate of his committee in Congress, much as the Joint Committee had sought control over space after Sputnik. In 1967 Holifield led Congress to revise the Atomic Energy Act to allow research not only in atomic energy but also in fields "relating to the protection of the public health and safety." The revised act, along with the passage of the Clean Air Act that year, signaled a federal commitment to environmental problems.[78]

Oak Ridge hence proposed to create an institute for environmental research, to include a strong component of social science. Although the institute failed to materialize, Oak Ridge obtained support in 1970 for a large environmental program, funded with $1.5 million by an interdisciplinary initiative of the National Science Foundation (NSF), which would last through the conversion to ERDA.[79] But Oak Ridge did not have to create an environmental program from scratch, in response to congressional initiatives. The health program at Clinton Labs during the war had started environmental monitoring of the local White Oak creek and lake, which drained into the Clinch River and into which the lab drained its radioactive effluents. After the war, the lab's Health Physics Division under Karl Morgan continued radioecological studies of the White Oak drainage as part of its waste disposal research. In the mid-1950s the lab expanded the program and added professional ecologists to bring an ecosystems approach to the problem of radioactive waste disposal. In 1970 Oak Ridge elevated the radiation ecology group to a new Ecological Sciences Division, which two years later became the Environmental Sciences Division.[80]

Like Oak Ridge, Argonne responded to the suggestion of environmental research by Holifield and the AEC, and likewise built on the lab's earlier entry into the field. Upon receipt of the AEC's letter of late 1966, lab director Albert Crewe established a study group on environmental pollution at Argonne. The group's report of February 1967 pointed out that Argonne had long experience with environmental problems, dating back to the wartime Met Lab, through its research in radiological physics, industrial safety, biomedical effects of nuclear energy, chemical engineering, and meteorology. Not surprisingly, it recommended the creation of a department of environmental studies. The nucleus for such a department soon emerged around the meteorology program and reactor engineers from the collapsing research reactor and nuclear rocket projects.[81]

The group's first project, funded by the AEC and the National Air Pollution Control Administration, involved modeling sulfur dioxide for the city of Chicago. The three-year effort capitalized on experience in systems analysis and modeling from reactor development and local weather data from the meteorology program; it developed an important method of diffusion modeling of air pollution. In December 1969 Argonne's director, Robert Duffield, formalized the environmental work in a Center for Environmental Studies under Leonard Link's direction. The center was explicitly multidisciplinary, mingling biologists and reactor engineers, chemists and meteorologists. Although the center started with funds from lab overhead, it then had to pay its own way from outside sponsors, which it found in the NSF, the Environmental Protection Agency, the Federal Aviation Administration, and several state and local agencies. The AEC eventually pitched in too, so that by 1973 Argonne's environmental work was thriving with a $5 million budget, over 5 percent of the total lab program. That year the lab absorbed the center into the mainstream of the lab program in a new Energy and Environmental Systems Division.[82]

The entry of Argonne and Oak Ridge into environmental research demonstrates how small science at the labs could provide the seed from which large programs would grow. An ecologist or two here at Oak Ridge, a meteorologist or two there at Argonne—and with time each lab had major departments, with dozens of scientists and multimillion-dollar budgets, dedicated to environmental research. And Argonne and Oak Ridge were not the only labs to enter the field. Brookhaven maintained a small radioecology program through its biomedical research, although Goldhaber rejected a proposed expansion in 1966 to include marine ecology. The following year, however, Brookhaven at the AEC's request added a program for atmospheric sampling of air pollution to its Meteorology Division.[83] Berkeley formed an environmental research group in 1970, encompassing seventy projects already under way or planned, ranging from atmospheric aerosols to diseases induced by pollution. The group would soon become a division, and then the largest division at the lab, consuming close to a quarter of the lab's effort; it fissioned in 1977 into separate divisions for Energy and Environment and for Earth Sciences.[84] Livermore also took up environmental research, working with Berkeley on the impact of supersonic planes on the earth's ozone budget and developing model "smog maps" of the San Francisco Bay area. Livermore had in place a $3.4 million program by 1972, most of it under the AEC but also for the state of California, Pacific Gas and Electric Company, and the NSF, Department of Transportation, and Environmental Protection Agency.[85] The AEC accommodated the environmental work of the labs in a reconstituted Division of Biomedical and Environmental Science.

The national laboratories could diversify, but they could not desert their original missions—or sponsors. Congressman Holifield in 1969 praised the increasing attention to environmental research in the national labs and elsewhere. But he also wished "to sound a note of concern about a growing tendency on the part of specialists in a particular discipline of science and technology to assume omniscience in every field of knowledge."[86] Weinberg had approached Senator Howard Baker of Tennessee, a member of the Senate Subcommittee on Air and Water Pollution, about the proposed environmental institute at Oak Ridge. At Baker's request a committee of Oak Ridge scientists had prepared a report on the subject, *The Case for National Environmental Laboratories.* The report provided the basis for a bill, introduced by Baker's committee in 1970, calling for the creation of six national environmental laboratories, including, coincidentally, the possibility of "the transfer of certain research function and facilities of existing national laboratories of the Atomic Energy Commission or any other Federal agency."[87]

Faced with the loss of a national lab to another agency, governed by another committee in Congress, Holifield and the Joint Committee warned the labs in general, and Oak Ridge in particular, of the limits to diversification:

> The Joint Committee wishes to sound a note of caution in connection with the activities of AEC's national laboratories. These laboratories are major national assets. They were created, they exist and they are needed for AEC's nuclear missions. In recognition of the potentially valuable assistance that these laboratories could provide in environmental and other health and safety problem areas, regardless of any nuclear relevance, the Congress, on the initiative of this Committee, changed the law in 1967. . . . During the discussion in the House [about this amendment] it was made clear that the added authority was not to be used for "empire building." . . .
>
> The Joint Committee sees signs that ambition to acquire new knowledge and expertise in fields outside the present competence and missions of an AEC national laboratory . . . is spurring at least one laboratory to solicit activities unrelated to its atomic energy programs and for which it does not now have special competence or talents.[88]

Any empire building by the labs, that is, had to aid the empire building of the Joint Committee by expanding, not reducing, its influence.

The AEC thence sent out guidelines to the lab directors reminding them that nuclear energy remained their primary mission, that any work for others had to go through the AEC and not originate in independent proposals to other agencies, and that the labs could undertake such work only if they already possessed expertise, not as a means to acquire it. The last ukase specified the social sciences as a field not covered by the competencies of the labs.[89]

ENERGY In 1965 a massive power failure plunged the northeastern United States into darkness. The blackout signaled the dependence of American society on reliable energy; power failures and brownouts in ensuing years underscored that demand was outstripping the supply of energy, culminating in 1973 in the oil embargo imposed on the United States by Arab nations as a consequence of their war with Israel.[90] New Yorkers living by candlelight, northeasterners shivering in winter without heat and sweltering in summer without air conditioning, and, most unthinkable, Americans unable to get gas for their cars—all these specters ensured that energy would become a national priority. Public and political attention to energy transformed the AEC into ERDA; the national labs would play a central role in the administrative transformation in particular and in the American response to the crisis in general.

The AEC's responsibility for nuclear energy did not require it to take on alternative energy sources as well. The AEC expanded its responsibility for energy through appeals to its recent environmental mandate. After President Nixon delivered an energy message to Congress in June 1971 emphasizing the need for a "clean" energy supply, Congress, again led by Representative Holifield and the Joint Committee, revised the Atomic Energy Act yet again to allow research under the AEC on "the preservation and enhancement of a viable environment by developing more efficient methods to meet the Nation's energy needs." The commission stressed to Congress that its "historic[!] concern with environmental considerations, and the scientific and technological expertise of its laboratories gained in developing nuclear energy, made it a logical contender for a strong technology development role in the new national energy policy."[91] Argonne and Berkeley reflected the awkward connection to environmentalism by combining energy and environmental research in a single department.

Many roads led to energy research. The environment was one. After the revision of the Atomic Energy Act in 1967, Argonne administrators encouraged the lab's chemical engineering division to initiate work in air pollution and released discretionary funds to get it started. A group led by Albert Jonke recognized that coal, a relatively abundant energy resource in the United States, contained much sulfur, which caused serious air pollution when coal was burned in power plants. Argonne's chemical engineers had decades of experience with the production of uranium hexafluoride for reactor fuels, which used limestone and a fluidized-bed technique to control the presence of sulfur dioxide. Jonke's group realized that they might apply the same process to coal combustion. Their fluidized-bed coal combustor burned a combination of finely powdered, suspended coal and limestone; promising results won support from the National Air Pollution Control Administration and then the Environmental Protection Agency and Office of Coal Research. The fluidized-bed project spawned a wide-ranging program of coal research in the 1970s, and Argonne sought to make it one of the major elements of the lab's mission.[92]

High-energy physics provided another route. In 1961 researchers at Bell Labs announced their achievement of a new type of superconducting material. Unlike normal or type-I superconductivity, which disappeared in the presence of small magnetic fields, their niobium compounds continued to conduct electricity with no resistance even in the presence of strong magnetic fields. Soon after the announcement of type-II superconductivity, scientists in the national labs began exploring the implications for high-energy physics—in particular, the possibility of winding electromagnets with superconducting wire. Brookhaven studied superconducting magnets for accelerators, and Argonne and Brookhaven designed superconducting magnets for their large bubble chambers. The possibilities also intrigued fusion scientists, and Livermore and Oak Ridge would later install superconducting magnets in fusion reactors. AEC funding of superconductivity research rose from $1.7 million in 1964 to $4.3 million in 1967; research in the national labs drew heavily on their burgeoning programs in materials and metallurgy.[93]

The transmission of electricity with no energy loss also promised to revolutionize the electrical power industry, even if the superconductors required refrigeration to the temperature of liquid helium, a few degrees above absolute zero. In the early 1970s Brookhaven parlayed its work on superconducting magnets for accelerators into a substantial project on superconducting power transmission under the NSF, and Oak Ridge and Los Alamos also undertook research on the subject.[94] Later, after the development of higher-temperature superconducting materials in the 1980s, the

Department of Energy would designate Argonne, Brookhaven, and the Ames Laboratory as the centers for a national initiative on superconductivity.[95]

Particle physics got Argonne into the solar energy business. Robert Sachs, appointed director of Argonne in 1973, came to the lab from the University of Chicago, where a colleague had developed a new technique for detecting Cerenkov radiation. The compound parabolic concentrator collected radiation from all directions, and thus got around the problem faced by solar collectors of tracking the sun across the sky. Argonne quickly obtained NSF support for a project on solar collectors, and the AEC and then ERDA would subsequently fund the work.[96] From another direction, the experience of Berkeley scientists with photochemistry, acquired in their pioneering explorations of photosynthesis in plants, led to one aspect of a solar energy program; the other aspect, in photovoltaics, derived from the thriving materials research program.[97]

Nuclear reactors also led to energy programs, besides the immediate problem of nuclear energy. In the early 1960s chemical engineers at Argonne were studying electrochemical reactions at high temperatures in molten-salt and liquid-metal reactor fuels. Their results suggested electrochemical cells that could be regenerated by heat, and a group under Elton Cairns pursued the possibility. Regeneration by heat seemed to require excessively high temperatures, so Cairns and his group turned to electrical regeneration, testing various materials for electrodes. They found that lithium at the positive electrode and molten sulfur at the negative, with molten salt as the electrolyte, provided a lighter, more efficient battery that was rechargeable, operating at 725°F. The program at its peak occupied forty researchers under AEC support; after the higher-priority breeder reactor forced it out of the AEC budget, Cairns obtained support from the National Air Pollution Control Administration, based on the potential of the battery for pollution-free electric cars. In 1972 the NSF took over support of the work, which Cairns redirected to develop battery banks for the storage of electricity from power plants during off-peak hours. The following year the AEC, now legally empowered in the energy field, resumed its support.[98] Once again, lab scientists demonstrated their initiative in pursuing interesting technical problems and their ability to anticipate national needs. The AEC did not have to build an energy program from the ground up in the early 1970s, for lab scientists had already laid the foundation on their own.

Implications for Systemicity

The establishment of new programs at the national labs represented the confluence of diverse interests, ranging from scientists to presidents. They also raised again the question of centralization versus competition, evident in talk of a "national environmental laboratory" and also in the program for desalination. Although AEC staff by the mid-1950s had discounted the potential of nuclear reactors as a source of heat for distillation of seawater, Los Alamos and Argonne had continued to explore the possibilities on a small scale. Oak Ridge thought bigger: after the first of the lab's Advanced Technology Seminars identified the promise of desalination, the lab proposed an ambitious program to the Department of the Interior that would make it a "central water laboratory." The department and the AEC in 1962 began a less ambitious program, which did not at first focus on nuclear power. Oak Ridge scientists followed AEC staff in doubting the economics of nuclear-powered desalination plants, and instead focused on basic research into the chemistry of desalination, based on the lab's experience with solution chemistry, separation processes, and corrosion. But the calculations of a Los Alamos reactor engineer, R. P. Hammond, suggested economies of scale that would make nuclear desalination systems feasible. Weinberg retained Hammond as a consultant using discretionary funds, and Oak Ridge soon returned with another ambitious proposal for a dual-purpose nuclear power plant that would desalinate water. President Kennedy touted the prospects of such "nuplexes" in a speech in September 1963, and President Johnson proclaimed a crash program in July 1964.[99]

Nuclear desalination thence grew to a $3 million program by 1968, about 10 percent of the total reactor effort at Oak Ridge. Although Oak Ridge served as the primary focus for nuclear-powered desalination systems, other national labs entered the program. Brookhaven, for example, parlayed previous reactor shielding research into a substantial project on concrete polymer materials for desalination plants.[100] Additional political support helped drive the further diversification of the program, even into the social sciences. In 1967, after the Six-Day War in the Middle East, former AEC commissioner Lewis Strauss and former President Eisenhower, both longtime promoters of peaceful uses of nuclear energy, took up desalination as a way to solve political problems in the Middle East. Nuclear reactor complexes could produce power to run desalination plants and pump the water to the desert. Construction of the massive plants, reservoirs, and pipelines would employ Arab refugees, who could then find work in the agriculture and industry that the water and power would make possible. The AEC responded by directing Oak Ridge to conduct a Middle East study, for which the lab hired economists and political scientists. The Strauss-Eisenhower plan eventually fizzled; Oak Ridge reactor scientists could not solve thousands of years of ethnic and religious conflict in the Middle East with a technological fix. But the possibility supported a substantial research program at the lab and led it into the social sciences—with the consent of the AEC, despite the commission's later opposition to social science research for environmental programs.[101]

The example of desalination indicates that programs in the national labs could wield strong influence on wider developments. Feedback effects then spurred further diversification; Brookhaven put its concrete polymer research at the service of the Department of Transportation for new road materials. Similarly, lab scientists made the limited test ban possible by ensuring that both verification of Soviet compliance and continued American testing underground would be feasible. The test ban in turn spurred refinement of underground testing methods, geophysics and atmospheric science to detect clandestine tests, and the development of sophisticated simulation techniques using flash x-ray linacs, lasers, computer hydrocodes, and conventional high explosives. In each case lab scientists did not merely react to social and political developments but instead helped to drive them; in both desalination and the test ban the scientific, technological, and programmatic interests of lab scientists merged with the interests of diverse actors, inside and outside the government, in the United States and abroad.

Research for outside sponsors sustained and stimulated lab scientists but brought its own problems. The pursuit of new sponsors by the national labs promoted the "job shop" view that they maligned, with lab scientists now accountable to diverse sponsors outside the AEC. The labs, however, were already job shops under the AEC, and it could matter little from the lab's perspective whether the lab program was spread among several AEC divisions or among several agencies.[102] AEC staff do not seem to have worried, as Shields Warren had in the early 1950s, about the diffusion of political support and loss of concentrated advocacy for the national labs. The comments of Spofford English on the directive of 1964 instead expressed concern over distractions from AEC programs and the possibility that other agencies would leave lab staff adrift with abrupt cancellation of programs. Neither seemed a pressing concern; work for others could instead benefit AEC programs, and most of the outside work, at Oak Ridge at least, had long-term support.[103] As before, if the labs had to choose between job shop treatment or no jobs at all, the choice was clear.

The national labs, abetted by the AEC and Joint Committee, did not spread into environmental and energy research unimpeded; other agencies, with their own political advocates, also staked a claim in these fields. The Environmental Protection Agency, created in 1970, emerged with the strongest stake in the environment; for energy, the departments of Interior, Transportation, and Housing and Urban Development, as well as the Environmental Protection Agency (EPA), exerted their interests. The GAC in 1972 noted that several of these agencies, especially the EPA, were seeking to establish their own central research labs for non–nuclear energy research.[104] As the national labs appealed to these agencies for support, their proposals entered a diverse institutional pool; instead of competing just with one another, or with other AEC labs or contractors, the national labs found themselves up against the research labs of the sponsoring agency as well as other would-be diversifiers. The resultant interactions further reduced the insularity of the national lab system.

Another consequence of diversified sponsors was a shift toward short-term applications instead of long-term research. The anti-science trend of the 1960s led to an increasing emphasis on applied science; if the federal government was supporting scientific research, taxpayers should expect to see some return besides, say, knowledge of another elementary particle. President Johnson asked scientists what they could do for Grandma and declared in 1966 that "a great deal of basic research has been done. . . . I think the time has now come to zero in on the targets by trying to get this knowledge fully applied."[105] The trustees of AUI perceived that Brookhaven's work for other agencies often sought quick results instead of basic research.[106] Brookhaven accommodated its new sponsors in a Department of Applied Science, created out of the old nuclear engineering department, and lab scientists learned to stress practical applications of their research in testimony to Congress.[107]

The initiative for diversification could spring from different levels in the hierarchy of the lab system. Weinberg drove the diversification of Oak Ridge from the director's desk, expressing his entrepreneurial activism in the seminars he called to explore alternative missions. At Argonne, by contrast, diversification emerged from senior scientists such as Leonard Link or those at lower levels, who identified promising topics on their own and found ways to pursue them.[108] Impetus for diversification came also from the congressional Joint Committee, eager as always for ways to expand its influence, and the AEC itself, which transmitted the suggestions of the Joint Committee and tolerated the transformations. Likewise, each level imposed limits to diversification: Brookhaven's director and trustees refused initial attempts to expand the ecology program; AEC staff tried to limit work for others to a small fraction of lab programs; and the Joint Committee warned the labs not to abandon atomic energy.

How extensive was work for others? For 1973 it ranged from 4 percent of the total lab operating budget at Argonne to 10 percent at Los Alamos; Oak Ridge had the most significant outside workload at 17 percent, although Pacific Northwest Laboratory had 42 percent. System-wide, the labs gave about 13 percent of their effort to outside work (see appendix). The amount is not trifling: 13 percent of half a billion dollars could buy a lot of research back then, and now. And the rate of growth was significant; Brookhaven, for example, increased its non-AEC work from $70,000 in 1966 to $900,000 in 1970, and would double that in 1971 and again in 1972.[109] But it suggests that the AEC accommodated much of the diversification at the labs within its own programs.

The national labs were not the only scientific institutions to diversify in response to fiscal pressure and new national priorities. As the space race sputtered after its first lap, NASA centers such as the Jet Propulsion Laboratory found new programs and sponsors in the 1970s—including energy, under Department of Energy sponsorship. Labs under the Department of Defense did likewise, such as MITRE, a center for electronic systems engineering at MIT. In 1969 the secretary of defense allowed and encouraged defense labs to take up work for other agencies, including transportation, housing and urban development, pollution control, and health and medicine. In response, MITRE in the first half of the 1970s more than doubled its outside work, so that a third of its program was non-defense; the General Accounting Office warned that MITRE was straying too far from its military mission.[110]

But these labs were following a trend set by the national labs of the AEC, whose diversification in the late 1960s and 1970s only extended a process that dated to their inception. Biomedical research—not an original mission of the labs—led to radioecology and then environmental research; high-energy physics, another add-on, led to superconductivity and then electrical power transmission. The multiprogram status of the national labs allowed and encouraged their diversification, as did their name; "national" labs, for both the scientists within them and their overseers, implied a national resource available to meet national needs. Hence attempts to appropriate the term for specific fields failed. Talk of "national space labs" in the late 1950s or "national environmental labs" a decade later missed the point. National labs encompassed a variety of fields and could acquire new ones to meet new national priorities.

9

CONCLUSION: STRATEGY AND STRUCTURE

The national laboratories have survived and multiplied over the fifty-odd years of the nuclear age. They remain among the most important institutions in American science and technology, although their character has changed to fit the context of the post–Cold War world. The history of the labs demonstrates that this process of adaptation began in the earliest days of the system, and that the structure of the system thenceforth encouraged it. The system itself, however, was not an inevitable development. Its origins in a wartime crash project under the military did not require continuation into peacetime under a civilian agency. The survival of wartime labs, especially Los Alamos and Oak Ridge, was not assured, nor did Berkeley and Brookhaven—or, later, Livermore—have to join the system. Lab scientists and AEC staff had to define the missions of the labs, the policies to manage them, and, most fundamentally, the balance between centralization and competition.

The history of the national laboratory system illuminates the interplay between strategy and structure in the evolution of large social institutions.[1] The strategies of the national lab system—that is, the determination of the goals of the labs and the allocation of resources to accomplish those goals—were negotiated by numerous individual actors within the labs and their political framework. The structure of the system, itself a product of collective decisions, constrained subsequent strategic options. The presence of several sites forced scientists at each lab and their overseers to formulate their programs with an eye toward present and future programs at the other sites. The systemic structure of the national labs had crucial consequences for the historical actors, and it stamped the labs as a new, distinctly American type of scientific institution.

THE ACTORS

A diverse assortment of individuals shaped the evolution of the national lab system. Scientists, lab directors, academic and industrial administrators, program managers, commissioners, science advisers, budget examiners, generals, legislators, and presidents: all maintained their own interests in the system and allied their interests with those of others to advance their own cause. The lab system evolved under the influence of complicated combinations of interest groups and competing interests within groups on particular issues. Amid this complex flux of interests, two groups deserve particular attention. The first, lab scientists, are the subject of much historical writing; the second, program managers, of virtually none.

The Independence of Scientists

A central question for historians of postwar American science, and of physics in particular, concerns the alignment of scientific research with the goals of national security. Paul Forman has advanced the thesis that American physicists in the Cold War became captives of military patronage, and "were now far more used *by* than using American society, far more exploited by than exploiting the new forms and terms of their social integration." The work of the national labs for national security has led some historians, including Forman, to categorize them as military institutions.[2] Lab scientists did devote much effort to national security programs and proved willing to orient their programs even more toward military needs in times of emergency. But they also provide evidence for the other side of the argument: that postwar physicists found "intellectually compelling areas of inquiry" whether relevant or not to national security, and that Forman's thesis assumes "some true basic physics" to exist from which postwar physicists were diverted.[3] Leaving aside the counterfactual and disciplinary components of the debate, detailed examination of the national labs demonstrates that, despite their extensive work for national security, lab scientists proved adept at attaining scientific independence.

In the late 1950s the Lab Directors' Club outlined competing conceptions of the labs. The institutional concept, they said, held that the labs should hire the best scientists available and let them determine what to do at the discretion of the lab director. The project concept, by contrast, framed the labs as a collection of projects assigned and controlled by the

AEC; the labs hired scientists to work on particular projects. In other words, under the institutional view the labs were a collection of people pursuing their own interests; under the project view, a collection of projects assigned by the government. Both concepts, however, allowed for the independence of scientists. To attract and retain top scientists, the labs adopted the scientists-on-tap model, which, for example, allowed Frederick Reines and Clyde Cowan at Los Alamos to take a break from weapons work for basic research that would detect the neutrino and win a Nobel Prize. Constant attention to the morale of lab scientists by the AEC and its advisers ensured careful treatment of their research proposals. Lab scientists could also justify their research as a contribution to the storehouse of knowledge, to the apprenticeship of visiting scientists, and to the training of new generations of scientists.

The project concept did not necessarily entail passivity on the part of lab scientists. The lab directors embraced the project perspective when it suited their purpose and satisfied lab scientists. Weinberg and the chemical and reactor engineers at Oak Ridge pushed the homogeneous reactor project as an interesting technical problem and persuaded the AEC to support it, which then helped stave off centralization of reactors. Brookhaven's nuclear engineers had to overcome the resistance of lab administrators to win approval of their liquid-metal-fueled reactor. Nuclear reactors certainly satisfied national interests; the resources committed to the reactor program attracted people to the field of nuclear engineering and supported them at the labs. But national interests did not just flow down to the labs from the federal government. Reactor technology itself derived from research predating the war, which scientists and engineers introduced to the government and thereafter developed with federal support. Reactors served scientific and technical interests, which helped to construct national priorities. Science, as Forman observes, is a socially integrated institution.[4]

Lab scientists did not merely trim their sails to prevailing social and political winds; they could influence the wind direction. The reactor projects at Brookhaven and Oak Ridge provided challenging research in chemistry, metallurgy, physics, and engineering and were actively pursued by, and not imposed upon, lab scientists. Similarly, Ernest Lawrence convinced the AEC that high-energy accelerators served the national interest and cultivated commissioners and congressmen to ensure that new designs from Berkeley received consideration; Berkeley scientists then used them to discover the denizens of the elementary particle zoo. The national security state of the early Cold War, though dominant, was not a monolith, and lab scientists and their advocates learned to play off competing interests within the government to define their missions. Lab directors could also use discretionary budgets to seed small programs for their labs, some of which became major projects. Both Weinberg and Bradbury used discretionary budgets to start research on controlled thermonuclear fusion, which had programmatic justifications but also led to important theoretical work in magnetohydrodynamics.

The entrepreneurial spirit of lab directors like Lawrence and Weinberg suffused the lower levels at each lab. Consider two examples, in different fields and at opposite ends of the time period. Alexander Hollaender inherited a moribund biology program at Oak Ridge, and through his initiative and political acumen built it into a centerpiece of the AEC's biomedical program by 1950, with large projects such as Megamouse and many smaller ones on basic research. At Los Alamos, Louis Rosen—"a promoter and a politician" and, in his own words, "not a conservative by nature"—put weapons work aside in the 1960s to pursue an accelerator for the lab. Rosen's "sales campaign" targeted congressmen and budget managers as well as advisory panels and AEC staff, and resulted in the Los Alamos Meson Physics Facility.[5] Although Rosen cited possible uses for weapons research to help win approval, the accelerator would further shift the Los Alamos program away from weapons.

The competitive nature of the national laboratory system and its political framework rewarded entrepreneurial and political talents in addition to scientific ability. References to "sales campaigns," which also appeared in Argonne's quest for a research reactor, illustrate the American spirit of enterprise. Scientists such as Hollaender and Rosen thrived in the lab system, while those who lacked their drive and acumen, such as the early leaders of biomedical and solid-state research at Brookhaven, often failed to win promotion or were weeded out.

But a new, entrepreneurial breed of "Laboratory Man" did not exclusively populate the system. Bench scientists pursuing small science flourished under the leadership of entrepreneurs, who ensured steady support of their work and sheltered them from interference from Washington. Lab directors and scientists, with the help of colleagues on advisory committees, persuaded the AEC to support certain fields, such as biomedicine and solid-state science, either for potential contributions to national security or as a humanitarian antidote to weapons programs. Within those fields, individual scientists had freedom to pursue interesting problems wherever they led, including into neighboring areas connected only remotely, if at all, with atomic energy. Their independence is evident in John Gofman's research on atherosclerosis at Berkeley, or the work of Peter Pringsheim at Argonne and Paul Levy at Brookhaven on color centers in alkali halide crystals. The enterprise with which lab scientists identified and diversified into fertile fields of science contradicts an image of them as passive tools of the military.

The Conflict of Interest of Managers

The trick for the AEC, as the GAC observed at the outset, lay in striking "the proper balance between freedom in the research laboratories to explore the boundaries of scientific knowledge and positive direction by the Commission."[6] The delicate balance between scientific autonomy and political accountability depended on the performance of program managers. The crucial role of program managers in postwar American science has yet to receive sufficient scrutiny from historians. The directors and staff of the AEC's divisions of Research, Military Application, Biology and Medicine, and Reactor Development played particularly important roles in the history of the national laboratory system. The institutional and project concepts identified by the lab directors had an administrative component: the institutional concept, favored by the lab directors, devolved administrative responsibility to the labs, while the project concept retained responsibility in AEC staff.

The gradual ascendance of the project perspective at the AEC illustrates the increasing importance of program managers, although their presence did not preclude scientific autonomy. Program managers were the highest level of administration to review the operating programs of each lab in detail. They therefore defined the boundaries of basic research, the limits to diversification, and the amount of freedom allowed lab scientists—for example, Shields Warren's determination that the AEC should support Gofman's program at Berkeley. Since operating programs for individual labs disappeared into divisional budgets, program managers provided consideration of the system as a whole. As the arbiters of systemicity, they refereed interlaboratory competition and enforced specialization when the labs could not attain it themselves.

Two types of conflict of interest complicated the difficult roles of program managers. The first derived from the hiring practices of the AEC, which looked to the labs to fill staff positions. Many program managers, especially in the Division of Research, were drawn from the ranks of lab employees. The directors of the division often were former lab scientists or had strong interactions with the labs: Kenneth Pitzer, a chemist from Berkeley, succeeded James Fisk, the first director; Thomas H. Johnson, the head of the physics department at Brookhaven, replaced Pitzer; John H. Williams, a physicist from Minnesota and a member of MURA in its negotiations with Argonne, served after Johnson. Divisional staff also came from the labs, such as Donald Stevens, a solid-state scientist from Oak Ridge who joined the metallurgy and materials branch of the Research Division in time to oversee the expansion of materials research in the late 1950s, and Spofford English, another Oak Ridge veteran who led the chemistry branch. After the reorganization of 1961, the AEC appointed English the first assistant general manager for research and development.

As a result, program managers had to resist the interests of former (and possible future) colleagues and friends while allocating resources throughout the system. There is little outright evidence of institutional bias in their actions, except perhaps for Williams's support of university institutes for materials research instead of expanded programs at the national labs. As the AEC realized, the labs provided a natural source of expertise on atomic energy and of potential staff. In this respect, the AEC differed little from other federal regulatory agencies forced to seek technical expertise in the very institutions they were supposed to oversee, amid the rise of the regulatory state in twentieth-century America.

The conflict of interest extended to the GAC and to the commission itself. A partial list of GAC members with ties to the labs would include Oppenheimer, Rabi, Seaborg, von Neumann, Wigner, and Teller, as well as Williams and Pitzer; Williams came to the GAC from the AEC, on which von Neumann had previously served and which Seaborg and Haworth would later join. Since GAC members kept their full-time jobs while they served, they faced a direct conflict of interest. We have seen that Rabi and Seaborg took sides in the early contest between Brookhaven and Berkeley over a large accelerator, and Rabi continued through the early 1950s to represent Brookhaven's interest in debates before the committee.[7] Wigner was one of the few to display qualms over possible conflicts in his service on the GAC. He recused himself from a debate over an accelerator at Princeton, his employer, and later refused to serve as a consultant on the reactor review committee at Argonne.[8] The latter action sparked discussion of the U.S. statutes prohibiting conflict of interest for government employees, and of the implications for the AEC if the law barred experts such as Wigner from service. The AEC chose to conclude "that employment with a national laboratory is not the sort of relationship which is contemplated by the cited statutes," although it did expect GAC members to recuse themselves from discussions concerning their employers.[9]

Underneath personal conflicts of interest lay another, wider conflict that stemmed from the position of program managers on the path of proposals. Program managers served as mediators between the labs and the commission, the rest of the executive branch, and Congress. That is, they translated scientific goals into political priorities, and political goals into scientific programs. The managers thus performed a dual role: as boosters of the labs, who promoted lab research and defended lab programs and budgets before Congress; and as responsible managers, who had to keep the labs under fiscal control and working together as a system to fulfill the AEC mission. The combination of partisan promotion and meticulous management reflected the competing forces of autonomy and accountability. The various program divisions took different approaches to this problem, which did not necessarily follow the amount of programmatic research involved. The Division of Reactor Development demanded the strictest bureaucratic accountability from its programs; the divisions of Research and Biology and Medicine were looser; Military Application the most tolerant of all. The personal experience of Research Division staff in the labs, in addition to their professional and intellectual sympathies as scientists, inclined them to defend autonomy and allow a good measure of independence to lab scientists.

THE SYSTEM: A NEW SPECIES

In 1949 Weinberg opined that "the large national laboratory has emerged as a new entity, a new experiment in the conduct of organized research"—in short, and in accordance with our ecological metaphor, "a new species."[10] Several characteristics of the labs help to distinguish them from other institutional inhabitants of the scientific landscape: their mixture of public and private sectors through contract operation; their blend of basic research and technological development, and through it their relationships with academia, industry, and the government; and the subjects of their programs and the implications of those subjects, including security restrictions and moral questions associated with their work on nuclear technology. Two traits in particular set the national labs apart from their antecedents: their size and their structure.

Scale and Scope

The national laboratory system exemplified a central characteristic of American science and technology: size. Visiting European scientists in the 1930s had been impressed by the American infatuation with the big machine: "Americans seem to work very well, only they obviously insist on making everything as big as possible."[11] In the national labs, big science and the technological sublime found their fullest expression, not just in the physical size of machines but also in the range of imagination: genetics experiments on 10,000 mice, accelerators with 10,000-ton magnets, designs for 10,000-megaton nuclear bombs, the use of nuclear explosions to propel spaceships and to dissipate hurricanes, and nuclear desalination complexes to make deserts bloom.

The scale of the labs themselves—their hundreds and thousands of scientists, their multimillion-dollar budgets, their mammoth machines—dwarfed most other scientific institutions in postwar America, with the possible exception of industrial labs.[12] The effects of size on the performance of research and development laboratories remains a matter of debate.[13] Size did confer particular capabilities on the national labs, however. It gave them the power to prosecute crash programs. The crash program, another legacy of the Manhattan Project, also became a typical American endeavor, exemplified by NASA's Apollo project. In the first decade of the Cold War, the staff and facilities on tap in the labs provided the principal means to pursue crash programs both successful, such as the hydrogen bomb and submarine reactor, and otherwise, such as aircraft and rocket propulsion and controlled fusion. The Strategic Defense Initiative in the 1980s, which derived from work in the national labs, continued the tradition.

The national labs operated not only on a large scale but also with a wide scope. The concatenation of disciplines afforded by the labs, and the large facilities around which they congregated, provided fruitful combinations of interdisciplinary research, such as biophysics, bio-organic chemistry, and solid-state science. Lab scientists came to recognize interdisciplinary research as a particular strength of the national labs and a feature that distinguished them from academic research. Unlike university scientists, confined to disciplinary departments and discouraged by tenure committees from venturing too far afield, lab scientists from diverse disciplines mingled around reactors and accelerators and were actively encouraged by lab administrators and advisers to cross disciplinary boundaries in their research. Hence, for example, Brookhaven sought ways to increase the cross-pollination between its departments of biology, medicine, physics, and chemistry, among which it considered overlapping or joint membership on visiting committees and inviting prominent scientists with their feet in multiple disciplines, such as Leo Szilard and Max Delbrück, to spend time at the lab.[14] In the process, lab scientists helped to propagate the concept of interdisciplinarity itself.

But the national labs did not do just big science. Their overall scale obscures the smallness of much of their science. From many lab benches within the system the view seemed not so different from that of university researchers: groups of a few scientists studying photosynthesis with radioisotopic tracers or alkali halide crystals with ultraviolet light. The fine structure of the laboratory system contradicts the common conception of the labs as homes only to big science.[15]

This fine structure allowed the labs to avoid the pitfalls of size. Although the organization of the individual labs differed, the common presence of small science kept them nimble and quick on their feet, and hence able to identify and pursue new fields of research on the cutting edge: for instance, solid-state science and molecular biology, which have grown to become central programs at the labs, with their own large facilities (the Advanced Photon Source at Argonne, Advanced Light Source at Berkeley, and National Synchrotron Light Source at Brookhaven) and crash projects (the Human Genome Project, for which Berkeley, Los Alamos, and Livermore served as the centers of the effort of the Department of Energy). Not coincidentally, lab scientists perceived both of these fields as exemplars of the sort of interdisciplinary research at which they excelled. In return, the institutional support of new fields by the labs helped to establish them as disciplines.

Systemicity

The nimbleness of the labs stemmed from their structure, which differentiated them from other scientific institutions. Many of the other features of the national labs had appeared as components of previous institutions.

States had a long history of supporting research in their interest, including the establishment of scientific academies, surveys, and observatories. Some of this resembled big science: for instance, Tycho Brahe's observatory on Uraniborg in the sixteenth century, which was sponsored by the Danish Crown; employed teams of calculators to make use of the data from its huge mural quadrant and had skilled artisans in its workshops to build new instruments; boasted a chemical laboratory in addition to its astronomical facilities; published some of its research results on its own printing press; and maintained a culture of secrecy and its own security system—a pair of large dogs at the gates.[16] Also, science had long been enlisted in the development of military technology: scientists in revolutionary France instigated the establishment of a secret weapons laboratory, a "distant historical ancestor of Los Alamos," to develop new explosives.[17]

The integration of science into the economic and military, and hence political, systems of industrialized and industrializing nations, including the United States, accelerated in the late nineteenth and early twentieth centuries. The rise of science-based industries in the late nineteenth century, especially in electricity and optics, led national governments to sponsor labs to generate and maintain accurate scientific standards in support of industry. Germany's Physikalisch-Technische Reichsanstalt (established in 1887), Britain's National Physical Laboratory (1900), and the National Bureau of Standards in the United States (1901) provided for basic research in support of their metrological mission, but became preoccupied with work of immediate industrial importance.[18] The national labs differed from the National Bureau of Standards in their pursuit of basic scientific research. Other possible precedents existed in the laboratories of military services and the National Advisory Committee for Aeronautics, but these too tended toward developmental projects and were operated directly by the government; by contrast, contract operation of the national labs introduced a new hybrid of public and private institution, albeit one dependent on the public purse.[19] The national labs also aimed to provide large facilities for use by visiting researchers, not just in-house staff; unlike similar big facilities, such as the 200-inch telescope on Mount Palomar sponsored by the Rockefeller Foundation for national use, the labs provided equipment for many scientific fields and did so at government expense, alongside programs of national security interest.

But the main novelty of the national labs lay in their systemic structure. Instead of a single, central lab established around a discipline, or an instrument, or a technology, the national labs duplicated disciplines, machines, and missions and distributed them geographically. There had been several earlier proposals for a central research lab in the United States, most notably one spurred by the application of science to national priorities in World War I. That plan, for a physical science lab under the National Research Council, entailed private funding. A counterproposal for several regional labs sparked debates over the merits of centralization, and after university scientists opposed the creation of new institutions the proposal came to naught. New patterns of government support and modes of scientific research after World War II realized the idea of national labs in systemic form.[20]

Systemicity added another dimension to the difference conferred on the labs by their size. Historians have noted the rise of big science and often point to the national labs as individual exemplars. Each lab expressed the large-scale, capital-intensive, multidisciplinary team research and hierarchical, bureaucratic organization typical of the genre, and as typically compared to an industrial environment. The decentralized arrangement of the national laboratories as a system, however, required an even higher degree of organization, and one closer to that of the modern industrial corporation, defined as "a collection of operating units, each with its own specific facilities and personnel, whose combined resources and activities are coordinated, monitored, and allocated by a hierarchy of middle and top managers."[21] The complex organization of the laboratory system stymied the best efforts of the AEC and lab scientists to master it, and required scientists to adapt to structures more familiar to the corporate boardroom than to science. The importance to the national laboratory system of program managers, whose coordinating role provided the counterpart to mid-level corporate managers, indicates the institutionalization in science of the complex organization that had come to characterize twentieth-century American society.

The national lab system, however, displayed interorganizational as well as intraorganizational dynamics. The individuality of the labs and their programmatic similarity forced them, unlike units within industrial corporations, to compete and negotiate with one another for programs, like individual industrial firms. Competition kept them on their toes and ready to jump disciplinary or programmatic fences into new fields, but also spurred them to differentiate their programs from those of their competitors. Specialization and diversification, the effects of systemicity, had important implications of their own for the evolution of the national labs.

Specialization spurred a shift from regional to "national" labs. Despite their name, the national labs were originally conceived to serve regions. The labs had just begun to recognize and fulfill their regional obligations in the late 1950s when specialization forced them into service for a national constituency. Instead of duplicating the facilities of the other sites in the system, each lab began to build specialized, unique facilities unavailable elsewhere. In the meantime, the advent of safe, fast air travel had made them accessible from the whole country, not just their own region. The example of research reactors in the late 1950s demonstrated the shift: from generalized research reactors at several sites to one specialized beam reactor at Brookhaven for physicists and a flux-trap reactor at Oak Ridge for chemists. The trend may also be traced through the history of accelerators in the ensuing decades. The rising costs of accelerators and their operation precluded a non-zero-sum game, or duplication of facilities. For the next generation of high-energy accelerator beyond Brookhaven's AGS, the AEC established Fermilab, a new "national" lab with a single, unique machine and single purpose and constituency.

Specialization and service of a single, nationwide constituency sparked proliferating claims to "truly national" status, a label applied to the Los Alamos Meson Physics Facility, to Brookhaven in general and the AGS in particular, and to an electron linac and the heavy-ion facility at Oak Ridge, in addition to Fermilab.[22] The location of the heavy-ion accelerator highlighted the lessons of specialization, which once again were lost on Argonne and learned by Oak Ridge. In 1971 the visiting committee for physics at Oak Ridge perceived that "the trend of the future is towards centers of specialization," and Weinberg defined the lab's proposal for a heavy-ion facility as "national rather than regional in character." By contrast, Argonne repeated its earlier mistake in research reactors and proposed a "regional accelerator facility" for heavy ions, which lost out to the "National Heavy Ion Laboratory" at Oak Ridge.[23] By the early 1970s the AEC had approved three new accelerators, one for each region of the energy spectrum and at different sites: the heavy-ion facility at Oak Ridge for low-energy nuclear physics, the meson factory at Los Alamos for medium energies, and Fermilab for high energies.[24]

While specialization redefined regional labs as truly national labs by expanding their constituencies, diversification turned atomic energy labs into national labs by broadening their mission. They diffused first into fields within the AEC's purview. The presence of large accelerators at Argonne, Oak Ridge, and Los Alamos indicates the erasure of the initial rough categories of weapons, reactor, and accelerator labs, evident also in the emergence of reactor work at Brookhaven, Livermore, and Los Alamos. By the late 1950s the initial missions in weapons and reactors seemed to hold little long-range potential, and the AEC was assigning new high-energy accelerators to single-purpose labs such as SLAC. The labs looked beyond atomic energy for programmatic sustenance and diversified into meteorology, astrophysics, molecular biology, and other fields tangential at best to AEC missions. Events in the 1960s then fostered further diversification, for instance, into environmental research and non–nuclear energy sources. Diversified programs brought diversified sponsors, after the AEC and then Congress expanded the mandate of the national labs to allow work for other agencies. The diversification beyond atomic energy received institutional expression in 1974 in the reconstitution of the AEC as the Energy Research and Development Administration, which soon became the Department of Energy.

NATIONAL LABS AND NATIONAL GOALS

What's in a name? The growth of the labs into their identity as "national labs" suggests an alignment with national priorities. Although pluralism and small science provided some autonomy to lab scientists, systemicity and the political framework forced them also to attend to political priorities. As perfect commercial competition is supposed to ensure responsiveness to the marketplace, so the competitive lab system responded to the priorities of the AEC, the rest of the executive branch, and Congress. Scientists at all of the labs worked for the national defense during the emergency of the early 1950s. After Eisenhower's speech on Atoms for Peace, the labs scrambled to propose international training schemes and pursue new power reactor designs. The labs entered the space race with plans for nuclear rocket propulsion, atmospheric research, and ways to contribute to science education. Similar examples abound for the environmental and energy crises of the 1960s and 1970s.

One school of thought in organizational theory has concluded from other studies that competition within government bureaucracies can kick decisions up to the policy level; if competing interests cannot settle claims to resources among themselves, policy makers may intervene and apply their own priorities.[25] The lab directors demonstrated their anticipation of the theory by colluding at times to avoid political decisions: thus Zinn at

Argonne and Weinberg at Oak Ridge circumvented the decision to centralize reactors, Livermore and Los Alamos divided up weapon design assignments, and Berkeley and Brookhaven agreed to play leapfrog with accelerator energies. It is no coincidence that collusion surfaced in programs that enjoyed bountiful resources. In less prosperous times or programs, competition prevailed and lab proposals went up the ladder for arbitration, where political priorities could take precedence over the programmatic priorities of the labs. Competition among the national labs thus ensured that they would answer to national priorities.

To attribute the responsiveness of the labs to systemicity does not, however, illuminate the sources of systemicity itself. For that we must look beyond immediate political goals. The national laboratory system expressed long-held American ideals of competition, decentralization, and diversity. Lab scientists, program managers, commissioners, and congressmen repeatedly appealed to laissez-faire principles of competition as the surest, most efficient route to scientific and technological progress. The AEC likewise supported both geographic and administrative decentralization of the labs as a way to devolve scientific resources and policy decisions away from central authority to local scientists and representatives, another long-held aim of Americans traditionally wary of distant central government. Decentralization, along with contractor operation, fostered the individuality of the labs; their adaptation to regional politics, geography, and scientific communities produced institutional diversity within the system and programmatic diversity within, and beyond, the AEC mission.

To highlight the expression of American themes in the national laboratories, consider some comparable institutions in other countries. Britain in the immediate postwar period expanded its nationalized sector. It did not inherit an atomic energy establishment assembled haphazardly under the pressure of war. Finally, its geography, and hence geographic politics, did not demand or foster a far-flung system of research labs, nor did Britain have the scientists to staff one or the money to fund it. Hence Britain established Harwell, a single research lab run by the government.[26] A representative of Brookhaven, after a trip to Harwell in 1952, noted, "Unlike this country, all work [in Britain] in the field of nuclear energy, both basic and applied, is concentrated at one location."[27] The French, with even less to build on, also nationalized industry after the war. The ruined French economy and a shortage of scientists reinforced the long history of centralization of French science, and the French Atomic Energy Commission planned to concentrate nuclear research at Saclay.[28] Security concerns in both Britain and France reinforced tendencies to centralize. The examples of Harwell and Saclay suggest that competition and decentralization do not necessarily find institutional expression in democracies.[29]

If these are uniquely American themes, however, the national labs in the United States reflected them imperfectly. We must distinguish description from causation; the merits of competition were generally invoked post facto in the establishment of the system, with the exception of Livermore. Appeals to competition did help sustain the system once it was in place; and competition for programs and facilities was quite real and intense, as were its effects, specialization and diversification. Competition, however, did not exactly obey laissez-faire principles—or, to follow our ecological metaphor, a Darwinian theory of evolution. Lab programs, such as aircraft nuclear reactors, died out, but the labs themselves never died. A benevolent federal government sustained them, from postwar demobilization through the decline of their original missions, and thus appeased lab scientists and regional universities, their congressional representatives, the military services, and other interest groups.

The preservation of the national labs and the managerial coordination of competition suggests another characteristic of the American context. In the twentieth century the United States implemented the regulatory state and social welfare programs in order to mitigate the effects of free market competition on individuals and the environment, and industry increasingly followed the visible hand of corporate managers. The Cold War elevated technocratic intervention to new heights at the same time that the ideological struggle between capitalism and communism put a premium on laissez-faire principles. Appeals to the advantages of competition in the national labs thus served a rhetorical purpose: confirmation that capitalistic competition, applied to the production of nuclear weapons or particle physics, ensured the eventual sociopolitical, scientific, and technological triumph of the American system over the Soviet alternative. Hence the rationale pronounced by a Brookhaven physicist for his lab's competition with Argonne for a massive bubble chamber: "The Soviet Union provided a good example of the one-way approach. The accelerators and other facilities visited there gave sorry evidence . . . of the lack of competition."[30]

Just as the lab system implemented imperfect competition, so did it fail to institutionalize the other American ideals completely. Scientists in the labs and on the GAC pointed to the accomplishments of Los Alamos and the MIT Rad Lab during the war and recommended centralization of reactors at Argonne and continued centralization of weapons work at Los Alamos. Politicians could question duplicative decentralization on fiscal grounds. Lab scientists also came to oppose administrative decentralization, which added another bureaucratic layer between them and program managers in Washington; instead they urged centralization of authority for the national labs. The subsequent diversification and growth of the labs elicited investigations by the AEC and Congress. The well-publicized Galvin committee of 1995 was only the latest in a long line of blue-ribbon panels called to consider the continued multiplication of the missions of the national labs.[31] Other characteristics of the labs—secrecy, subsidies, government intervention, collusion to circumvent political decisions—run counter to professed American ideals, although not to American history. We may attribute the imperfect implementation of some American themes to two other noble American traditions: pork barrel politics and log-rolling. But if general goals were implemented in the labs for less than ideal reasons, they still supported systemicity and thus encouraged the labs to satisfy more specific national goals. Both the representation of general ideals and the responsiveness to specific priorities stemmed from the situation of the national laboratory system within the social and political context of Cold War America.

APPENDIX 1

APPENDIX 2

ABBREVIATIONS USED IN THE NOTES

NOTES

INDEX

APPENDIX 1

Laboratory Operating Budgets, 1948–1966, in Millions of Current Dollars

FY	Weapons	Reactors	Research	Biomed	Other	Total
		Argonne National Laboratory				
1948		NA	2.3	NA	NA	NA
1949		NA	4.1	NA	NA	NA
1950		6.0	4.6	NA	NA	NA
1951		9.0	4.8	2.6	2.3	18.7
1952		8.0	5.5	2.8	4.3	20.6
1953		8.1	6.1	2.4	2.1	18.7
1954		8.5	6.2	2.6	0.3	17.6
1955		10.5	6.1	2.7	0.4	19.7
1956		12.0	6.8	2.8	0.6	22.2
1957		14.4	8.4	2.8	0.5	26.1
1958		17.9	9.2	2.9	1.0	31.0
1959		19.6	11.6	3.4	0.8	35.4
1960		22.3	13.3	3.8	0.8	40.2
1961		21.9	15.6	4.2	1.0	42.7
1962		21.3	19.2	4.8	1.1	46.4
1963		23.7	22.5	5.3	1.1	52.6
1964		27.3	27.8	5.8	1.2	62.1
1965		27.8	32.0	6.1	1.2	67.1
1966		31.7	34.8	6.8	1.0	74.3
		Brookhaven National Laboratory				
1948		—	6.4	—	—	6.4
1949		—	4.3	0.8	—	5.1
1950		0.2	5.9	1.4	—	7.5
1951		0.5	5.1	1.8	0.1	7.5

FY	Weapons	Reactors	Research	Biomed	Other	Total
1952		1.2	5.3	2.2	—	8.7
1953		1.7	5.5	2.2	—	9.4
1954		1.7	5.6	2.4	—	9.7
1955		1.9	5.8	2.6	—	10.3
1956		2.5	6.2	2.7	0.1	11.5
1957		3.7	7.0	2.9	0.1	13.7
1958		5.1	8.3	3.6	1.0	18.0
1959		4.4	9.4	4.2	0.3	18.3
1960		3.4	11.8	4.6	0.5	20.3
1961		3.2	16.6	4.8	0.7	25.3
1962		3.9	18.9	5.2	0.9	28.9
1963		4.7	24.3	5.8	1.3	36.1
1964		5.3	27.8	6.2	1.4	40.7
1965		5.7	30.2	6.6	1.4	43.9
1966		5.9	32.7	7.1	1.5	47.2
		Lawrence Radiation Laboratory (Berkeley)				
1948		—	3.8	1.9	—	5.7
1949		0.1	3.5	1.3	0.1	5.0
1950		—	4.2	1.3	—	5.5
1951		—	4.7	1.5	0.2	6.4
1952		—	4.8	1.6	3.4	9.8
1953		—	4.7	1.7	—	6.4
1954		—	4.6	1.7	—	6.3
1955		—	5.0	1.7	—	6.7
1956		—	10.4	1.8	—	12.2
1957		—	12.5	1.6	—	14.1
1958		—	17.0	1.9	—	18.9
1959		—	19.9	2.0	—	21.9
1960		—	17.6	2.2	—	19.8
1961		0.2	20.9	2.5	—	23.6
1962		0.2	23.6	2.6	—	26.4
1963		0.1	25.4	3.0	—	28.5
1964		—	27.9	3.2	—	31.1
1965		—	30.1	3.3	—	33.4
1966		—	32.1	3.8	—	35.9
		Lawrence Radiation Laboratory (Livermore)				
1953	8.8	1.2	0.3	—	—	11.3
1954	12.5	1.5	0.7	—	—	14.7
1955	NA	2.5	2.2	—	—	NA
1956	20.5	1.5	—	—	—	22.0

FY	Weapons	Reactors	Research	Biomed	Other	Total
1957	26.4	3.3	—	—	—	29.7
1958	35.4	2.9	—	—	—	38.3
1959	41.9	6.9	—	—	—	48.8
1960	42.8	13.9	6.0	—	3.3	66.0
1961	47.0	18.7	6.1	—	3.7	75.5
1962	71.0	19.4	6.5	—	3.6	100.5
1963	70.7	21.7	7.0	0.2	4.1	103.7
1964	83.5	11.4	5.6	1.8	6.2	108.5
1965	83.5	3.1	5.9	3.0	5.9	101.4
1966	85.8	3.5	5.9	3.4	6.8	105.4
Los Alamos Scientific Laboratory						
1948	NA	NA	NA	NA		NA
1949	NA	NA	NA	NA		NA
1950	NA	NA	NA	NA		NA
1951	NA	—	0.1	0.7		NA
1952	NA	—	0.1	0.6		NA
1953	48.0	—	0.1	0.7		48.8
1954	50.4	—	0.3	0.7		51.2
1955	NA	NA	1.0	0.7		NA
1956	40.2	4.3	1.1	0.8		46.4
1957	32.9	8.1	1.8	0.9		43.7
1958	37.9	10.9	2.3	0.9		52.0
1959	38.5	15.4	3.0	1.0		57.9
1960	36.4	18.9	2.2	0.9		58.4
1961	36.7	26.2	2.0	1.0		65.9
1962	43.4	33.0	1.9	1.0		79.3
1963	42.8	36.3	2.0	1.0		82.1
1964	47.4	35.1	1.8	1.1		85.4
1965	49.0	31.8	1.8	1.3		83.9
1966	54.2	34.3	4.1	1.4		94.0
Oak Ridge National Laboratory						
1948		NA	5.7	1.6	NA	14.4
1949		NA	7.9	1.7	NA	14.5
1950		NA	6.4	1.7	NA	15.0
1951		8.0	4.8	1.9	3.5	18.2
1952		11.2	6.0	2.2	5.0	24.4
1953		11.0	7.5	2.3	7.9	28.7
1954		11.4	7.8	2.3	7.5	29.0
1955		19.4	8.5	2.4	2.4	32.7
1956		26.5	8.6	3.8	2.0	40.9

FY	Weapons	Reactors	Research	Biomed	Other	Total
1957		35.3	9.8	4.2	2.2	51.5
1958		27.5	13.3	3.7	3.6	48.1
1959		29.0	14.9	4.6	4.0	52.5
1960		27.7	17.7	5.3	5.8	56.5
1961		30.5	21.1	5.8	5.6	63.0
1962		32.3	23.8	6.5	3.7	66.3
1963		32.0	25.2	7.6	4.6	69.2
1964		31.2	25.9	8.6	5.0	70.7
1965		31.3	27.3	9.4	6.1	74.1
1966		31.1	30.1	10.2	6.6	78.0

Sources: *The Future Role* [for FY1951–1955]; "Material for Lab Directors Meeting," Ames, May 1966 (GM 5625/15) [for FY1956–1966]. JCAE, 85th Cong., 2d sess., *Physical Research Program* (Feb 1958), 738 [research budgets for FY1948–1950]. There are several, mostly small discrepancies among the three sources.

Categories refer to the sponsoring AEC division: Military Application, Reactor Development, Research, and Biology and Medicine. "Other" includes production of nuclear materials and radioisotopes, and training, education, and information. NA signifies data were not available.

Additional sources for Brookhaven: Operating costs for 1948 were apparently assigned to Research. Biomedical for 1949 from C. Wilson to W. Kelley, 16 Apr 1948 (DBM, 20/MH&S: BNL). Reactors for 1949 extrapolated to zero, and "other" for 1949 also assumed nil. Figures for 1950 from Haworth to Fackenthal, 16 Jan 1950 (BNL-DO, I-8).

Additional sources for Berkeley: Budget for 1948 from budget summary, contract W-7405-eng-48, 19 Mar 1947, ORO-1397; for 1949, from A. Tammaro to E. O. Lawrence, 18 Nov 1948 (DC, 5/Rad Lab: Budget); biomedical for 1950 from Tammaro to R. Underhill, 9 Jan 1950 (DC, 7/Finance-Admin: Budget). "Other" for 1949 refers to production of weapons materials, and for 1951 and 1952 apparently to Livermore.

Additional sources for Livermore: Weapons budgets for 1953 and 1954 from AEC 99/16 (EHC, PLB&L-7: Los Alamos). Totals include operating costs for weapons tests; excluding tests gives weapons budget of $6.1 million for 1953, $8.5 million for 1954. "Other" includes Plowshare. *The Future Role* lists Sherwood (fusion) under physical research, but data for Livermore from Ames meeting apparently include it under weapons.

Additional sources for Los Alamos: Weapons budgets for 1953 and 1954 from AEC 99/16. Excluding weapons tests gives weapons budget of $38.7 million for 1953, $41.3 million for 1954. It is not clear whether weapons budgets for subsequent years include weapons tests operations. Reactor budgets assumed nil before 1955. Sherwood assigned to Research.

Additional sources for Oak Ridge: Figures for 1948–1950 estimated from "Tentative Cost and Personnel Schedule," 22 Mar 1948, ORNL/CF-48-3-336; budget estimates, 1949 and 1951, ORO-1476; and "Budget and Personnel Comparison Survey," FY1949–50 (ORNL-DO, Budget-Instructions and procedure). "Other" refers mostly to production program.

APPENDIX 2

Laboratory Operating Budgets for 1973, in Millions of Current Dollars

Laboratory	Weapons	Reactors	Research	Biomed/ Env	CTR	Plowshare	Work for others	Total
Argonne	—	42	34	9	—	—	3.7	92.9
Brookhaven	—	—	37.9	8.3	—	—	2.9	49.1
Berkeley	—	—	29.0	4.1	—	—	2.4	35.5
Livermore	107.4	—	—	3.3	7.8	5.2	10.9	135.6
Los Alamos	77.4	9.3	9.4	2.8	5.6	—	8.9	113.4
Oak Ridge	—	27	32	14	—	—	16.2	93.2

Source: AEC, Technical Information Center, *Atomic Energy Commission Research and Development Laboratories: A National Resource,* TID-26400 (Sep 1973). Categories refer to the sponsoring AEC division: Military Application, Reactor Development, Physical Research, Biomedical and Environmental Research, Controlled Thermonuclear Research, and Civilian Application of Nuclear Explosives; work for others refers to other federal agencies. Not included are generally small programs for production of nuclear materials, radioisotopes, training and education, and regulation.

ABBREVIATIONS

ACBM x   Advisory Committee for Biology and Medicine, meeting number x, in DBM

AIP   Niels Bohr Library, Center for History of Physics, American Institute of Physics, College Park, Maryland

AMW   Alvin M. Weinberg papers, Children's Museum, Oak Ridge, Tennessee

ANL-DO   Argonne National Laboratory, Director's Office Project Files, 1955–1970, AEC records, RG 326, NARA-GL

ANL-PAB   Argonne National Laboratory, Records of Policy Advisory Boards, 1957–1967, AEC records, RG-326, NARA-GL

AUI   Associated Universities, Inc., minutes of Board of Trustees and Executive Committee (abbreviated as Ex Comm) meetings, Brookhaven National Laboratory

BAS   *Bulletin of the Atomic Scientists*

BNL-DO   Brookhaven National Laboratory, Director's Office Files, in History Office; microfilm at AIP (Roman numerals indicate series number)

BNL-HA   Brookhaven National Laboratory, Historical Archives, History Office

BOB   Office of Management and Budget Subject Files, Records of the Military Division re: Budgetary Administration in Certain Independent Agencies, RG 51, series E-57, NARA II

Crease   Robert P. Crease, *Making Physics: A Biography of Brookhaven National Laboratory, 1946–1972* (Chicago, 1999)

DBM   Division of Biomedical and Enviromental Research [formerly Division of Biology and Medicine], Central Subject File, RG 326, NARA II

DBM-DOE Division of Biology and Medicine files (Record Group 326–1132), DOE

DC Donald Cooksey files, LBL

DOE Department of Energy, History Division, Germantown, Maryland

EHC Energy History Collection, DOE

EOL Ernest O. Lawrence papers, The Bancroft Library, UC Berkeley

ES records on epidemiology studies, DOE Public Reading Room, Oak Ridge Operations Office

The Future Role U.S. Congress, Joint Committee on Atomic Energy, 86th Cong., 2d sess., *The Future Role of the Atomic Energy Commission Laboratories* (Washington, D.C., 1960)

GAC x General Advisory Committee meeting number x, AEC, minutes and reports of the GAC, 1947–1974, RG 326, series E-70, NARA II; declassified copies for most meetings in EHC and some in JCAE

GM AEC, records of the General Manager, DOE

Heilbron, Seidel, and Wheaton J. L. Heilbron, Robert W. Seidel, and Bruce R. Wheaton, *Lawrence and His Laboratory: Nuclear Science at Berkeley, 1931–1961* (Berkeley, 1981)

HEP 57–64 AEC, Correspondence Relating to High Energy Physics, 1957–1964, RG 326, series E-39, NARA II

Hewlett and Anderson Richard G. Hewlett and Oscar E. Anderson, Jr., *A History of the United States Atomic Energy Commission,* vol. 1, *The New World, 1939–1946* (University Park, Pa., 1962)

Hewlett and Duncan Richard G. Hewlett and Francis Duncan, *A History of the United States Atomic Energy Commission,* vol. 2, *Atomic Shield, 1947–1952* (University Park, Pa., 1969)

Hewlett and Holl Richard G. Hewlett and Jack M. Holl, *Atoms for Peace and War, 1953–1961: Eisenhower and the Atomic Energy Commission* (Berkeley, 1989)

Holl Jack M. Holl, *Argonne National Laboratory, 1946–96* (Urbana, Ill., 1997)

HREX Human Radiation Experiments database, available through DOE, Office of Human Radiation Experiments Web site, http://www.ohre.doe.gov

HSPS *Historical Studies in the Physical and Biological Sciences*

HSPT Human Studies Project Team, released documents, LANL

JCAE records of the Joint Committee on Atomic Energy, unclassified general subject file, RG 128, NARA I

JCAE III records of JCAE, Appendix III: Declassified Records from Classified General Subject File, RG 128, NARA I

Johnson and Schaffer Leland Johnson and Daniel Schaffer, *Oak Ridge National Laboratory: The First Fifty Years* (Knoxville, Tenn., 1994)

KP Kenneth Pitzer papers, The Bancroft Library, UC Berkeley

LANL Los Alamos National Laboratory, Archives

LASL-DO Los Alamos Scientific Laboratory, Director's Office files of Norris Bradbury (B-9 files), LANL

LBL Lawrence Berkeley National Laboratory, Archives and Records

NARA I National Archives and Records Administration I, Washington, D.C.

NARA II National Archives and Records Administration II, College Park, Maryland

NARA-GL National Archives and Records Administration, Great Lakes Region, Chicago

ORNL/CF Oak Ridge National Laboratory, Central Files, Laboratory Records

ORNL-DO Oak Ridge National Laboratory, Director's Office Files, Laboratory Records

ORO records on human radiation experiments, DOE Public Reading Room, Oak Ridge

RRC-UT Radiation Research Collection, Special Collections, Hoskins Library, University of Tennessee, Knoxville

- MS-652 and MS-1261 Alexander Hollaender papers
- MS-982 Arnold H. Sparrow papers
- MS-1067 AEC records
- MS-1167 Charles L. Dunham papers
- MS-1709 ORNL Biology Division records

Sec'y 47–51 AEC, records of the Secretariat, 1947–1951, RG 326, series E-67A, NARA II

Sec'y 51–58 AEC, records of the Secretariat, 1951–1958, RG 326, series E-67B, NARA II

Sec'y 58–66 AEC, records of the Secretariat, 1958–1966, DOE

TBL The Bancroft Library, University of California, Berkeley

UC Regents University of California Regents, Committee on AEC Projects, minutes, University of California Office of the President, Oakland

UCh-VPSP Office of Vice President for Special Projects, 1940–1969, records, Regenstein Library, Special Collections, University of Chicago

Weinberg Alvin Weinberg, *The First Nuclear Era: The Life and Times of a Technological Fixer* (New York, 1994)

NOTES

The notes refer to the location of records in 1996–1998. Some of them may have been moved since. For notes citing a series of documents, material quoted in the text is cited in the order in which it appears, unless otherwise indicated. Citations follow the form: source, box/folder.

INTRODUCTION

1. Lowell Wood and John Nuckolls. "The Development of Nuclear Explosives," in Hans Mark and Lowell Wood, eds., *Energy in Physics, War, and Peace* (Boston, 1988), 311–317, on 316.
2. *The Future Role,* 3. Lab figures totaled from individual lab budgets. Physical science here excludes engineering.
3. ACBM 45, 25–26 Jun 1954; AEC minutes, 22 Sep 1955 (Sec'y 51–58, 176/BAF-2: 1957, vol. 1).
4. E.g., Du Pont in 1956 spent $140 million on R&D, including technical support of production and sales. David A. Hounshell and John Kenly Smith, Jr., *Science and Corporate Strategy: Du Pont R&D, 1902–1980* (Cambridge, 1988), 328, 335. Compare the staff sizes of Du Pont, Bell Labs, GE, Westinghouse, Eastman Kodak, and other large labs in *Industrial Research Laboratories of the United States,* 9th ed. (Washington, D.C., 1950), 10th ed. (Washington, D.C., 1956), and 11th ed. (Washington, D.C., 1960).
5. Hewlett and Anderson; Hewlett and Duncan; Hewlett and Holl.
6. On accelerators: Daniel S. Greenberg, *The Politics of Pure Science* (New York, 1967); Leonard Greenbaum, *A Special Interest: The AEC, Argonne National Laboratory, and the Midwestern Universities* (Ann Arbor, 1971); Zuoyue Wang, "The Politics of Big Science in the Cold War: PSAC and the Funding of SLAC," *HSPS,* 25:2 (1995), 329–356; Catherine Westfall, "The First 'Truly National Laboratory': The Birth of Fermilab" (Ph.D. diss., Michigan State University, 1988); Elizabeth Paris, "Lords of the Ring: The Fight to Build the First

U.S. Electron-Positron Collider," *HSPS*, 31:2 (2001), 355–380; Lillian Hoddeson and Adrienne W. Kolb, "The Superconducting Super Collider's Frontier Outpost, 1983–1988," *Minerva*, 38 (2000), 271–310. On the weapons program: Sybil Francis, "Warhead Politics: Livermore and the Competitive System of Nuclear Weapon Design" (Ph.D. diss., MIT, 1996); Richard Rhodes, *Dark Sun: The Making of the Hydrogen Bomb* (New York, 1995). On fusion: Joan Lisa Bromberg, *Fusion: Science, Politics, and the Invention of a New Energy Source* (Cambridge, Mass., 1982).

7. Heilbron, Seidel, and Wheaton; Robert W. Seidel, "Accelerating Science: The Postwar Transformation of the Lawrence Radiation Laboratory," *HSPS*, 13:2 (1983), 375–400; Johnson and Schaffer; Holl; Crease. See also Joanne Abel Goldman, "National Science in the Nation's Heartland: The Ames Laboratory and Iowa State University, 1942–1965," *Technology and Culture*, 41:3 (2000), 435–459; Necah Furman, *Sandia National Laboratories: The Postwar Decade* (Albuquerque, 1989), and Leland Johnson, *Sandia National Laboratories: A History of Exceptional Service in the National Interest*, ed. Carl Mora, John Taylor, and Rebecca Ullrich (SAND07–1029, Sandia National Laboratory, 1997).
8. Robert W. Seidel, "A Home for Big Science: The AEC's Laboratory System," *HSPS*, 16:1 (1986), 135–175.
9. Nathan Reingold, "Vannevar Bush's New Deal for Research: or, The Triumph of the Old Order," *HSPS*, 17:2 (1987), 299–344; David M. Hart, *Forged Consensus: Science, Technology, and Economic Policy in the United States, 1921–1953* (Princeton, 1998); James G. Hershberg, *James B. Conant: Harvard to Hiroshima and the Making of the Nuclear Age* (New York, 1993); Daniel J. Kevles, *The Physicists: The History of a Scientific Community in Modern America* (New York, 1978), 324–366; Peter Galison, *Image and Logic* (Chicago, 1997), 239–311; Peter Galison and Barton Bernstein, "In Any Light: Scientists and the Decision to Build the Superbomb, 1952–1954," *HSPS*, 19:2 (1989), 267–347; Roger L. Geiger, *Research and Relevant Knowledge: American Research Universities since World War II* (New York, 1993); Rebecca S. Lowen, *Creating the Cold War University: The Transformation of Stanford* (Berkeley, 1997); Michael Aaron Dennis, "'Our First Line of Defense': Two University Laboratories in the Postwar American State," *Isis*, 85:3 (1994), 427–455; Jessica Wang, *American Science in an Age of Anxiety: Scientists, Anticommunism, and the Cold War* (Chapel Hill, 1999); Walter A. McDougall, *The Heavens and the Earth: A Political History of the Space Age* (New York, 1985).
10. Paul Forman, "Behind Quantum Electronics: National Security as Basis for Physical Research," *HSPS*, 18:1 (1987), 149–229; Stuart W. Leslie, *The Cold War and American Science: The Military-Industrial Complex at MIT and Stanford* (New York, 1993); Daniel J. Kevles, "Cold War and Hot Physics: Science, Security, and the American State, 1945–1956," *HSPS*, 20:2 (1990), 239–264;

Roger L. Geiger, book review of Leslie, in *Technology and Culture,* 35:3 (1994), 629–631.

11. Arnold Kanter, *Defense Politics: A Budgetary Perspective* (Chicago, 1979); Francis, "Warhead Politics," 25–27; William M. Evan, *Organization Theory: Structures, Systems, and Environments* (New York, 1976), 126.
12. Alfred D. Chandler, *Strategy and Structure: Chapters in the History of the Industrial Enterprise* (Cambridge, Mass., 1962), 13–17, on 14.
13. Louis Galambos, "The Emerging Organizational Synthesis in Modern American History," *Business History Review,* 44:3 (1970), 279–290, and "Technology, Political Economy, and Professionalization: Central Themes of the Organizational Synthesis," *Business History Review,* 57 (1983), 471–493; Robert H. Wiebe, *The Search for Order, 1877–1920* (New York, 1967); Alfred D. Chandler, *The Visible Hand: The Managerial Revolution in American Business* (Cambridge, Mass., 1977); David A. Hounshell, "Hughesian History of Technology and Chandlerian Business History: Parallels, Departures, and Critics," *History and Technology,* 12 (1995), 205–224; Robert D. Cuff, "An Organizational Perspective on the Military-Industrial Complex," *Business History Review,* 52:2 (1978), 250–267.
14. Alan Brinkley, quoted in Brian Balogh, "Reorganizing the Organizational Synthesis: Federal-Professional Relations in Modern America," *Studies in American Political Development,* 5 (1991), 119–172, on 120; Cuff, "Organizational," 258–259; see also James Q. Wilson, *Bureaucracy: What Government Agencies Do and Why They Do It* (New York, 1989).
15. Wesley Shrum and Robert Wuthnow, "Reputational Status of Organizations in Technical Systems," *American Journal of Sociology,* 93:4 (1988), 882–912.
16. Thomas P. Hughes, *Networks of Power: Electrification in Western Society, 1880–1930* (Baltimore, 1983); Wiebe E. Bijker, Thomas P. Hughes, and Trevor Pinch, eds., *The Social Construction of Technological Systems: New Directions in the Sociology and History of Technology* (Cambridge, Mass., 1987).
17. AEC, *Fifth Semiannual Report to Congress* (Jan 1949), 53, 67.
18. AEC, *Second Semiannual Report to Congress* (Jul 1947), 2–3, and *Fifth Semiannual Report to Congress* (Jan 1949), 50, 77.
19. Wayne Brobeck to W. Borden, 20 Oct 1949 (JCAE III, 37/LRL); Seaborg, in GAC 47, 13–15 Dec 1955, EHC.
20. Groves to Oppenheimer, 29 Jul 1943, in Alice Kimball Smith and Charles Weiner, eds., *Robert Oppenheimer: Letters and Recollections* (Cambridge, Mass., 1980), 262–263.
21. John Warner, in GAC 66, 28–30 Oct 1959, EHC: "If Brookhaven had been located only a mile or two from LaGuardia airport, many more students would be desirous of visiting Brookhaven, and thereby the past buildup of permanent staff at the BNL would have been partially unnecessary"; Zinn, quoted in Holl, 52; site selection criteria, Nov 1965, in Westfall, "Fermilab," 426; T. A.

Heppenheimer, *Turbulent Skies: The History of Commercial Aviation* (New York, 1995).

22. Norris Bradbury, "Los Alamos—The First 25 Years," in J. O. Hirschfelder, H. P. Broida, and L. Badash, eds., *Reminiscences of Los Alamos* (Dordrecht, 1980), 161–175, on 161; Weinberg, 67–71; Holl, 173–174. Two other exceptions: Eugene Wigner directed Clinton through the summer of 1947, and I. I. Rabi's active interest as a trustee of Brookhaven advanced its status.
23. GAC 2, 2–3 Feb 1947, EHC, seeking "a man of recognized standing in physics who would give style and inspiration" to Los Alamos; GAC, Report of Subcommittee on Research, in Oppenheimer to Lilienthal, 3 Apr 1947, GAC 3, EHC, presenting an argument for a central reactor lab: "It may be just possible to find one man fully qualified as director; it is very uncertain that several can be found for several effective laboratories." The GAC later worried about the inexperience of York at Livermore: Hewlett and Holl, 180.
24. See, e.g., Robert O. Keohane, "Hegemonic Leadership and U.S. Foreign Economic Policy in the 'Long Decade' of the 1950s," in William P. Avery and David P. Rapkin, eds., *America in a Changing World Political Economy* (New York, 1982), 49–76, on 49.
25. *New York Times,* 17 Mar 1954, 1.
26. I. I. Rabi, GAC 41, 14 Jul 1954, in Hewlett and Holl, 180.
27. Strauss, 9 Feb 1949, in Seidel, "Home," 136–137.
28. James R. Newman and Byron S. Miller, "The Socialist Island," *BAS,* 5:1 (1949), 13–15, on 13; Edward Teller with Allen Brown, *The Legacy of Hiroshima* (New York, 1962); V. W. Hughes in AUI, Ex Comm, 20 September 1963; A. V. Crewe to M. Goldhaber, 4 Feb 1965 (BNL-HA, 33/Lab directors' meetings).
29. David A. Hollinger, "Free Enterprise and Free Inquiry: The Emergence of Laissez-Faire Communitarianism in the Ideology of Science in the United States," *New Literary History,* 21 (1990), 897–919.

1. ORIGINS

1. Bush had in 1940 established the National Defense Research Committee, or NDRC, which retained an advisory role in OSRD. Daniel J. Kevles, *The Physicists: The History of a Scientific Community in Modern America* (New York, 1978), 296–301.
2. Holl, 6.
3. Holl, 7–9; Arthur Holly Compton, *Atomic Quest* (New York, 1956), 80–81; Hewlett and Anderson, 41–43, 53–56; Richard Rhodes, *The Making of the Atomic Bomb* (New York, 1986), 397–400.
4. Hewlett and Anderson, 71–83; Holl, 7–13; Rhodes, *Atomic Bomb,* 424–428.
5. Compton, *Atomic Quest,* 110–111.

6. Holl, 13–14.
7. Hewlett and Anderson, 76–77.
8. Hewlett and Anderson, 91, 185–193; Johnson and Schaffer, 16–18.
9. Hewlett and Anderson, 198, 207, 212; Johnson and Schaffer, 18–21; Holl, 22–24, 31.
10. Hewlett and Anderson, 227–230; Oppenheimer to Robert Bacher, 10 Jun 1942, in Alice Kimball Smith and Charles Weiner, eds., *Robert Oppenheimer: Letters and Recollections* (Cambridge, Mass., 1980), 225; Rhodes, *Atomic Bomb,* 447–451.
11. David Hawkins, *Project Y: The Los Alamos Story,* vol. 1., *Toward Trinity* (Los Angeles, 1983), 3–5; Oppenheimer to J. H. Manley, 12 Oct 1942, in Smith and Weiner, *Oppenheimer,* 231; U.S. Atomic Energy Commission, *In the Matter of J. Robert Oppenheimer* (Cambridge, Mass., 1971), 12, 28; Lillian Hoddeson et al., *Critical Assembly: A Technical History of Los Alamos during the Oppenheimer Years, 1943–1945* (Cambridge, 1993), 57–58.
12. Smith and Weiner, *Oppenheimer,* 239.
13. Oppenheimer to Manley, 12 Oct 1942.
14. AEC, *In the Matter,* 12; Oppenheimer to Conant, 1 Feb 1943, in Smith and Weiner, *Oppenheimer,* 247–248; Hewlett and Anderson, 230–232; Rhodes, *Atomic Bomb,* 448, 452–455.
15. Holl, 6, 24.
16. Johnson and Schaffer, 27; Weinberg, 45.
17. George Everson to Underhill, 17 Feb 1944 (EOL, 29/38).
18. Bush to OSRD staff, 8 Aug 1944 (EOL, 29/38).
19. Holl, 32–33; Compton, in Robert W. Seidel, "Accelerating Science: The Postwar Transformation of the Lawrence Radiation Laboratory," *HSPS,* 13:2 (1983), 375–400, on 379.
20. Hewlett and Anderson, 322–323; Holl, 29.
21. Jeffries report, 18 Nov 1944, excerpts in app. A of Alice Kimball Smith, *A Peril and a Hope: The Scientists' Movement in America, 1945–47* (Chicago, 1965), 539–559; Holl, 32; Hewlett and Anderson, 323–325.
22. Groves to George L. Harrison, 19 Jun 1945, in Seidel, "Accelerating," 378; R. C. Tolman to E. O. Lawrence, 16 Sep 1944 (EOL, 29/37).
23. Nichols, Jan 1945, in Holl, 33; Hewlett and Anderson, 337–338, 344–345, 366. The other subcommittees included one under James Franck on social and political implications of atomic energy, which produced the famous Franck report.
24. Holl, 28.
25. T. R. Hogness to W. Bartky, 13 Jun 1945 (EOL, 29/36); A. J. Dempster to Bartky, 28 Jun 1945, ibid.
26. W. H. Zinn to Bartky, 15 Jun 1945 (EOL, 29/36).

27. Dempster to Bartky, 28 Jun 1945, and Zinn to Bartky, 15 Jun 1945 (EOL, 29/36).
28. M. D. Whitaker et al., "Organization of the National Nucleonics Program," n.d. [ca. Jun 1945] (EOL, 29/36).
29. F. H. Spedding to Hilberry, 16 Jun 1945 (EOL, 29/36); J. C. Stearns memo, 14 Jun 1945, ibid.
30. H. J. Curtis to A. H. Compton, 16 Jun 1945, and J. G. Hamilton to Hilberry, 13 Jun 1945 (EOL, 29/36). Hamilton directed the Crocker Lab (a subsidiary of Lawrence's Radiation Lab), which would soon rely on federal support; Curtis would later direct Brookhaven's biology department.
31. Bartky, "Preliminary Report," 16 Jun 1945 (EOL, 29/36).
32. Hewlett and Anderson, 367–369.
33. Seidel, "Accelerating," 378. By September 1944 Bush was planning to disband OSRD upon the defeat of Germany; Nathan Reingold, "Vannevar Bush's New Deal for Research: or, The Triumph of the Old Order," *HSPS,* 17:2 (1987), 299–344, on 319.
34. K. Priestley to MED Area Engineer, 22 Jul 1946 (EOL, 32/1); Lawrence to Groves, 13 Jul 1945 (EOL, 29/38); Seidel, "Accelerating," 379.
35. Hawkins, *Project Y,* 484–485.
36. Oppenheimer to Division and Group Leaders, 20 Aug 1945 (LANL A-84–019, 60/10).
37. Hewlett and Anderson, 627; Johnson and Schaffer, 27.
38. Norris Bradbury, "Los Alamos—The First 25 Years," in J. O. Hirschfelder, H. P. Broida, and L. Badash, eds., *Reminiscences of Los Alamos* (Dordrecht, 1980), 161–175, on 162; Norris Bradbury, "The First 20 Years at Los Alamos," *LASL News,* 1 Jan 1963.
39. Bradbury to Groves, 3 Nov 1945 (LANL WWII Director's Files, 19/4).
40. Bradbury, "First 20 Years"; Edith C. Truslow and Ralph Carlisle Smith, *Project Y: The Los Alamos Story,* vol. 2, *Beyond Trinity* (Los Angeles, 1983), 271.
41. "Some Comments on the Results of the Questionnaire," 14 Mar 1946 (LANL A-84–019, 31/5); Leslie R. Groves, *Now It Can Be Told* (New York, 1962), 378–379.
42. Weinberg, 48.
43. Johnson and Schaffer, 29.
44. Clinton Labs Research Council, minutes, 13 Feb 1946, ORNL/CF-46-2-13.
45. "History of the Activities of the Manhattan District Research Division," 15 Oct 1945–31 Dec 1946, ORO-1050, chap. 12.
46. Oppenheimer to Division and Group Leaders, 20 Aug 1945 (LANL A-84–019, 60/10).
47. Groves to Bradbury, 4 Jan 1946 (LANL WWII Director's Files, 19/4).

48. Bradbury interview by Arthur L. Norberg, 1976, TBL, 48; Bradbury, "First 20 Years"; Bradbury, "First 25 Years."
49. Minutes of Crossroads meeting, 26 Jan 1946, and Bradbury to Vice Adm. W. H. P. Blandy, 7 Jan 1946 (LANL WWII Director's Files, 23/7).
50. B. Dauben to J. H. Lawrence, 28 Jun 1946 (John H. Lawrence papers, TBL, 4/28).
51. Lawrence to Groves, 13 Jul 1945 (EOL, 29/38); Lawrence to Monroe Deutsch, 21 Aug 1945 (EOL, 20/11); Lawrence, "Recommended Program," 15 Sep 1945 (EOL, 22/10).
52. Groves to Lawrence, 15 Oct 1945 (EOL, 22/10); K. Priestley to Lawrence, 1 Nov 1945 (EOL, 22/4).
53. Groves to Lawrence, 28 Dec 1945 (EOL, 22/10); W. B. Reynolds to W. Norton, 12 Jan 1946, and Lawrence to R. Sproul, 31 Jan 1946 (EOL, 22/4).
54. Lawrence telex [to Cooksey], 29 Mar 1946 (EOL, 22/4).
55. Hewlett and Anderson, 627; Weinberg, 56–57.
56. "History of Activities," chap. 12; Holl, 39–40.
57. Hewlett and Anderson, 633.
58. F. Daniels, "Proposed Program for the National Nucleonics Laboratory at Argonne," n.d. [probably 21 Feb 1946], ONRL/CF-46-3-123.
59. John S. Rigden, *Rabi: Scientist and Citizen* (New York, 1987), 184.
60. G. B. Pegram to Groves, 17 Jan 1946, in Crease, 14.
61. Allan A. Needell, "Nuclear Reactors and the Founding of Brookhaven National Laboratory," *HSPS,* 14:1 (1983), 93–122, on 94, 97.
62. Ibid., 98; Crease, 14–15.
63. Pegram to Groves, 3 Mar 1946 (BNL-HA, AUI Planning Committee).
64. Crease, on 15 and 16.
65. "History of Activities," chap. 12, 4.
66. Hewlett and Anderson, 633–637.
67. Holl, 40.
68. "History of Activities," chap. 12, 4.
69. Pegram to northeastern institutions, 5 Mar 1946, in Robert W. Seidel, "A Home for Big Science: The AEC's Laboratory System," *HSPS,* 16:1 (1986), 135–175, 139n7.
70. L. A. DuBridge to Lt. Col. Stanley Stewart, 31 Dec 1946; Stewart to A. V. Peterson, 31 Dec 1946; Carroll Wilson to DuBridge, 9 Jan 1947; and DuBridge to Wilson, 13 Jan 1947 (Sec'y 47–51, 60/Nuclear Science Lab for West Coast). The initial impetus for the proposal seems to have come from Lawrence, who proposed a reactor for Berkeley in late 1945 but, according to DuBridge, suggested to Groves building it in southern California. It may also have stemmed from talk at Los Alamos of moving the weapons lab to southern California; see Groves, *Now It Can Be Told,* 378–379.

71. U.S. House, 80th Cong., 1st sess., *Independent Offices Appropriation Bill for 1948*, 1477.
72. E. U. Condon to Nichols, 21 Feb 1946, and Columbia University, "Nuclear Science Research Program," 2 Mar 1946, ORNL/CF-46-3–123.
73. Crease, 15; N. F. Ramsey, "Summary of Meeting at Columbia University," 16 Feb 1946 (BNL-HA, AUI Planning Committee).
74. Clinton Labs, Research Council, minutes, 6 Feb 1946, ORNL/CF-46-2-113.
75. C. A. Thomas to Nichols, 11 Feb 1946, ORNL/CF-46–2–170; Clinton Research Council, minutes, 13 Feb 1946, ORNL/CF-46-2-13.
76. K. T. Compton to Pegram, 19 Mar 1946 (BNL-HA, AUI Planning Committee).
77. Clinton Steering Committee minutes, 4 Feb 1946, ORNL/CF-46-2-29.
78. Clinton Steering Committee minutes, 15 Feb 1946, ORNL/CF-46-2-209.
79. A. Hunter Dupree, *Science in the Federal Government* (Baltimore, 1986), 275; Rexmond C. Cochrane, *Measures for Progress: A History of the National Bureau of Standards* (Washington, D.C., 1966), 68; Kevles, *Physicists*, 66–67, 81, 189–190.
80. Cochrane, *Measures*, 332, 357–364, 377–388; Hewlett and Anderson, 25, 32, 66, 168–169, 233.
81. Deficiency Appropriation Act of 3 Mar 1901 and Joint Resolution of 12 Apr 1892, in Cochrane, *Measures*, app. C, 539–540.
82. Cochrane, *Measures*, 224–225.
83. Warren Weaver diary, 27 Oct 1939, in J. L. Heilbron and Robert W. Seidel, *Lawrence and His Laboratory: A History of the Lawrence Berkeley Laboratory*, vol. 1 (Berkeley, 1989), 475.

2. INDIVIDUALITY

1. Daniel J. Kevles, *The Physicists: The History of a Scientific Community in Modern America* (New York, 1978), 298–301.
2. Hewlett and Duncan, 19.
3. Holl, 8.
4. Hewlett and Anderson, 192–193; Holl, 21; Johnson and Schaffer, 18.
5. J. L. Heilbron and Robert W. Seidel, *Lawrence and His Laboratory: A History of the Lawrence Berkeley Laboratory*, vol. 1 (Berkeley, 1989), 114, 509.
6. Hewlett and Anderson, 230.
7. David Hawkins, *Project Y: The Los Alamos Story*, vol. 1., *Toward Trinity* (Los Angeles, 1983), 29–50; R. M. Underhill, interview by Arthur Norberg, 1976, TBL, 39.
8. Underhill to David Dow, 21 Dec 1944, and Lansdale to Oppenheimer, 30 Dec 1944 (LANL, WWII Director's Files, 1/8).

9. Underhill interview, 2, 18.
10. JCAE, 81st Cong., 1st sess., *Investigation into the United States Atomic Energy Project,* 304.
11. Johnson and Schaffer, 29.
12. Holl, 22, 33, 40–41, 49.
13. Report of Advisory Committee for Selection of ANL Management Agency, 16 Apr 1951, and Howard Brown to Medford Evans, 31 Jul 1950 (DBM, 17/MH&S-21: ANL).
14. Lawrence to Sproul, 17 Jan 1944 (EOL, 20/22); Robert W. Seidel, "Accelerating Science: The Postwar Transformation of the Lawrence Radiation Laboratory," *HSPS,* 13:2 (1983), 375–400, on 377–384.
15. Contract no. W-7405-eng-36, 20 Apr 1943 (LANL, WWII Director's Files, 1/7); Underhill interview, 21.
16. Gregg Herken, "The University of California, the Federal Weapons Labs, and the Founding of the Atomic West," in Bruce Hevly and John M. Findlay, eds., *The Atomic West* (Seattle, 1998), 119–135, on 121.
17. Hewlett and Anderson, 314–315.
18. Underhill to Sproul, 9 Mar 1946 (LANL A-83-0033, 15/8).
19. Oppenheimer to Monroe Deutsch, 24 Aug 1945; to Lawrence, 30 Aug 1945; to R. G. Sproul, 29 Sep 1945; and to R. T. Birge, 29 Sep 1945, in Alice Kimball Smith and Charles Weiner, eds., *Robert Oppenheimer: Letters and Recollections* (Cambridge, Mass., 1980), 295, 301, 306, 307. Oppenheimer's antagonism toward the university contributed to his decision not to return to its faculty after the war.
20. Underhill interview, 59, 64–65.
21. Underhill to Bradbury, 3 Jan 1946; Bradbury to Groves, 11 Jan 1946; Bradbury to Nichols, 5 Mar 1946 (LANL, WWII Director's Files, 1/8).
22. Underhill to Sproul, 9 Mar 1946 (LANL A-83-0033, 15/8).
23. Underhill interview, 78, 76.
24. C. F. Dunbar to E. C. Shoup, 13 Oct 1948 (BNL-DO, II-1).
25. Mervin Kelly, testimony in JCAE, *Investigation,* 813.
26. Atomic Energy Act of 1946, sec. 1(b)(3), sec. 3(b), in Hewlett and Anderson, app. 1.
27. Underhill, memo of phone conversation with Carroll Wilson, 10 Sep 1947 (EOL, 32/3).
28. Oppenheimer to Lilienthal, GAC 3, 3 Apr 1947, EHC.
29. J. R. Coe et al. to Lilienthal, 28 May 1947 (RRC-UT, MS-652, 10/40).
30. JCAE, *Investigation;* Hewlett and Duncan, 355–361. The investigation alarmed the University of California, whose Regents unanimously conveyed their "expressed desire that steps be taken to withdraw from the Los Alamos contract at as early a date as it is practicable to do so." UC Regents, 10 Jun 1949.

31. AEC press seminar, 14–15 Mar 1949 (BNL-DO, II-5).
32. Director of Production, "Direct Operations by the Atomic Energy Commission," draft Sep 1949 (DBM, 31/O&M-3).
33. Ibid.
34. T. Keith Glennan to AEC and AEC staff, 29 Dec 1950 (EHC, folder "Management of National Labs").
35. Dunbar to Shoup, 13 Oct 1948 (BNL-DO, II-1).
36. J. M. Knox to Haworth, 14 Oct 1949 (BNL-DO, VII-32).
37. The subcommittee started out as the Special Committee on the Los Alamos Project and then changed its name in 1948.
38. Dunbar to Shoup, 13 Oct 1948 (BNL-DO, II-1); Bethe to P. M. Morse, 6 Nov 1947 (BNL-DO, I-4).
39. Carol Gruber, "The Overhead System in Government-Sponsored Academic Science: Origins and Early Development," *HSPS,* 25:2 (1995), 241–268.
40. AUI, Board of Trustees, minutes, 25 Oct 1947.
41. Lawrence to Sproul, 10 Jul 1947 (EOL, 23/19).
42. Clark Kerr to Lawrence, 2 Mar 1953 (EOL, 32/10); Kerr to UC deans et al., 27 Oct 1952 (EOL, 19/15); and UC organization charts, ca. 1954 (EOL, 19/5).
43. Peter J. Westwick, "Abraded from Several Corners: Medical Physics and Biophysics at Berkeley," *HSPS,* 27:1 (1996), 131–162, on 151–153.
44. R. T. Birge to [Dean] A. R. Davis, 12 Jul 1950, and Birge to Sproul, "Non-Academic Employees" and "Expense and Equipment," 22 Nov 1947 (Records of the Department of Physics, 1920–1962, UC Berkeley, TBL, 1/11, 1/7). The other labs, however, could envy the cheap (or free) labor of graduate students (AUI, Ex Comm, 15 Jun 1951).
45. Minutes, UC Regents, Committee on AEC Projects, 21–22 Mar 1952 (EOL, 19/37).
46. AEC, *Tenth Semiannual Report* (Jul 1951), 63, and *Eleventh Semiannual Report* (Jan 1952), 44.
47. Heilbron and Seidel, *Lawrence,* 103–113.
48. Underhill to W. B. Reynolds, 10 Aug 1949 (DC, 7/Administrative-Finance: Overhead); Underhill to A. Tammaro, 22 Jan 1952 (LANL A-83–0033, 9/1); Committee on AEC Projects, UC Regents, minutes, 21–22 Mar 1952 (EOL, 19/37); UC Regents, 25 Jul 1952, 21 Nov 52, 30 Jan 1953, and 26 Jun 1953.
49. Gruber, "Overhead," 243; Underhill interview, 33; Heilbron and Seidel, *Lawrence,* 511–512.
50. AEC, *Seventh Semiannual Report* (Jan 1950), 158.
51. Gruber, "Overhead," 265.
52. "Direct Operations."
53. Fletcher Waller to Wilson, 3 Apr 1950, and D. B. Langmuir to C. G. Worthington, 10 May 1950 (DBM, 34/Research Committee).
54. In late 1941 Columbia's bursar found it "necessary for us to try again to 'edu-

cate' Dean Pegram as to what is meant by 'overhead.'" Pegram helped found Brookhaven. Gruber, "Overhead," 260.

55. K. Priestley to A. P. Pollman, 26 Feb 1948 (DC, 5/Rad Lab: Budget); UCRL budget request, 21 Apr 1948 (DC, 5/Rad Lab: Budget request); C. E. Larson to A. H. Holland, 16 May 1950 (ORNL-DO, Budget-Costs).
56. Report for Selection of ANL Management.
57. AEC, "Management Fee for AUI," Jan 1952 (DBM, 18/MH&S-21: BNL).
58. Paul Green to Underhill, 18 Jul 1949, and Underhill to W. B. Harrell and J. Campbell, 20 Jul 1949 (DC, 7/Admin-Finance: Overhead).
59. AEC Research Committee, "Notes on Discussion of National Laboratory Problems," 14 Nov 1949 (EHC, Sec'y 47–51, Mgt. of national labs); AEC 317, 1 May 1950, and AEC 317/5, 21 Dec 1950 (Sec'y 47–51, 67/635.12-BNL); AEC, "Management Fee."
60. Report for Selection of ANL Management, and "Negotiations with the University of Chicago," 13 Dec 1951 (DBM, 17/MH&S-21: ANL).
61. Sumner Pike, in GAC 4, 30 May–1 June 1947, EHC.
62. Hewlett and Anderson, 66–67, 76–79, 103–104.
63. "Statement by Research Directors of C.N.L.," n.d. (RRC-UT, MS-652, 1/26).
64. Johnson and Schaffer, 50–51; Hewlett and Anderson, 122–126; Weinberg, 68–69.
65. M. D. Peterson to C. N. Rucker, 14 Jul 1948 (ORNL-DO, Research-General).
66. AUI, Ex Comm, minutes, 16 Nov 1951.
67. Glennan to AEC.
68. U.S. House, 80th Cong., 2d sess., *Supplemental Independent Offices Appropriation Bill for 1949,* 805–806; Oral Rinehart, budget instructions, May 1949 (ORNL-DO, Budget-Instructions and procedures); C. E. Larson to A. H. Holland, 16 May 1950 (ORNL-DO, Budget-Costs); Hewlett and Duncan, 420–421.
69. Harold Orlans, *Contracting for Atoms* (Washington, D.C., 1967), 55–56; Frank K. Edmondson, *AURA and Its National Observatories* (Cambridge, Mass., 1997); Dominique Pestre, "The First Suggestions, 1949–June 1950," in Armin Hermann et al., eds., *History of CERN,* vol. 1, *Launching the European Organization for Nuclear Research* (Amsterdam, 1987), 63–95, on 89.
70. Holl, 153–174, 182–189.
71. Orlans, *Contracting,* 13–15.
72. AEC, *Ninth Semiannual Report to Congress* (Jan 1951), 41, 55–69.
73. M. Boyer to Division and Office Directors, 30 Jun 1953 (Sec'y 51–58, 174/BAF-2: 1954); Topnotch II agenda item 7, 24–28 Sep 1953 (Sec'y 51–58, 65/O&M-6: Topnotch II); AEC, response to Fay-Carpenter report, May 1954, ORO-1763.
74. Paul Foster to AEC staff, 22 Oct 1958 (Sec'y 58–66, 1423/7).

75. UC Regents, 14 Jul 1955 and 21 Nov 1958.
76. W. B. McCool to Starbird, 12 Jan 1959 (Sec'y 58–66, 1423/7).
77. UC Regents, 24 Jun 1955.
78. Orlans, *Contracting,* 4–5. The AEC did operate a few small labs itself, for instance, the Health and Safety Lab near Brookhaven in New York.
79. AEC, *Ninth Semiannual Report* (Jan 1951), 41.
80. The exceptions were Hanford and Sandia, which were expanding their research efforts.
81. R. Birge to R. Sproul, 15 Dec 1945, in Westwick, "Abraded," 138; UC Regents, 23 Apr 1954.
82. Seidel, "Accelerating," 384–385.
83. Holl, 8–9, 22.
84. Organization charts for Los Alamos, ca. Aug 1943 and 22 Sep 1944, and Oppenheimer to Coordinating Council, 20 Apr 1944 (LANL, WWII Director's Files, 6/1); Hawkins, *Project Y,* 29–34.
85. Johnson and Schaffer, 55; Weinberg, 71.
86. M. E. Day to R. A. Nelson, 3 Jun 1948 (DC, 5/Personnel: Professional); Seidel, "Accelerating," 386–387, 390–391. See also Peter Galison, *Image and Logic* (Chicago, 1997), 250–255, 346.
87. Seidel, "Accelerating," 385–386; Robert W. Seidel, "From Mars to Minerva: The Origins of Scientific Computing in the AEC Labs," *Physics Today* (Oct 1996), 33–39.
88. ORNL organization charts for 1 Jul 1953 and 1 Jan 1962; Johnson and Schaffer, 70.
89. L. G. Hawkins to A. E. Dyhre, 30 Mar 1949 (LANL A-83–0033, 7/3); Necah Furman, *Sandia National Laboratories: The Postwar Decade* (Albuquerque, 1989), 277, 312–339; Leland Johnson, *Sandia National Laboratories: A History of Exceptional Service in the National Interest,* ed. Carl Mora, John Taylor, and Rebecca Ullrich (SAND07–1029, Sandia National Laboratory, 1997), 18–31.
90. Johnson and Schaffer, 63–64.
91. ORNL Policy Committee, minutes, 21 Dec 1955 (ORNL-DO, Meetings-Policy Committee).
92. P. Sandidge, "Study of Clinton Laboratories Departments and Personnel," 1 May 1947, ORNL/CF-47–5–76; comparative analyses of AEC laboratories, FY1953 (ORNL-DO, Budget Jul–Dec 1954); AEC, *Fifteenth Semiannual Report* (Jan 1954).
93. M. S. Livingston to Haworth, 20 Jan 1948 (BNL-DO, IV-19); Policy Committee minutes (ORNL-DO, Meetings-Policy Committee).
94. Knox to Haworth, 14 Oct 1949 (BNL-DO, VII-32).
95. W. P. Leber to A. V. Peterson, 7 Mar 1947, ORO-1643.
96. Seidel, "Accelerating," 387n32; Johnson and Schaffer, 36.

97. Seidel, "Accelerating," 387.
98. Oppenheimer to J. H. Manley, 12 Oct 1942, in Smith and Weiner, *Oppenheimer,* 231–232; Oppenheimer to S. K. Allison, 7 Jun 1943; Allison to Oppenheimer, 19 May 1943; Oppenheimer to Groves, 13 and 27 May 1944; Nichols to Oppenheimer, 29 May 1944; Conant to Oppenheimer, 20 Nov 1942 (LANL, WWII Director's Files, 4/10).
99. Col. G. W. Beeler to L. DuBridge, 9 May 1946 (BNL-HA, University-Laboratory Liaison); Allan A. Needell, "Nuclear Reactors and the Founding of Brookhaven National Laboratory," *HSPS,* 14:1 (1983), 93–122, 102.
100. Needell, "Nuclear Reactors," 105; Morse to Wigner, 10 Dec 1946 (BNL-DO, I-1).
101. Morse to Wigner, 10 and 13 Dec 1946, and to L. R. Thiesmeyer, 17 Oct 1946 (BNL-DO, I-1).
102. Leslie R. Groves, *Now It Can Be Told* (New York, 1962), 382.
103. Haworth to F. W. Loomis, 20 Nov 1947 (BNL-DO, I-6); H. A. Winne to Carroll Wilson, 26 Mar 1948, Wilson to Morse, 31 Mar 1948, and Morse to Wilson, 12 Apr 1948 (BNL-DO, I-5).
104. R. P. Johnson to C. N. Rucker, 9 Sep 1948 (ONRL-DO, Program-ORNL).
105. MED, Advisory Committee on Research and Development, minutes, 15 Jun 1946, ONRL/CF-46-6-229.
106. AEC, *Sixth Semiannual Report,* 113.
107. Morse to Thiesmeyer, R. A. Patterson, and L. B. Borst, 7 Mar 1947, and Morse to James H. Lum, 14 Mar 1947 (BNL-DO, I-2).
108. Westwick, "Abraded," 141.
109. Raemer E. Schreiber, interview by Arthur L. Norberg, 1976, TBL, 15, 29–30.
110. S. A. Goudsmit, memo to R. L. Cool et al., 18 May 1956, and Lawrence to M. Oliphant, 4 Apr 1956, in John L. Heilbron, "Creativity and Big Science," *Physics Today* (Nov 1992), 42–47, on 44; N. Hilberry, "Some Comments on the Objectives of the Commission's National Laboratories," 3 Jan 1958, quoted in Robert W. Seidel, "A Home for Big Science: The AEC's Laboratory System," *HSPS,* 16:1 (1986), 135–175, on 164–165.
111. William H. Whyte, Jr., *The Organization Man* (New York, 1957), 235.
112. Hewlett and Duncan, 66, 35.
113. Gale Young to Joseph Brewer, 19 Jun 1947, and Weinberg to A. Hollaender, 24 Dec 1947 (RRC-UT, MS-652, 10/83).
114. Crease, 26.
115. Crease, 33, 32.
116. C. P. Rhoads to R. D. Conrad, 11 Oct 1948 (BNL-DO, V-49); AUI, Ex Comm, minutes, 18 Jun 1948.
117. Walter W. Stagg to J. R. Abersold, 22 Jan 1946 (LANL, WWII Director's Files, 4/11).

118. Darol Froman, interview by Arthur L. Norberg, 1976, TBL, 44; Schreiber interview, 26.
119. Cabell Phillips, "Scientists Ponder Jobs on the Atomic Project," *New York Times,* 19 Sep 1948, E-7.
120. Mervin Kelly testimony in JCAE, *Investigation,* 813; M. S. Livingston to Haworth, 20 Jan 1948 (BNL-DO, IV-19).
121. I. I. Rabi, "Report of Subcommittee on Personnel Policy," 16 Apr 1946 (BNL-HA, AUI Planning Committee).
122. M. Kuper to Ad Hoc Committee on Tenure, 31 Mar, 7 Apr, and 13 May 1948; J. B. H. Kuper to Policy and Program Committee, 9 Jun 1948; Charles Dunbar to Haworth, 11 Aug 1948; and Haworth to scientific staff, 1 Sep 1948 (BNL-DO, VI-21).
123. Hollaender to Personnel Committee, 26 Jul 1949, Summary of activities of the Personnel Committee, 19 Aug 1949, and Personnel Committee minutes, 9 Nov 1949 (ORNL-DO, Meetings-Personnel Committee); ORNL Research Council minutes, 10 Nov 1949 (ORNL-DO, Meetings-Research Council).
124. Underhill interview, 65.
125. Sproul to Birge, 27 Apr 1950 (EOL, 23/20).
126. Lincoln Constance to Glenn Seaborg, 17 Dec 1959, in Westwick, "Abraded," 153.
127. Lawrence to Birge, 9 Jun 1950, Birge to Sproul, 12 Jun 1950, and Sproul to Bradbury, 17 Jul 1950 (EOL, 23/20); Raymond T. Birge, *History of the Physics Department,* vol. 5 (UC Berkeley, Physics Library, 1966), XVIII:47.
128. Report for Selection of ANL Management.
129. W. C. Parkinson to G. W. Beadle, 6 Jul 1961 (ANL-PAB, 9/Physics).
130. AUI trustees, minutes, 20 Jan 1956 and 19 Jan 1962; graphs of age of ANL staff in GAC 48, 12–13 Jan 1956; Haworth, "The Future Role of Brookhaven," 5 Dec 1955, and Weinberg, "Some Problems in the Development of the National Laboratories," Dec 1955 (BNL-DO, III-21); ANL Policy Advisory Board, minutes, 8 May 1961 and 16 Oct 1963 (ANL-PAB, boxes 3 and 4).
131. Loomis et al. to Nichols, 19 Mar 1946 (LANL A-84–019, 49/5).
132. Wilson, "Appointment of Scientific Personnel Committee," 20 May 1947 (LANL A-83–0033, 7/5).
133. Report of the Committee on Scientific Personnel (3d draft), 29 Jul 1947 (UCh-VPSP, 32/4).
134. Bradbury to Nichols, 5 Mar 1946 (LANL, WWII Director's Files, 1/8); AUI Ex Comm, minutes, 24 Oct 1946; A. E. Dyhre, "History—Business Office," 3 Jun 1948 (LANL, A-83–0033, 7/3); Los Alamos, Personnel Administrative Panel, minutes, 28 Jun 1948 (LANL A-83-0033, 8/1); Bradbury to Sproul, 26 Jan 1949 (LANL, A-83-0033, 9/1); see ORNL-DO, Meetings-Policy Committee.
135. E. R. Jette to Bradbury, 31 Jan 1947 (LANL A-84-019, 49/5); Froman to

Underhill, 5 Jun 1950 (LANL A-83-0033, 14/4); Bradbury to Sproul, 10 Nov 1955 (EOL, 19/37).

136. Lawton Geiger to W. B. Harrell, table 2-A, 5 Sep 47 (UCh-VPSP, 33/4).
137. Geiger to Harrell, ibid.; conference on wage and salary increases, 23 Dec 1948 (LANL A-83-0033, 8/5); Bradbury to division leaders, 25 Apr 1949 (LANL A-83-0033, 8/4); Bradbury, interview by Arthur L. Norberg, 1976, TBL 6.
138. Westwick, "Abraded," 141.
139. Bradbury to Sproul, 26 Jan 1949 (LANL A-83-0033, 9/1).
140. AUI Ex Comm, 15 Jun 1951, 20 May 1955, 21 Sep 1956.
141. W. Reynolds to E. C. Shute, 31 Aug 1956 (EOL, 22/9).
142. AEC 811/3, 10 Oct 1956 (Sec'y 51–58, 62/O&M-6: Meetings and conferences, vol. 2); GAC 51, 29–31 Oct 1956, EHC; AUI Ex Comm, 13 Dec 1956.
143. GAC 24, 4–6 Jan 1951; GAC 36, 17–19 Aug 1953, EHC; ACBM 47, 3–4 Dec 1954, and G. Failla to Strauss, 7 Mar 1955, ACBM 47.
144. GAC, "Recommendations Relative to Administrative Policy for the AEC Research Laboratories," 8 Jan 1954, GAC 38; T. H. Johnson to managers, AEC Operations Offices, 19 Feb 1954 (BNL-DO, III-21).
145. GAC 76, 19–21 Oct 1961, and K. Pitzer to Seaborg, 21 Oct 1961 (EHC, Sec'y 58–66, O&M-7: GAC minutes).
146. ORNL, "Justification—FY1950 Budget," 25 Aug 1948, ORO-1206. The founders of Fermilab helped popularize claims to "truly national" status; see Lillian Hoddeson, "Establishing KEK in Japan and Fermilab in the US: Internationalism, Nationalism and High Energy Accelerators," *Social Studies of Science,* 13 (1983), 1–48, on 17; Catherine Westfall, "The First 'Truly National Laboratory': The Birth of Fermilab" (Ph.D. diss., Michigan State University, 1988).
147. GAC 15, 14–15 Jul 1949.
148. AEC, *Sixth Semiannual Report* (Jul 1949), 153.
149. BNL, Policy and Program Committee, 25 Nov 1947 (BNL-DO, IV-16).
150. AUI, Board of Trustees, 19 Oct 1951; Haworth to E. L. Van Horn, 8 Jan 1952 (BNL-DO, I-10); G. F. Tape to Van Horn, 26 Jun 1952 (BNL-Central Records, DO-87).
151. GAC 47, 13–15 Dec 1955, EHC, 15–16, 28; Seidel, "Home," 158; Holl, 72, 90–92.
152. E.g., H. Urey to Morse, 17 Mar 1947, and Fermi to Morse, 31 Mar 1947 (BNL-DO, I-2); Morse to John von Neumann, 18 Apr 1947, and to Oppenheimer, 16 May 1947 (BNL-DO, I-3).
153. Morse to Thiesmeyer, 18 Oct 1946 (BNL-DO, I-1); Morse to Shoup, 3 Nov 1947 (BNL-DO, I-4); AUI, Ex Comm, 21 May 1948.
154. Clinton Steering Committee, minutes, 4 Feb 1946, ORNL/CF-46-2-29; Lilienthal, in U.S. Senate, 80th Cong., 2d sess., *Supplemental Independent Offices Appropriation Bill for 1949,* 94–95.

155. ORNL Policy Committee, 11 Feb 1952 (ORNL-DO, Meetings-Policy Committee); Haworth to Van Horn, 22 Mar 1954 (BNL-DO, I-12); Schreiber, in Richard Rhodes, *Dark Sun: The Making of the Hydrogen Bomb* (New York, 1995), 278; GAC 37, 4–6 Nov 1953, EHC; T. H. Johnson to managers, AEC Operations Offices, 19 Feb 1954 (BNL-DO, III-21).
156. ANL Policy Advisory Board, 21 Oct 1959 (ANL-PAB, box 2); ANL Physics Review Committee, 2 Oct 1959; Hilberry to R. W. Harrison, 10 Oct 1960; and Physics Review Committee, 20–22 May 1965 (ANL-PAB, 9/Physics).

3. INTERDEPENDENCE

1. David Edge, "Competition in Modern Science," in Tore Frangsmyr, ed., *Solomon's House Revisited* (Canton, Mass., 1990), 208–232.
2. Leonard S. Reich, *The Making of American Industrial Research: Science and Business at GE and Bell, 1876–1926* (Cambridge, 1985), 110, 186–191; Eliot Marshall, "Ethics in Science: Is Data-Hoarding Slowing the Assault on Pathogens?" *Science,* 275 (7 Feb 1997), 777–780; Stephen Hilgartner, "Data Access Policy in Genome Research," in Arnold Thackray, ed., *Private Science: Biotechnology and the Rise of the Molecular Sciences* (Philadelphia, 1998), 202–218.
3. Charles Coulston Gillispie, "Science and Secret Weapons Development in Revolutionary France, 1792–1804: A Documentary History," *HSPS,* 23:1 (1992), 35–152.
4. Spencer R. Weart, "Scientists with a Secret," *Physics Today,* 29 (1976), 23–30.
5. Hewlett and Anderson, 110, 170, 227–229; Holl, 25–26, 30–31.
6. K. D. Nichols to D. Cooksey, n.d. [ca. Aug 1945] (EOL, 21/18); Groves to Underhill, 14 Aug 1945, and W. B. Harrell to Underhill, 15 Aug 1945 (LANL A-83-0033, 15/8).
7. Henry DeWolf Smyth, *Atomic Energy for Military Purposes* (Stanford, 1989); Hewlett and Anderson, 368, 400–401, 406–407.
8. Hewlett and Anderson, 1–2, 647; Col. W. S. Hutchison, "The Manhattan Project Declassification Program," *BAS,* 2 (1 Nov 1946), 14–15.
9. Morse to Spedding, 12 Mar 1948; Spedding to F. Waller, 3 Nov 1947 (BNL-DO, I-5); Morse to W. Kelley, 17 Dec 1947 (BNL-DO, I-4).
10. Loomis et al. to Nichols, 19 Mar 1946 (LANL A-84-019, 49/5).
11. R. M. Underhill to Bradbury, 28 Apr 1948; Bradbury to Underhill, 6 May 1948; and Underhill, minutes of Special [Regents'] Committee on the Los Alamos Project, 27 Apr 1948 (EOL, 19/36).
12. Hewlett and Duncan, 23–26; Fields to Wilson, 25 Apr 1947 (Sec'y 47–51, Lab Directors' meetings, EHC); C. F. Dunbar to E. C. Shoup, 13 Oct 1948 (BNL-DO, II-1).
13. Louis N. Ridenour, "Secrecy in Science," *BAS,* 1 (1 Mar 1946), 3, 8; Edward

Teller, "Scientists in War and Peace," ibid., 10–11; cf. Edward Teller, "The First Year of the Atomic Energy Commission," *BAS*, 4 (Jan 1948), 5–6, insisting now that atomic information "is much too dangerous to be allowed to circulate in an uncontrolled manner."

14. Hewlett and Duncan, 88–95; Holl, 62.
15. Sumner T. Pike, "A Commissioner Speaks," *BAS*, 4 (Jan 1948), 15–17; AUI, Ex Comm, 19 Dec 1947; T. H. Davies, "'Security Risk' Cases—A Vexed Question," and Stephen White, "Report on Oak Ridge Hearings" and "The Charges Presented in Oak Ridge Cases," *BAS*, 4 (Jul 1948), 193–196, on 196.
16. AUI, Ex Comm, 19 Dec 1947; Dunbar to Shoup.
17. Cooksey to J. J. Flaherty, 26 Oct 1948 (EOL, 33/1).
18. GAC 15, 14–15 Jul 1949; Pitzer to Lawrence, 21 Jul 1949, and Lawrence to Pitzer, 26 Jul 1949 (EOL, 32/6).
19. GAC 16, 22–23 Sep 1949, EHC; AUI, Ex Comm, 19 May 1950, and Board of Trustees, 21 Jul 1950 ("prime importance").
20. JCAE, 81st Cong., 1st sess., *Investigation into the United States Atomic Energy Project*, 14, 2, 787; Hewlett and Duncan, 355–361; Holl, 81–84.
21. David P. Gardner, *The California Oath Controversy* (Berkeley, 1967), 248; R. G. Sproul to Bradbury, 5 Jul 1950 (LANL A-83-0033, 9/1); G. Chew to Birge (quote), 24 Jul 1950, in Raymond T. Birge, *History of the Physics Department*, vol. 5 (Physics Library, UC Berkeley), XIX:46; David Kaiser, "Democracy Takes a Triple Turn: Geoffrey Chew, the Theoretical Community, and Nuclear Democracy," paper delivered at annual meeting of History of Science Society, Kansas City, 1998; Peter J. Westwick, "Abraded from Several Corners: Medical Physics and Biophysics at Berkeley," *HSPS*, 27:1 (1996), 131–162, on 143. John Manley, associate director at Los Alamos and also the secretary to the GAC, did apparently protest, although he remained in both positions until 1951.
22. AEC, *Fourth Semiannual Report to Congress* (Jul 1948), 51.
23. GAC 15, 14–15 Jul 1949.
24. U.S. Senate, 83rd Cong., 1st sess., *Second Independent Offices Appropriations for 1954*, 11.
25. Serber in Lawrence to Tammaro, 4 Nov 1949 (EOL, 32/6); J. G. Beckerley, "Secrecy in Nuclear Engineering," *Nucleonics*, 10:1 (1952), 36–38, and "Declassification Problems in Power Reactor Information," *Nucleonics*, 11:1 (1953), 6–8.
26. Hewlett and Duncan, 332–334; Robert W. Seidel, "A Home for Big Science: The AEC's Laboratory System," *HSPS*, 16:1 (1986), 135–175, on 146.
27. AEC Program Council minutes, 24 Jun 1948 (Sec'y 47–51, 33/Program Council).
28. AEC, *Fifth Semiannual Report* (1949), 110–111; JCAE, 82d Cong., 2d sess.,

Amending the Atomic Energy Act (1952), 39; Harold Green, "The Unsystematic Security System," *BAS,* 11 (Apr 1955), 118–122, 164.

29. Oppenheimer to Dean, 30 Apr 1952, GAC 30; AEC, "Response to Fay-Carpenter Report," Mar 1954, ORO-1763.
30. Hewlett and Holl, 43–72, 119–143.
31. Ralph Lapp, "The Lesson of Geneva," *BAS,* 11 (Oct 1955), 275, 308; cf. Frederick Seitz and Eugene Wigner, "On the Geneva Conference: A Dissenting Opinion," *BAS,* 12 (Jan 1956), 23–24.
32. "News Roundup," *BAS,* 11 (Mar 1955), 102–103; Edward Shils, "Security and Science Sacrificed to Loyalty," *BAS,* 11 (Apr 1955), 106–109, 130; S. A. Goudsmit [quote], "The Task of the Security Officer," *BAS,* 11 (Apr 1955), 147.
33. Lawrence to H. A. Fidler, 20 Oct 1954; P. L. Schiedermayer memo, 20 Oct 1954; Strauss to Lawrence, 15 Nov 1954; Cooksey to Clark Kerr, 14 Dec 1954 (EOL, 32/11).
34. GAC 39, 31 Mar and 1–2 Apr 1954, EHC; Haworth to Van Horn, 2 Aug 1955 (BNL-DO, I-13); "News Roundup," *BAS,* 12 (Feb 1956), 62.
35. G. Failla to Strauss, 7 May 1955, in ACBM 50.
36. AEC minutes, 2 Apr 1952 (Sec'y 51–58, 141/R&D-7: CTR); Joan Lisa Bromberg, *Fusion: Science, Politics, and the Invention of a New Energy Source* (Cambridge, Mass., 1982), 30–31.
37. GAC 37, 4–6 Nov 1953, EHC.
38. AEC 532/4, 8 May 1953, and AEC 532/15, 15 Jan 1954 (Sec'y 51–58, 141/R&D-7: CTR); Bromberg, *Fusion,* 38–39.
39. R. R. Wilson to T. H. Johnson, 17 Aug 1953, and AEC 532/10, 24 Sep 1953 (Sec'y 51–58, 141/R&D-7: CTR).
40. J. G. Beckerley to R. Thornton, 2 Mar 1953, and G. A. Kolstad to Thornton, 13 Mar 1953 (EOL, 32/10); GAC 37, 3–6 Nov 1953, EHC. The Greek patent was sealed but was scheduled to be opened by the Greek patent office on 1 Jan 1954; congressional staff urged the AEC to get the envelope away from the Greeks. Corbin Allardice to Strauss, 5 Nov 1953 (Sec'y 51–58, 141/R&D-7: CTR).
41. AEC minutes, 25 May 1955, and AEC 532/19, 18 Mar 1955 (Sec'y 51–58, 141/R&D-7: CTR). For the debates over declassification see Sec'y 51–58, 113/Reactor Development-1: CTR.
42. Bromberg, *Fusion,* 71–78, 89–90.
43. AEC 532/10, 24 Sep 1953 (Sec'y 51–58, 141/R&D-7: CTR).
44. "Technical Meeting, Research Staffs," 17–19 Jun 1946, ONRL/CF-46-6-247; W. P. Leber to M. D. Whitaker, 29 May 1946, ORNL/CF-46-5-507; Bradbury to Groves, 15 Jun 1946 (EOL, 19/36); Edith C. Truslow and Ralph Carlisle Smith, *Project Y: The Los Alamos Story,* vol. 2, *Beyond Trinity* (Los Angeles, 1983), 270, 430–431.

45. Handwritten notes [apparently by Lawrence], 16 Oct 1946; Wigner to A. V. Peterson, 17 Oct 1946; and Lawrence to Peterson, 29 Nov 1946 (EOL, 32/27).
46. See list of information meetings (EOL, 32/27); information meeting agendas for Chicago, 21–23 Apr 1947, ORNL/CF-47-4-346; Clinton, 13–15 Oct 1947, ONRL/CF-47-10-216; Argonne, 18–20 Oct 1948, ES-423; and Los Alamos, 5–7 May 1949, ORNL/CF-49-5-127; A. Hollaender to A. H. Holland, Jr., 12 Mar 1948, ORO-328; Hollaender to Warren, 29 Mar 1948, ORO-319.
47. AEC, *Seventh Semiannual Report to Congress* (Jan 1950), 185; JCAE, *Investigation,* 821.
48. Fields to Wilson, 25 Apr 1947 (Sec'y 47–51, Lab Directors' meetings, EHC).
49. AEC, *Seventh Semiannual Report to Congress* (Jan 1950), 164–167.
50. John R. Munkirs, *The Transformation of American Capitalism* (Armonk, N.Y., 1985), 107–118.
51. Morse to Scientific Advisory Committee, 6 Jan 1947 (BNL-DO, I-2); R. D. Conrad to Morse, 15 Sep 1947 (BNL-DO, II-15); Morse to E. Reynolds, 22 Sep and 3 Dec 1947 (BNL-DO, I-4); BNL, Policy and Program Committee, 28 Oct 1947 (BNL-DO, IV-16). Another non-AUI scientist joined one subcommittee for a total of twelve members.
52. Report of subcommittee of AUI Trustees, "Proposed AUI Visiting Committees to BNL," 15 Jul 1949, and AUI, "Visiting Committees," 8 Nov 1949 (BNL-DO, II-16); Haworth to department heads, 1 Sep 1949 (BNL-DO, I-7).
53. R. W. Dodson to Weinberg, 11 Oct 1954 (BNL-DO, V-14); Visiting Committee for Chemistry to C. E. Larson, 29 Oct 1954 (ORNL-DO, Advisory Committees-Chemistry); ORNL Policy Committee minutes, 3 Nov 1954 (ORNL-DO, Meetings-Policy Committee); Haworth to department chairs, 23 Oct 1957 (BNL-DO, I-15); T. H. Johnson in GAC 47, 13–15 Dec 1955, EHC, 25.
54. ANL Policy Advisory Board, minutes, 17 Jul 1957 (ANL-PAB, box 2).
55. See, e.g., AUI Board of Trustees, 15 Jul 1955.
56. ORNL, Research Committee, minutes, 4 Jun 1958 (ORNL-DO, Committees-Research Committee).
57. AEC, *Eleventh Semiannual Report to Congress* (Jan 1952), 46.
58. Donald J. Keirn to Fields, 14 Oct 1954; V. G. Huston (for Fields) to Bradbury and York, 22 Oct 1954; Darol Froman to Huston, 16 May 1955, in AEC 855, 19 Aug 1955 (Sec'y 51–58, 52/MR&A-5: Rocket propulsion); Raemer Schreiber, "What Happened to LASL?" draft Nov 1991, LANL VFA-1240.
59. Schreiber to Bradbury, 12 May 1955, in AEC 855, 19 Aug 1955.
60. Hewlett and Duncan, 20.
61. AEC, *Fifth Semiannual Report to Congress* (Jan 1949), 50; GAC 7, 21–23 Nov 1947, and GAC 12, 3–5 Feb 1949.
62. Nathan Reingold, "Vannevar Bush's New Deal for Research: or, The Triumph of the Old Order," *HSPS,* 17:2 (1987), 299–344, on 308, 328.

63. AEC, *Second Semiannual Report* (July 1947), 3, 5; Hewlett and Duncan, 18–21.
64. AEC, *First Semiannual Report* (Jan 1947), 6–7.
65. Alfred D. Chandler, *The Visible Hand: The Managerial Revolution in American Business* (Cambridge, Mass., 1977), 99–109.
66. Hewlett and Duncan, 33–42.
67. Report by Conant, DuBridge, and Hartley Rowe, in GAC 10, 4–6 Jun 1948, EHC.
68. Carroll Wilson to managers of directed operations, 5 Aug 1948, ORO-1547.
69. Hewlett and Duncan, 197–201; AEC, *Sixth Semiannual Report* (July 1949).
70. Hewlett and Duncan, 112–114, 251–252; AEC, *Third Semiannual Report* (Jan 1948), 34; Wilson, GM-32, "Division of Biology and Medicine," 15 Sep 1948 (RRC-UT, MS-1067, folder 2).
71. Howard Brown, "A Report on AEC Management Practices in Dealing with National Laboratories," 10 Mar 1950 (Sec'y 47–51, 60/Operation and management of national labs).
72. Ibid.; Wilson, draft GM bulletin, "Laboratory Coordinators," 9 Aug 1950 (DBM, 31/O&M-3).
73. GAC 10, 4–6 Jun 1948, EHC.
74. Hewlett and Duncan, 96–101; Paul Boyer, *By the Bomb's Early Light: American Thought and Culture at the Dawn of the Atomic Age* (New York, 1985), 107–121, 291–302; Spencer R. Weart, *Nuclear Fear: A History of Images* (Cambridge, 1988), 158–160.
75. GAC 10, 4–6 Jun 1948, EHC.
76. GAC 24, 4–6 Jan 1951; AUI, Ex Comm, 16 Nov 1951.
77. AEC 283/37, 19 Feb 1954, and Nichols to principal staff, 8 Mar 1954 (Sec'y 51–58, 57/O&M-2: General Manager).
78. GAC 39, minutes, 31 Mar and 1–2 Apr 1954, EHC; AEC, *Sixteenth Semiannual Report* (Jul 1954), ix.
79. Fred Schuldt to William D. Carey, 29 Sep 1954 (BOB, 8/Scientific Research-General).
80. Hewlett and Duncan, 336–337; R. M. Underhill, interview by Arthur Norberg, 1976, TBL, 8–9.
81. Luis Alvarez, quoted in Daniel S. Greenberg, *The Politics of Pure Science* (New York, 1967), 132; Robert W. Seidel, "Accelerating Science: The Postwar Transformation of the Lawrence Radiation Laboratory," *HSPS*, 13:2 (1983), 375–400, on 383.
82. Shoup, in BNL, Policy and Program Committee, minutes, 24 Jul 1947 (BNL-DO, IV-16); Morse to Malcolm R. Warnock, 16 Jul 1947 (BNL-DO, I-3).
83. Hewlett and Duncan, 334–336.
84. AEC GM-Bulletin 14, 14 Jul 1948 (ORNL-DO, Budget-General).

85. "Instructions for Preparation and Submission of Project Proposal and Authorization Forms," 10 Dec 1951 (DC, 11/Finance-Admin: Budget).
86. Carroll Wilson to Managers of Directed Operations, 5 Aug 1948, ORO-1547.
87. Brown, "Report."
88. Walter J. Williams to Strauss, 5 Oct 1953 (Sec'y 51–58, 58/O&M-2: Military Application).
89. JCAE, *Investigation,* 604–616, 637–667.
90. Shoup to files, 1 Dec 1948 (BNL-DO, II-1).
91. James Grahl to files, 17 May 1950 (BOB, 1/Field trips: '47–'55).
92. Wayne Brobeck to Bill Borden, 20 Oct 1949 (JCAE III, 37/LRL).
93. Grahl to Schuldt, 29 Sep 1949 (BOB, 1/Field trips: '47–'55).
94. Williams to Strauss.
95. Haworth, in AUI, Board of Trustees, 20 Jan 1950.
96. AEC Research Committee, minutes, 10 Apr 1950 (Sec'y 47–51, 60/Oper. and mgt. of national labs); Grahl to Schuldt. The labs did little or no research for the other AEC divisions.
97. Peter B. Natchez and Irvin C. Bupp, "Policy and Priority in the Budgetary Process," *American Political Science Review,* 67 (1973), 951–963.
98. AUI, Ex Comm, 20 Mar 1953; AEC, *Ninth Semiannual Report to Congress* (Jan 1951), 64.
99. Hewlett and Anderson, 410, 438; Brian Balogh, *Chain Reaction: Expert Debate and Public Participation in American Commercial Nuclear Power, 1945–1975* (Cambridge, 1991), 51–56.
100. Balogh, *Chain Reaction,* 52.
101. BOB, "Atomic Energy Section," 7 Feb 1956 (BOB, 1/AEC Admin.-General).
102. AEC 533/15, 29 Sep 1952, and 533/20, 17 Dec 1952 (Sec'y 51–58, 174/BAF-2:1954); AEC 625/22, 16 Jun 1953 (Sec'y 51–58, 174/BAF-2:1955); Schuldt, "FY 54 Work program-Atomic Energy Unit," 13 Oct 1952 (BOB, 1/Work program).
103. Natchez and Bupp, "Policy and Priority"; see also Aaron Wildavsky, *The New Politics of the Budgetary Process* (Glenview, Ill., 1988), 77–79.
104. Mr. Shapley, "Research and Development," draft 1 May 1953 (BOB, 8/Research-FY53).
105. Schuldt to W. F. Schaub, draft 13 Jan 1954, BOB (15/Budget, General-FY55).
106. Schuldt to Carl Tiller, 7 Aug 1953 (BOB, 15/Budget, General-FY55).
107. ORNL Research Council, 21 Oct 1955 (ORNL-DO, Meetings-Research Council).
108. George T. Mazuzan and J. Samuel Walker, *Controlling the Atom: The Beginnings of Nuclear Regulation, 1946–1962* (Berkeley, 1984), 59–92.
109. "Atomic Energy Section," 7 Feb 1956 (BOB, 1/AEC Admin.-General); Harold P. Green and Alan Rosenthal, *Government of the Atom: The Integration of Powers* (New York, 1963), 76.

110. Hewlett and Anderson, 504–513, 529–530.
111. "Atomic Energy Section."
112. BOB, Military Division, "Inventory of the Organization and Management of the Atomic Energy Commission," 29 Sep 1952 (BOB, 2/Organization and personnel-FY53).
113. U.S. House, 80th Cong., 1st sess., *Independent Offices Appropriation Bill for 1948,* 1505; U.S. House, 80th Cong., 2d sess., *Independent Offices Appropriation Bill for 1949,* 820; U.S. House, 81st Cong., 1st sess., *Independent Offices Appropriation Bill for 1950,* 1081.
114. U.S. Senate, 80th Cong., 2d sess., *Supplemental Independent Offices Appropriations Bill for 1949,* 90; Wildavsky, *New Politics,* 99.
115. U.S. House, 80th Cong., 1st sess., *Independent Offices Appropriation Bill for 1948,* 1505.
116. Rep. Howard Smith, *Congressional Record,* 102 (24 Jul 1956), 14246, and Rep. Ben Jensen, *Congressional Record,* 104 (22 Jul 1958), 14655; Green and Rosenthal, *Government of Atom,* 78n15. See also Rep. Jensen, quoted in Wildavsky, *New Politics,* 74: "I am not schooled in the art"; and Rep. W. Sterling Cole, in JCAE hearings with T. H. Johnson in 1956: "Now I know what you are talking about but I do not understand it," quoted in Greenberg, *Politics of Pure Science,* 222.
117. U.S. Senate, 82d Cong., 2d sess., *Independent Offices Appropriations, 1953,* 661–664; JCAE, 84th Cong., 2d sess., *Authorizing Legislation,* 17.
118. U.S. House, 82d Cong., 1st sess., *Independent Offices Appropriation Bill for 1952,* 816. Gore should not be confused with his son, the former vice president.
119. Rep. Henry M. Jackson, "Joint Committee on Atomic Energy—A New Experiment in Government," *Nucleonics,* 10:8 (1952), 8–9, on 9; Gore in U.S. House, 82d Cong., 2d sess., *Independent Offices Appropriation for 1953,* 1095.
120. BOB Military Division to BOB Director, 12 Jun 1958 (BOB, 16/Post-submission-FY59); C. W. Fischer to Schaub, draft 19 Sep 1958 (BOB, 8/Research programs-FY59).
121. Green and Rosenthal, *Government of Atom,* 85.
122. W. Hamilton and K. Mansfield to Borden, 26 Jul 1950, memo of meeting of Lawrence, McMahon, Borden, Mansfield, and Hamilton, 27 Jul 1950, Borden to files, 7 Aug 1950, and J. S. Walker to files, 19 Nov 1951 (JCAE-III, 1/Accels); Walker to files, 13 Nov 1951 (JCAE-III, 38/LASL).
123. AEC minutes, 28 Feb 1958 (Sec'y 51–58, 177/BAF-2:1959 vol. 3); Green and Rosenthal, *Government of Atom,* 83–85; Wildavsky, *New Politics,* 193–195. The Joint Committee joined a trend in Congress toward annual authorization by committees for agencies in the 1950s.
124. Green and Rosenthal, *Government of Atom,* 85.
125. AEC minutes, 5 Jun 1952 and 8 Jul 1952 (Sec'y 51–58, 173/BAF-2:1953); "The

Budget Processes of the Atomic Energy Commission," 4 Nov 1958 (Sec'y 58–66, 1318/2).

126. AUI Ex Comm, 21 May 1948 and 20 Mar 1953; ORNL Policy Committee, 12 Aug 1953 (ORNL-DO, Meetings-Policy Comm.), and Research Council minutes, 25 Apr 1952 (ORNL-DO, Meetings-Research Council); Reynolds to Pollman, 22 Dec 1948 (DC, 5/Rad Lab: Budget); AEC Research Committee, "Third Meeting to Discuss National Laboratory Problems," 17 Oct 1949 (DBM, 34/Research Committee); AEC Research Committee, minutes, 12 Jun 1950 (EHC, folder Sec'y 47–51, 635.123: Mgt. of national labs).
127. ORNL Policy Committee, 7 Sep 1955 (ORNL-DO, Meetings-Policy Committee).
128. Pollman, notes on telephone conversation with Mr. Shute, 25 Jul 1949, and Reynolds to Fidler, 27 Jul 1949 (DC, 7/Finance: Budget-1949); Reynolds to Pitzer, 24 May 1950 (DC, 7/Finance-Admin: Budget-1950); C. E. Andressen to Reynolds, 6 Nov 1950 (DC, 7/Finance-Admin: Budget-Internal corres.).
129. Luis Alvarez, quoted in Greenberg, *Politics of Pure Science,* 131; Schreiber, "What Happened to LASL?"; Norris Bradbury, "Los Alamos—The First 25 Years," in J. O. Hirschfelder, H. P. Broida, and L. Badash, eds., *Reminiscences of Los Alamos* (Dordrecht, 1980), 161–175, on 166; Darol K. Froman, interview by Arthur L. Norberg, 1976, TBL, 52; conversation with Alvin Weinberg, 18 June 1997, Oak Ridge.
130. Johnson and Schaffer, 43.
131. Thiesmeyer to Shoup, 16 Jun 1947 (BNL-Central Records, DO-87).
132. Roy Nelson to Priestley, 20 May 1947; Priestley to Nelson, 21 May 1947; Pollman to Priestley, 18 Jun 1947; Priestley to Pollman, 29 Jul 1947 (DC, 2/AEC: Berkeley).
133. Fields to Wilson, 25 Apr 1947 (EHC, Sec'y 47–51, Lab Directors meetings).
134. Conversation with Tom Row, 10 Jun 1997, ORNL.
135. AUI Ex Comm, 14 Apr and 23 Sep 1949; AUI Board of Trustees, 15 Apr and 21 Oct 1949.
136. Reynolds to D. Saxe, 12 Apr 1950; Saxe to Reynolds, 3 May 1950; and P. M. Goodbread to Fidler, 13 Dec 1950 (DC, 7/Finance-Admin: Budget 1950).
137. Andressen to Reynolds, 6 Nov 1950 (DC, 7/Finance-Admin: Budget 1950).
138. John Bugher to Oscar Smith, 11 Sep 1953 (DBM, 31/O&M-1: 1953).
139. Response to Fay-Carpenter.
140. F. C. Vonderlage to C. E. Larson, 11 Jan 1951 (ORNL-DO, Budget-General).
141. AEC, "Additional Personnel Requirements FY57," 30 Apr 56 (Sec'y 51–58, 176/BAF-2: 1957); U.S. House, 80th Cong., 2d sess., *Supplemental Independent Offices Appropriations Bill for 1949,* 781–782; U.S. Senate, 83d Cong., 1st sess., *Second Independent Offices Appropriations for 1954,* 31.
142. Hewlett and Duncan, 42; Dedication of AEC headquarters, 8 Nov 1957 (Sec'y 51–58, box 92).

143. "Biology and Medicine Program," draft, 20 Mar 1950, and AEC, Program and Budget Committee memo no. 10, 5 Apr 1950 (ORNL-DO, Budget-General).
144. U.S. House, 82d Cong., 1st sess., *Independent Offices Appropriation for 1952,* 822.
145. Hezz Stringfield to ORNL Policy Committee, 12 Aug 1953, and T. H. Johnson to Operations Office Managers, 22 Jul 1953 (ORNL-DO, Budget-General).
146. C. E. Center to S. R. Sapirie, 8 Feb 1955 (ORNL-DO, Budget-General).
147. Don Burrows to Sapirie, 14 Jan 1955, and Burrows to Managers of Operations and Division Directors, 21 Jan 1955 (ORNL-DO, Budget-General).
148. BNL, Policy and Program Committee, 2 Nov 1949 (BNL-DO, IV-16); AUI, Ex Comm, 17 Mar and 20 Apr 1950.
149. Weinberg sent copies of his speeches to Schuldt at the Bureau of the Budget, with some flattery: "In my opinion you know more about atomic energy than anybody else in the United States." Weinberg to Schuldt, 9 Nov 1961, and Schuldt to Weinberg, 6 Jul 1961 (AMW, Schuldt, F. W.).
150. Schuldt, "Draft Work Program," 30 Jan 1956, and "Summary Work Program-Atomic Energy Section," 7 Jan 1957 (BOB, 1/Work Program).
151. AEC press release, 26 Mar 1955, and AEC 868/2, 5 Dec 1955 (Sec'y 51–58, 71/O&M-7: McKinney Panel).
152. Sapirie to Center, 12 Jan 1953 (ORNL-DO, Budget-General).
153. Froman to Weinberg, 20 Dec 1960 (AMW, Lab Directors Meetings); AUI, Ex Comm, 18 Nov 1960. Froman's complaint stemmed in part from the increasing work at Los Alamos for other divisions, including Reactor Development.
154. James Q. Wilson, *Bureaucracy: What Government Agencies Do and Why They Do It* (New York, 1989), 241–244, 258–260, 366–367.
155. U.S. House, 80th Cong., 2d sess., *Supplemental Independent Offices Appropriations Bill for 1949,* 831.
156. Response to Fay-Carpenter.
157. "Notes for Meeting with Mr. McCone," 5 Sep 1958 (BOB, 17/AEC FY60 General Budget).
158. Sybil Francis, "Warhead Politics: Livermore and the Competitive System of Nuclear Weapon Design" (Ph.D. diss., MIT, 1996), 149.
159. AUI Trustees, 19 Apr 1957.
160. Lilienthal to Lawrence, 7 May 1947; J. B. Fisk to Lawrence, 8 May 1947; and Lawrence to Lilienthal, 13 May 1947 (EOL, 32/2); Hewlett and Duncan, 101, 107–109.
161. See penciled list of meetings on inside of folder "D.O., Lab Directors' Meetings" (BNL-HA, box 33).
162. GAC 47, 13–15 Dec 1955, EHC, 15, 21–24.
163. AEC 811/3, 10 Oct 1956 (Sec'y 51–58, 62/O&M-6: Meetings and conferences).
164. Weinberg to Haworth, 12 Jan 1959 (BNL-DO, III-21).

165. A. Weinberg, untitled draft, ca. Nov 1957, and Weinberg to Haworth, 6 Nov 1957 (BNL-DO, III-21); Haworth to Bradbury, to Weinberg, to Hilberry, to Cooksey, to Spedding, and to York, 13 Dec 1957 (BNL-DO, VII-31); Haworth to Hilberry and to Weinberg, 7 Jan 1958, and to Lawrence, 18 Jan 1958 (BNL-DO, I-16).
166. "Relationships between the Atomic Energy Commission and the National Laboratories," draft to K. E. Fields, 7 Jan 1958 (BNL-DO, III-21). See also Weinberg, untitled draft, ca. Nov 1957, and N. Hilberry, "Some Comments on the Objectives of the Commission's National Laboratories," draft, 3 Jan 1958 (BNL-DO, III-21).
167. "Relationships," draft.
168. Ibid.
169. Ibid.
170. Haworth to Lawrence, 18 Jan 1958, and Haworth to Harry S. Traynor, 18 Jan 1958 (BNL-DO, I-16).
171. Traynor and John G. Adams, "AEC Organization Study," 15 Apr 1958, 73–81, and AEC meetings 1362 and 1363, minutes, 25 Apr 1958 (Sec'y 51–58, 54/O&M-1: General Policy).
172. Loomis et al. to Nichols, 19 Mar 1946 (LANL A-84–109, 49/5); "Report of the Advisory Board on Relationships of the Atomic Energy Commission with Its Contractors," 30 Jun 1947 (BNL-DO, III-20).
173. BOB, "Inventory."
174. Weinberg to Haworth, 12 Jan 1959 (BNL-DO, III-21).
175. Weinberg, "The National Laboratories and the Atomic Energy Commission," draft, Jan 1961 (BNL-DO, III-22; copy of second draft in BNL-HA).
176. Bradbury to Weinberg, 30 Jan 1961 (AMW, Lab Directors Meetings).
177. GAC 71, 24–26 Oct 1960; Pitzer to Seaborg, 2 May 1961, in GAC 74; GAC 75, 13–15 Jul 1961, EHC.
178. AEC press release, 11 Aug 1961 (BNL-HA, Director's Office Misc. A).
179. Charles Dunbar to AUI Trustees, 18 Aug 1961 (BNL-HA, Director's Office Misc. A); Weinberg to A. V. Crewe, 22 Jun 1962 (AMW, Lab Directors Meetings).

4. COLD WAR WINTER

1. Carroll Wilson, testimony in U.S. House, 80th Cong., 1st sess., *Independent Offices Appropriation Bill for 1948,* 1481; GAC 2, 2–3 Feb 1947, EHC.
2. Richard Rhodes, *Dark Sun: The Making of the Hydrogen Bomb* (New York, 1995), 282–284; David Alan Rosenberg, "U.S. Nuclear Stockpile, 1945 to 1950," *BAS,* 38 (May 1982), 25–30.
3. Oppenheimer to Lilienthal, 30 Oct 1949, GAC 17, EHC; Peter Galison and

Barton Bernstein, "In Any Light: Scientists and the Decision to Build the Superbomb, 1952–1954," *HSPS*, 19:2 (1989), 267–347; Herbert F. York, *The Advisors: Oppenheimer, Teller, and the Superbomb* (San Francisco, 1976), 1–74.

4. GAC 23, 30–31 Oct and 1 Nov 1950, EHC.
5. Holl, 95; York, *Advisors*, 76; Johnson and Schaffer, 84–85; Oppenheimer to Lilienthal, 3 Dec 1949, GAC 18.
6. Weinberg to C. N. Rucker, 6 Jan 1949 (ORNL-DO, Radiological warfare).
7. AUI, Board of Trustees, 21 Oct 1949 and 21 Apr 1950.
8. Haworth to Van Horn, 12 Jul 1951 (BNL-HA, Haworth declassified files, box 36).
9. G. F. Tape to G. B. Collins, 29 Jun 1951, and Van Horn to Haworth, 3 Aug 1951 (BNL-DO, V-1); Tape to Van Horn, 7 Aug 1951 (BNL-Central Records, DO-87); P. W. McDaniel to M. Boyer, 14 Aug 1951 (Sec'y 51–58, 139/R&D-6: BNL).
10. Tape to Van Horn, 26 Sep 1951 (BNL-Central Records, DO-87); AEC Neutron Cross-Sections Advisory Group, minutes, 8–10 Oct 1951 (BNL-HA, Haworth declassified files, box 36); Haworth to Van Horn, 17 Dec 1951 (BNL-DO, V-1).
11. Haworth to Van Horn, 6 Jul 1950 (BNL-DO, I-8); Haworth to Van Horn, 18 Apr 1951 (BNL-HA, Haworth declassified files, box 36); BNL, "Annual Report," 1 Jul 1951; GAC 26, 8–10 May 1951, EHC; AUI, Ex Comm, 10 May, 15 Jun, and 21 Sep 1951, and Board of Trustees, 19 Oct 1951.
12. Rabi, 11 Oct 1949, in AEC, *In the Matter of J. Robert Oppenheimer* (Cambridge, Mass., 1971), 778; Alvarez, 14 Oct 1949, in Robert W. Seidel, "A Home for Big Science: The AEC's Laboratory System," *HSPS*, 16:1 (1986), 135–175, on 152–153.
13. AEC, *In the Matter*, 779; Heilbron, Seidel, and Wheaton, 63–64.
14. AEC, *In the Matter*, 782, 784–785; Seidel, "Home," 151; Hewlett and Duncan, 428–430, 552–553.
15. Heilbron, Seidel, and Wheaton, 65–75; AEC, *In the Matter*, 785; W. Hamilton and K. Mansfield to W. Borden, 26 Jul 1950, memo of meeting of Lawrence, McMahon, Borden, Mansfield, and Hamilton, 27 Jul 1950, and Borden to files, 7 Aug 1950 (JCAE-III, 1/Accels.).
16. GAC 21, 1–3 Jun 1950; GAC 22, 10–13 Sep 1950; A. H. Holland, Jr., to C. E. Larson, 7 Jun 1950, ORNL/CF-50-6-54; Rabi to Strauss, 3 Jun 1954, GAC 40 (JCAE-III, 33/GAC); Willard Libby to Clinton Anderson, 19 Apr 1955 (JCAE-III, 1/Accels.); Chronology, AEC San Francisco Operations Office (EOL, 32/33); Heilbron, Seidel, and Wheaton, 69–75.
17. Paul Forman, "Behind Quantum Electronics: National Security as Basis for Physical Research," *HSPS*, 18:1 (1987), 149–229, on 150; Daniel J. Kevles, "Cold War and Hot Physics: Science, Security, and the American State, 1945–

1956," *HSPS,* 20:2 (1990), 239–264; James G. Hershberg, *James B. Conant: Harvard to Hiroshima and the Making of the Nuclear Age* (New York, 1993), 492–494.
18. York, *Advisors,* 78, 125–127.
19. Hewlett and Duncan, 414–417, 438–441, 527–528, 535–537.
20. Sybil Francis, "Warhead Politics: Livermore and the Competitive System of Nuclear Weapon Design" (Ph.D. diss., MIT, 1996), 36; Hewlett and Duncan, 541.
21. Francis, "Warhead Politics," 37.
22. GAC 27, 11–13 Oct 1951, EHC.
23. Oppenheimer to Gordon Dean, 13 Oct 1951, and Libby to Dean, 13 Oct 1951, in GAC 27, EHC.
24. GAC 28, 12–14 Dec 1951, EHC; Hewlett and Duncan, 568–571.
25. Francis, "Warhead Politics," 39, 43.
26. Dean, diary entry for 16 Apr 1952, in Roger M. Anders, ed., *Forging the Atomic Shield: Excerpts from the Office Diary of Gordon Dean* (Chapel Hill, 1987), 212–213.
27. UCRL, "Status Report on MTA," quoted in Barton Bernstein, "Lawrence, Teller, and the Quest for the Second Lab," unpublished manuscript, 20 (I thank Barton Bernstein for a copy of his manuscript); York, *Advisors,* 130–131; Herbert F. York, *Making Weapons, Talking Peace: A Physicist's Odyssey from Hiroshima to Geneva* (New York, 1987), 62–66; Hewlett and Duncan, 582; Edward Teller with Allen Brown, *The Legacy of Hiroshima* (New York, 1962), 60; Francis, "Warhead Politics," 61–62. Cf. Oppenheimer querying Alvarez and McMillan on the connection between MTA and Berkeley's interest in weapons; they asserted there was "no strong interaction, different people are involved." GAC 30, 27–29 Apr 1952, EHC.
28. GAC 30, 27–29 Apr 1952, EHC; Oppenheimer to Dean, 30 Apr 1952, GAC 30.
29. Dean, memo for file, 1 Apr 1952, in Dean diary, 208.
30. AEC minutes, 8 Sep 1952, Lawrence Livermore National Lab, Archives.
31. Francis, "Warhead Politics," 62, 63–64; York, *Making Weapons,* 67–68.
32. GAC 33, 5–7 Feb 1953, EHC.
33. Francis, "Warhead Politics," 65–66.
34. Bernstein, "Quest," 29D; see also Gregg Herken, "The University of California, the Federal Weapons Labs, and the Founding of the Atomic West," in Bruce Hevly and John M. Findlay, eds., *The Atomic West* (Seattle, 1998), 119–135.
35. Francis, "Warhead Politics," chaps. 4 and 7.
36. Ibid., 149. The chair in 1958, Senator Clinton Anderson of New Mexico, had a political reason to question, implicitly, the lab at Livermore in order to defend his constituents at Los Alamos.

37. Teller, *Legacy*, 65.
38. James R. Shepley and Clay Blair, Jr., *The Hydrogen Bomb: The Men, the Menace, the Mechanism* (New York, 1954); Gordon Dean, review of Shepley and Blair, *BAS*, 10 (Nov 1954), 357, 362; *BAS*, 10 (Sep 1954), 283, 286; LASL press release, 24 Sep 1954, and Strauss to Bradbury, 22 Sep 1954 (LASL-DO, 310.1: P-Div. History).
39. Hewlett and Duncan, 44.
40. Oppenheimer to Lilienthal, 3 Apr 1947, GAC 3, EHC.
41. Ibid.
42. Ibid.
43. Hewlett and Duncan, 44–46, 66–68.
44. GAC 4, 30 May–1 June 1947, EHC.
45. Ibid.
46. GAC 5, 28–29 Jul 1947, EHC.
47. Johnson and Schaffer, 51; Hewlett and Duncan, 126.
48. J. C. Franklin to Prescott Sandidge, 31 Dec 1947 (RRC-UT, MS-652, 10/67).
49. Weinberg to Oppenheimer, 6 Jan 1948 (RRC-UT, MS-652, 10/83).
50. GAC, special meeting, 29–30 Dec 1947, EHC.
51. Wilson to Franklin, 3 Feb 1948 (ORNL-DO, Program-ORNL).
52. C. N. Rucker to Franklin, 22 Mar 1948, ORNL/CF-48-3-336; Fisk to Franklin, n.d. [ca. Apr 1948] (ORNL-DO, Program-ORNL).
53. Weinberg to Rucker, 22 Apr 1948, ORNL/CF-48-4-331.
54. F. H. Belcher to R. W. Cook, 2 Jun 1948 (ORNL-DO, Program-ORNL); ORNL, Laboratory Research Council, minutes, 3 Jun 1948 (ORNL-DO, Meetings-Research Council).
55. Weinberg to Rucker, 16 Jun 1948, ORNL/CF-48-6-196; W. P. Bigler to Zinn, 13 May 1948 (ORNL-DO, Program-ORNL); Hewlett and Duncan, 193–197; Holl, 67–71.
56. GAC 4, 30 May–1 June 1947, EHC.
57. Zinn to Fisk, 23 Jul 1948, ORNL/CF-48-8-13.
58. Ibid.
59. Ibid.
60. Hewlett and Duncan, 197–201.
61. Weinberg to Rucker, 17 Aug 1948, ORNL/CF-48-8-212; Weinberg to Teller, 30 Aug 1948, ORNL/CF-48–8–367; Weinberg, "Proposal for ONRL Research Reactor," 1 Oct 1948, ORNL/CF-48-10-25; Hewlett and Duncan, 201–205.
62. Hewlett and Duncan, 205, 214–219, 419; Weinberg, 79.
63. Hewlett and Duncan, 500; Holl, 116–117; C. E. Larson to J. H. Roberson, 16 Aug 1950, ORNL/CF-50-8-78.
64. Hewlett and Duncan, 208–209; Weinberg, "Research Program at ORNL," 22 Mar 1949, ORO-1727.

65. "Meeting to Discuss National Laboratory Problems," 3 and 17 Oct 1949 (DBM, 34/Research Committee).
66. Richard G. Hewlett and Francis Duncan. *Nuclear Navy, 1946–1952* (Chicago, 1974).
67. Hewlett and Duncan, *Atomic Shield,* 72–74, 190, 208, 211–212, 419–420.
68. Larson to A. H. Holland, Jr., "Brief History of the Aircraft Nuclear Propulsion Project at ORNL," 16 Jun 1950, ORNL/CF-50-6-74.
69. L. R. Hafstad, "Atomic Power for Aircraft," *BAS,* 5 (Nov 1949), 309–312, on 312; James Grahl to Fred Schuldt, 29 Sep 1949 (BOB, 1/Field trips).
70. Hewlett and Duncan, 211–212; Harold P. Green and Alan Rosenthal, *Government of the Atom: The Integration of Powers* (New York, 1963), 19, 242–247.
71. Larson, "Policy Statement Concerning ORNL," 10 Jul 1950 (ORNL-DO, Program-ORNL).
72. Topnotch II agenda item 5A, 24–28 Sep 1953 (Sec'y 51–58, 65/I&M-6: Topnotch II).
73. Hafstad to R. W. Cook, 14 Sep 1950, ORNL/CF-50-9-162.
74. GAC 25, 15–17 Mar 1951, EHC.
75. Fisk to Franklin, n.d. [ca. Apr 1948], and F. H. Belcher (quote) to Cook, 2 Jun 1948 (ORNL-DO, Program-ORNL).
76. Robert W. Seidel, "Accelerating Science: The Postwar Transformation of the Lawrence Radiation Laboratory," *HSPS,* 13:2 (1983), 375–400, on 392; Hewlett and Anderson, 633–637.
77. Allan A. Needell, "Nuclear Reactors and the Founding of Brookhaven National Laboratory," *HSPS,* 14:1 (1983), 93–122, on 93–104; Crease, 15–16, 20, 110.
78. Hewlett and Duncan, 79–80; Daniel J. Kevles, *The Physicists: The History of a Scientific Community in Modern America* (New York, 1978), 352–356
79. Glenn T. Seaborg, *Journal,* 29 Jul 1947 (Office for History of Science and Technology, UC Berkeley); Hewlett and Duncan, 79–84, 107–112; Seidel, "Accelerating," 393–394.
80. Seidel, "Accelerating," 394; Crease, 123; P. M. Morse to E. L. Van Horn, 5 Mar 1947 (BNL-DO, I-2).
81. Livingston memo, 20 Oct 1947, in Crease, 125.
82. GAC 7, 21–23 Nov 1947.
83. Fisk to Lawrence, 1 Dec 1947 (EOL, 32/3).
84. GAC 8, 6–8 Feb 1948.
85. GAC 8, 6–8 Feb 1948.
86. John Blewett, quoted in Crease, 128.
87. AUI, Ex Comm, 20 Feb 1948.
88. Memorandum for record, 22 March 1948 (EOL, 26/14); Seidel, "Accelerating," 396–397.

89. U.S. House, 81st Cong., 1st sess., *Independent Offices Appropriation Bill for 1950,* 1271.
90. Seidel, "Accelerating," 397.
91. AEC, *Fourth Semiannual Report to Congress* (July 1948), 39.
92. Oppenheimer to Lilienthal, 3 Dec 1949, GAC 18.
93. "Notes on a Meeting to Discuss Countermeasures against Atomic Bombs," 21 Jan 50, in Sumner Pike to Brien McMahon, 8 Mar 1950 (JCAE III, 63/Weapons).
94. R. T. Coiner to Lawrence, 9 Jun 1950, and Lawrence to Coiner, 14 Jun 1950 (EOL, 32/7).
95. AEC 603, 1 Dec 1952 (Sec'y 51–58, 138/R&D-6: Particle accels.). See also Jane Hall to H. A. Bethe et al., 14 Jan 1953 (LASL-DO, 410: Accelerators).
96. AEC 603/2, 26 Feb 1953 (EHC, R&D-6: Particle accels.). See also Robert W. Seidel, "The Postwar Political Economy of High-Energy Physics," in Laurie M. Brown, Max Dresden, and Lillian Hoddeson, eds., *Pions to Quarks: Particle Physics in the 1950s* (Cambridge, 1989), 497–507, on 503.
97. AEC minutes, 27 Feb and 23 Apr 1953, Dean to R. LeBaron, 12 May 1953 and AEC, status of decisions, AEC 603/2, 22 Jun 1953 (Sec'y 51–58, 138/R&D-6: Particle accels.); T. H. Johnson to K. D. Nichols, 12 Jan 1954 (Sec'y 51–58, 139/R&D-6: Particle accels.).
98. Stuart M. Feffer, "Atoms, Cancer, and Politics: Supporting Atomic Science at the University of Chicago, 1944–1950," *HSPS,* 22:2 (1992), 240–242.
99. Brookhaven bought a Van de Graaff from G.E., Brookhaven and Argonne bought 60-inch cyclotrons from Collins Radio, and Oak Ridge ordered a betatron from G.E. and considered buying a 60–70-inch cyclotron from the same company.
100. Holl, 156; Feffer, "Atoms"; Fisk to Lawrence, 1 Dec 1947 (EOL, 32/3).
101. Clinton Research Council, minutes, 6 Feb 1946, ORNL/CF-46-2-113; "Proposed Program" for FY1947, ORNL/CF-46-5-294; AEC, *Sixth Semiannual Report to Congress* (July 1949), 126.
102. Weinberg to Rucker, 22 Apr 1948, ORNL/CF-48-4-331; ORNL Research Council, 20 Apr 1948 (ORNL-DO, Meetings-Research Council).
103. Johnson and Schaffer, 65–69.
104. Roberson to Larson, 7 Sep 1950 (ORNL-DO, Program-ORNL); GAC 25, 15–17 Mar 1951, EHC.
105. Larson to Roberson, 19 Oct 1950, ORNL/CF-50-10-101; Johnson and Schaffer, 68–69.
106. AEC 603/8, 19 Aug 1953, and 603/10, 19 Nov 1953 (Sec'y 51–58, 139/R&D-6: Particle accels.); GAC 37, 4–6 Nov 1953, EHC.
107. R. S. Livingston to G. B. Rossi, 25 Nov 1953; Rossi to Livingston, 14 Dec 1953; and Lawrence et al., "Acceleration of Heavy Ions by UCRL Cyclotrons," ca.

1953 (EOL, 5/18); Rossi, "Review of Literature," 12 Jul 1955 (EOL, 24/16); Seaborg, *Journal,* 4 Aug and 24–28 Aug 1953.

108. Hewlett and Duncan, 245–247, 434–435.
109. AEC 603/8, 19 Aug 1953, and 603/10, 19 Nov 1953 (Sec'y 51–58, 139/R&D-6: Particle accels.); GAC 37, 4–6 Nov 1953, EHC.
110. AEC 603/10 and GAC 37, 4–6 Nov 1953, EHC.
111. GAC 38, 6–8 Jan 1954, EHC; Seaborg, *Journal,* Dec 1952 and 1953 passim; cf. Holl, 176–181.
112. GAC 38, 6–8 Jan 1954, EHC; Rabi to Strauss, 9 Jan 1954, GAC 38; AEC 603/14, 15 Jan 1954, and AEC minutes excerpt, 20 Jan 1954 (Sec'y 51–58, 139/R&D-6: Particle accels.); Seaborg, *Journal,* 7 Jan 1954.
113. AEC 728/3, 3 Sep 1954 (Sec'y 51–58, 174/BAF-2: 1956[BP]).
114. BNL Scientific Steering Committee, minutes, 13 May 1947 (BNL-DO, IV-19); John Wheeler to Morse, 5 Jun 1947 (BNL-DO, I-4); Morse to Brig. Gen. T. C. Rives, 13 Jul 1948 (BNL-DO, I-5); Haworth to Air Surgeon, Mitchell Field, 21 Mar 1950 (BNL-DO, V-63).
115. A. Hollaender to J. S. Putnam, 7 Oct 1947 (RRC-UT, MS-652, 9/7); "Mammalian Genetics Program," 27 Nov 1947, DOE Opennet, accession NV0707352.
116. GAC 8, 6–8 Feb 1948.
117. T. H. Johnson to Schuldt, 20 Sep 1955 (BOB, 8/Research programs-FY56).
118. "Meeting," 17 Oct 1949 (DBM, 34/Research Committee).
119. "Proposed Program for the Northeast Laboratory," 28 May 1946, in Needell, "Nuclear Reactors," 103.
120. AEC, *Second Semiannual Report to Congress* (July 1947), 9–10; U.S. Senate, 81st Cong., 1st sess., *Independent Offices Appropriation Bill for 1950,* 1216, 1217; Bradbury testimony to JCAE, 81st Cong., 1st sess., *Investigation into the U.S. Atomic Energy Project,* 816; U.S. Senate, 80th Cong., 2d sess., *Supplemental Independent Offices Appropriation Bill for 1949,* 83–84. See also Vannevar Bush, *Science, the Endless Frontier* (Washington, D.C., 1945), 2, 13–14: "Basic scientific research is scientific capital" and "the fund from which the practical applications of knowledge must be drawn."
121. GAC 4, 30 May–1 Jun 1947.
122. Ibid.
123. Oppenheimer to Lilienthal, 10 Oct 1947, in GAC 6.
124. "Plans for Clinton National Laboratory," in Wilson to Franklin, 3 Feb 1948, and Fisk to Franklin, ca. Apr 1948 (ORNL-DO, Program-ORNL).
125. Memo on Los Alamos policy, 9 Aug 1946 (LANL, A-84-019, 37/5); Farrington Daniels, "Proposed Program for the National Nucleonics Laboratory at Argonne," ca. Feb 1946, ORNL/CF-46-3-123; Thomas to Nichols, 11 Feb 1946, ORNL/CF-46-2-170.
126. "Meeting," 17 Oct 1949 (DBM, 34/Research Committee).

127. Louis Rosen to Darol Froman, 15 May 1958 (LASL-DO, 310.1: P-Div. History).
128. AEC, *Third Semiannual Report to Congress* (Jan 1948), 24, and *Fifth Semiannual Report to Congress* (Jan 1949), 67; U.S. House, 80th Cong., 1st sess., *Independent Offices Appropriation Bill for 1948,* 1490–92; U.S. House, 80th Cong., 2d sess., *Supplemental Independent Offices Appropriation Bill for 1949,* 767–768.
129. See, e.g., AEC, *Fifth Semiannual Report* (Jan 1949).
130. BNL, Scientific Steering Committee, minutes, 14 Apr 1947 (BNL-DO, IV-19); GAC 1, 3–4 Jan 1947, EHC; Atomic Energy Act of 1946, Sec. 1(b)(5), in Hewlett and Anderson, 715, app. 1.
131. Oppenheimer to Lilienthal, 15 Jul 1949, and minutes, GAC 15, 14–15 Jul 1949.
132. Oppenheimer to Dean, 11 May 1951, GAC 26; GAC to Dean, 30 Apr 1952, in GAC 30; GAC 30, 27–29 Apr 1952, EHC.
133. GAC 15, 14–15 Jul 1949; AEC, *Seventh Semiannual Report* (Jan 1950), 51–54; *Ninth Semiannual Report* (Jan 1951), 11; *Eleventh Semiannual Report* (Jan 1952), 29; and *Thirteenth Semiannual Report* (Jan 1953), 29–30.
134. ORNL, gray books for FY1957, 2 Jun 1955, ORNL/CF-55-3-3 and ORNL/CF-55-3-4.
135. T. H. Johnson in U.S. Senate, 82d Cong., 2d sess., *Independent Offices Appropriations, 1953,* 93.
136. Frederick Reines, "The Neutrino: From Poltergeist to Particle," in *Les Prix Nobel: The Nobel Prizes, 1995* (Stockholm, 1996), 96–115; Peter Galison, *Image and Logic* (Chicago, 1997), 460–463; cf. Trevor Pinch, *Confronting Nature: The Sociology of Solar-Neutrino Detection* (Dordrecht, 1986), 50–57.
137. AUI, Board of Trustees, 21 Oct 1949 and 21 Jul 1950; Haworth to Shields Warren, 12 May 1950 (BNL-DO, I-8); Tape to Van Horn, 28 Dec 1950 (BNL Central Records, DO-87); Pitzer to R. W. Cook, 12 Jan 1951, in Larson to Weinberg et al., 19 Jan 1951 (ORNL-DO, Research-General).
138. AEC, *Seventh Semiannual Report* (Jan 1950), 54.
139. Cf. Forman, "Quantum Electronics," 158n14. Instead of the 4 percent Forman's formula would provide, including biomedical and reactor research gives 12 percent of the total AEC budget for 1950 given to research, not including research under the weapons program. Forman's wider claim is that little if any of this work qualifies as basic research.
140. Kevles, *Physicists,* 358–359.
141. Appropriations Committee, hearings on NSF, 17 Feb and 24 Apr 1953, in C. Grobstein, "Federal Research and Development: Prospects, 1954," *BAS,* 9 (Oct 1954), 299–304, on 303; Smyth in U.S. House, 83d Cong., 1st sess., *Second Independent Offices Appropriations for 1954,* 401.

142. Draft Executive Order, 9 Jul 1953 (BOB, 8/Scientific research-general); AEC 183/5, 15 Sep 1953; AEC 183/6, 20 Nov 1953; Strauss to Dodge, 11 Dec 1953; and Executive Order 10521, 17 Mar 1954 (Sec'y 51–58, 77/O&M-12: NSF).
143. K. Pitzer to A. Tammaro, 25 Jan 1951 (DC, 11/Finance-Admin.: Budget).
144. Kirk McVoy to S. A. Goudsmit, in Goudsmit to Haworth, 20 Oct 1958 (BNL-DO, V-64).
145. C. E. Center, "Policy Statement Concerning ORNL," 10 Jul 1950 (ORNL-DO, Program-ORNL).
146. Johnson, "Research in Physics," Jan 1951 (BNL-DO, V-63).
147. GAC 4, 30 May–1 Jun 1947, EHC; Hewlett and Duncan, 82; Kevles, *Physicists,* 69, 82–83, 191.
148. Center, "Policy Statement."
149. AEC, *Ninth Semiannual Report* (Jan 1951), 55.
150. York to J. J. Flaherty, Livermore program for CY54–FY55, 18 Dec 1953 (EHC, PLB&L-7: Los Alamos); GAC 38, 6–8 Jan 1954, EHC. Cf. Forman, "Quantum Electronics," 158n14.

5. FALSE SPRING

1. Walter A. McDougall, *The Heavens and the Earth: A Political History of the Space Age* (New York, 1985), 58, 158.
2. Cf. ibid.; Brian Balogh, *Chain Reaction: Expert Debate and Public Participation in American Commercial Nuclear Power, 1945–1975* (Cambridge, 1991), 171n4.
3. Hewlett and Holl, 65–67, 71–72, 209–211; George T. Mazuzan and J. Samuel Walker, "Developing Nuclear Power in an Age of Energy Abundance, 1946–1962," *Materials and Society,* 7 (1983), 307–319, on 309.
4. David E. Lilienthal, *The Journals of David E. Lilienthal,* vol. 2, *The Atomic Energy Years, 1945–1950* (New York, 1964), entry for 14 Feb 1950; Hewlett and Duncan, 98–101, 185–221; Paul Boyer, *By the Bomb's Early Light: American Thought and Culture at the Dawn of the Atomic Age* (New York, 1985), 109–121, 291–302.
5. Rebecca S. Lowen, "Entering the Atomic Power Race: Science, Industry, and Government," *Political Science Quarterly,* 102:3 (1987), 459–479, on 471–472.
6. Statements by Sens. Cole and Hickenlooper, and AEC chairman Dean, in JCAE, 83d Cong., 1st sess., *Atomic Power Development and Private Enterprise,* 2, 6, 64; Mazuzan and Walker, "Developing," 308–309; Lowen, "Entering the Race," 474–475; Hewlett and Holl, 113–143, 183–208.
7. Hewlett and Holl, 439; Daniel Yergin, *The Prize: The Epic Quest for Oil, Money, and Power* (New York, 1991), 488.

8. GAC 47, 13–15 Dec 1955, EHC.
9. Johnson and Schaffer, 89; Hewlett and Holl, 410, 428–429, 504–505; ORNL budget figures in *The Future Role,* 51.
10. Hewlett and Holl, 209–237.
11. AUI, Ex Comm, 17 Dec 1954 and 25 Feb 1955; Jerome D. Luntz, "An International Agency," *Nucleonics,* 12:2 (1954), 7; A. F. Andresen and J. A. Goedkoop, "Neutrons in the Netherlands and Scandinavia," in G. E. Bacon, ed., *Fifty Years of Neutron Diffraction* (Bristol, 1987), 62–71, on 66. The description of Brookhaven as a "Mecca" appears in all three sources.
12. "Atoms for Peace in the UN," *BAS,* 11 (Jan 1955), 24–27; G. F. Tape to Dept. Chairs, 7 Feb 1955 (BNL-DO, VII-1); Haworth to Van Horn, 9 Apr 1958 (BNL-DO, I-16); Clarke Williams to Haworth, 21 Jan 1959 (BNL-DO, V-59); Andresen and Goedkoop, "Neutrons"; Haworth to D. I. Blokhintsev, 16 Aug 1957, and Haworth to Dept. Chairs, 17 Apr 1957 (BNL-DO, I-15).
13. Holl, 135–137; "News Roundup," *BAS,* 12:1 (Jan 1956), 29.
14. Haworth to Van Horn, 21 Mar 1956 (BNL-DO, I-14); BNL, "Study for an Asian Regional Nuclear Center," 15 Nov 1956 (BNL-HA). The nations involved were Burma, Cambodia, Ceylon, India, Indonesia, Japan, Laos, Malaya, Nepal, Pakistan (West and East), the Philippines, Thailand, and Vietnam.
15. Glenn T. Seaborg, *Journal,* 18 Apr 1961 (Office for History of Science and Technology, UC Berkeley); AUI, Ex Comm, 20 Apr 1961.
16. Dominique Pestre, "The First Suggestions, 1949–June 1950," in Armin Hermann et al., eds., *History of CERN,* vol. 1, *Launching the European Organization for Nuclear Research* (Amsterdam, 1987), 63–95, on 89. The acronym stood for the European Council (later Organization) for Nuclear Research.
17. Seaborg to Lawrence, 15 Mar 1955, Earl Hyde memo, n.d. [ca. 1955], and Lawrence to Fidler, 24 Mar 1955 (EOL, 32/12); Haworth to Van Horn, 13 Jun 1955 (BNL-DO, I-13).
18. Haworth to Van Horn, 11 Jun 1954 (BNL-DO, I-12).
19. Weinberg, "State of the Laboratory—1955" (BNL-HA, Haworth declassified files, box 36); Zinn, quoted in Holl, 138–139; International Conference on the Peaceful Uses of Atomic Energy, Geneva, 8–20 Aug 1955, *Proceedings,* 16 vols. (New York, 1956).
20. Heilbron, Seidel, and Wheaton, 60; Johnson to Morse, 12 Jan 1948, quoted in Robert W. Seidel, "Accelerating Science: The Postwar Transformation of the Lawrence Radiation Laboratory," *HSPS,* 13:2 (1983), 375–400, on 397n60.
21. George B. Collins to Haworth, 29 Jan 1952 (BNL-DO, V-5).
22. M. Stanley Livingston, *Particle Accelerators: A Brief History* (Cambridge, Mass., 1969), 60–67; Crease, 141–147.
23. AUI, Ex Comm, 16 Sep 1952.

24. AUI Ex Comm, 16 Sep 1952, 20 Mar 1953, and 18 Sep 1953; AEC 603, 1 Dec 1952 (Sec'y 51–58, 138/R&D-6: Particle accels.); Haworth to John and Hildred Blewett, 19 Oct 1953 (BNL-DO, I-11).
25. AEC 603/9, 8 Oct 1953 (Sec'y 51–58, 139/R&D-6: Particle accels.); Hewlett and Holl, 258.
26. AUI, Ex Comm, 14 Nov 1955; Lawrence to Clark Kerr, 15 Dec 1955 (EOL, 22/8); notes on meeting of AEC and MURA representatives, 8 Nov 1955 (Sec'y 51–58, 102/PLB&L-49: Midwestern Univs.); Holl, 168–169.
27. S. K. Allison et al. to T. H. Johnson, 30 Jan 1953, and Johnson to Zinn, 16 Feb 1953 (Sec'y 51–58, 138/R&D-6: Particle accels.); Daniel S. Greenberg, *The Politics of Pure Science* (New York, 1967), 209–269; Leonard Greenbaum, *A Special Interest: The AEC, Argonne National Laboratory, and the Midwestern Universities* (Ann Arbor, 1971); Holl, 152–174.
28. AEC 827/7, 21 Nov 1955; R. W. Cook to K. E. Fields, 15 Dec 1955; Zinn to Fields, 12 Dec 1955; Fields to J. J. Flaherty, 29 Dec 1955; Zinn to Flaherty, 5 Jan and 26 Jan 1956; Flaherty to Zinn, 17 Jan 1956 (Sec'y 51–58, 102/PLB&L-49: Midwestern Univs.).
29. JCAE, 84th Cong., 2d sess., *Authorizing Legislation* (17 Feb 1956), 27.
30. V. F. Weisskopf, "International Conference on High Energy Physics," *BAS*, 12 (Sep 1956), 259; T. H. Johnson to Tammaro and Fields, 22 Jul 1957 (GM, 5673/9).
31. *The Future Role*, 4.
32. Robert W. Seidel, "A Home for Big Science: The AEC's Laboratory System," *HSPS*, 16:1 (1986), 135–175, on 171; JCAE, 86th Cong., 1st sess., *Stanford Linear Electron Accelerator*, 1–8.
33. AEC 603/53, 28 Nov 1958, and McCool to J. H. Williams, 18 Dec 1958 (Sec'y 58–66, 1424/1).
34. Schuldt, "Draft Work Program," 30 Jan 1956, and BOB, "Summary Work Program," 7 Jan 1957 (BOB, 1/Work program).
35. GAC 61, 5–7 Jan 1959, and GAC 64, 4–6 May 1959, EHC; Southern Regional Accelerator Committee, minutes ("pittance"), 27 Jul 1959 (ORNL-DO, Southern Regional Accel.).
36. Rep. Melvin Price in JCAE, 86th Cong., 1st sess., *Stanford Linear Electron Accelerator*, 36.
37. Lanfranco Belloni, "The Italian Scenario," in Hermann et al., *History of CERN*, 353–382, on 361; see also, e.g., Dominique Pestre, "The Fusion of the Initiatives," ibid., 111–112.
38. V. F. Weisskopf, "A Theoretical Physicist at the Geneva Conference," *BAS*, 11 (Oct 1955), 278.
39. Armin Hermann, "Some Aspects of the History of High-Energy Physics, 1952–66," in Hermann et al., *History of CERN*, 41–94, on 61–63.

40. "News Roundup," *BAS,* 12 (June 1956), 229.
41. Zuoyue Wang, "The Politics of Big Science in the Cold War: PSAC and the Funding of SLAC," *HSPS,* 25:2 (1995), 329–356, on 337–340.
42. R. B. Brode to Haworth, 17 Aug 1959, and minutes, International Accelerator Study Committee, 15 Sep 1959 (BNL-DO, VIII-3).
43. John McCone and V. S. Emelyanov, memo, 24 Nov 1959 (BNL-DO, VIII-3); Hewlett and Holl, 531–536.
44. C. E. Falk, "Report on Escort of Russian High Energy Exchange Team," 5–8 Jul 1960, and "Report to U.S. and Soviet Atomic Energy Authorities," 16 Sep 1960 (BNL-DO, VIII-3); AUI Ex Comm, 23 Sep 1960.
45. AUI Ex Comm, 23 Sep 1960; George Kolstad to J. B. Adams, 26 Sep 1960 (BNL-DO, VIII-3); AUI Trustees, 27 Oct 1961; Seaborg, *Journal,* 12 Apr and 27 Jun 1961.
46. Keith Symon to Glenn Frye, 3 Jun 1960 (Sec'y 58–66, 1424/2); AUI, Ex Comm, 27 Oct 1961, 20 Sep 1963, and 15 Nov 1963; Goldhaber to Paul McDaniel, 24 Nov 1963, L. C. L. Yuan to Falk, 11 Nov 1963, and R. L. Cool to Falk, 11 Nov 1963 (BNL-HA, 31/Accelerator Dept.); Seidel, "Home," 172.
47. Weisskopf, "International," 260.
48. Wang, "PSAC and SLAC," 339.
49. Haworth to Harry S. Traynor, 11 Jul 1960 (BNL-DO, I-18).
50. W. K. H. Panofsky to Haworth, 22 Sep 1959 (BNL-DO, VIII-3).
51. Panofsky, "International Collaboration in High-Energy Physics," 26 Nov 1962 (HEP 57–64, 3/GAC-PSAC); Haworth to Jerome Wiesner, 18 May 1961 (Sec'y 58–66, 1424/2).
52. Goldhaber to McDaniel, 24 Nov 1963 (BNL-HA, 31/Accelerator Dept.).
53. Joan Lisa Bromberg, *Fusion: Science, Politics, and the Invention of a New Energy Source* (Cambridge, Mass., 1982), 13–37.
54. AEC minutes, 3 Sep and 21 Sep 1953, and AEC 532/10, 24 Sep 1953 (Sec'y 51–58, 141/R&D-7: CTR); Bromberg, *Fusion,* 37–38; Hewlett and Holl, 260–261.
55. Bromberg, *Fusion,* 30.
56. In 1959 Teller proposed the use of the fast electron injector to Livermore's Astron device for ballistic missile defense. Around 1953, however, no such uses had been suggested for fusion reactors. GAC 66, 28–30 Oct 1959, EHC.
57. AEC 852/160, 16 Oct 1956 (Sec'y 51–58, 121/R&D-1: Problems and policy).
58. AEC 532/39, 15 Mar 1957 (Sec'y 51–58, 114/Reactor Dev.-1: CTR).
59. Bromberg, *Fusion,* 74–75, 81–86; Weinberg, "The State of the Laboratory, 1957," ORNL/CF-57-12-127.
60. Bromberg, *Fusion,* 78, 83.
61. Ibid., 86–88.
62. Edward Creutz, ibid., 44.

63. J. M. B. Kellogg, notes attached to Kellogg to Bradbury, 1 Nov 1956 (LASL-DO, 410: Accelerators).
64. Kellogg to Bradbury, ibid.
65. AEC, *Eighth Semiannual Report* (Jul 1950), 149–153.
66. "Radiation Instrument Does $20-Million Business in 1952," *Nucleonics,* 11:2 (1953), 78.
67. B. J. Moyer to D. Cooksey, 10 Oct 1949, and J. S. Norton to Cooksey, 21 Oct 1949 (EOL, 32/6); U.S. House, 83d Cong., 2d sess., *Independent Offices Appropriations for 1955,* 2637, 2814–17; AEC, *Sixteenth Semiannual Report* (Jul 1954), 67–72; "Where AEC's Instrument Dollar Goes," *Nucleonics,* 12:9 (1954), 85.
68. Weinberg, "On the State of the Laboratory—1952," 30 Dec 1951, ORNL/CF-51-12-186.
69. AEC 152/23, 8 Apr 1952, and AEC minutes, 28 May 1952 (Sec'y 51–58, 139/R&D-6).
70. Topnotch II agenda item 5a, 24–28 Sep 1953 (Sec'y 51–58, 65/O&M-6: Topnotch II); Zinn, "Comment on National Laboratories," 6 Dec 1955 (EOL, 32/12).
71. GAC 27, 11–13 Oct 1951, EHC.
72. Allan A. Needell, "Nuclear Reactors and the Founding of Brookhaven National Laboratory," *HSPS,* 14:1 (1983), 93–122. The reactor cost $25 million, ten times the original, optimistic estimate of 1946 and $10 million more than the $15 million estimated in December 1947.
73. J. D. Jameson to E. C. Shoup, 22 Sep 1949 (BNL-DO, II-5).
74. AUI, Ex Comm, 18 Jan and 18 May 1951.
75. Van Horn to Tape, 7 Jul 1952, and Clarke Williams to C. R. Russell, 29 Sep 1952 (BNL-DO, V-83).
76. "Justification for an Accelerated LMFR Program," draft, ca. 25 Jul 1955, and Tape to Van Horn, 4 Aug 1955 (BNL-DO, V-83).
77. AUI, Ex Comm, 25 Feb and 21 Oct 1955.
78. Weinberg, "Some Problems in the Development of the National Laboratories," Dec 1955, and Haworth, "The Future Role of Brookhaven National Laboratory," 5 Dec 1955 (EOL, 32/12).
79. "New Industrial Association Formed to Foster Growth of Atomic Energy Industry," *Nucleonics,* 11:5 (1953), 70; Hewlett and Holl, 27.
80. Tammaro to Luedecke, 18 Sep 1959, and H. A. Stanwood to C. L. Dunham, 30 Nov 1959 (DBM-DOE, 3374/16); Francis McCune to Luedecke, 22 Oct 1959 (Sec'y 58–66, 1404/1).
81. "Wanted: Reactor Engineers," *Nucleonics,* 10:2 (1952), 10–13; Johnson and Schaffer, 74–75; Weinberg, 51–54.
82. AUI Board of Trustees, 16 Jan 1953.
83. AEC 496/23, 14 Feb 1955 (Sec'y 51–58, 69/O&M-7: JCAE).

84. Glenn Seaborg and Daniel M. Wilkes, *Education and the Atom* (New York, 1964), 5.
85. Haworth to Dunham, 23 Dec 1955 (BNL-DO, I-13), and Haworth, "Future Role."
86. Hewlett and Holl, 254, and Johnson and Schaffer, 74–75.
87. R. L. Thornton, "Long Range Objectives of the Radiation Laboratory," 30 Nov 1955 (EOL, 22/8).
88. Graduate students at UCRL [Rad Lab], 12 Aug 1958 (EOL, 19/27).
89. AUI, Ex Comm, 20 Mar 1959.
90. AEC Division of Research, "The Contract-Research Program," 10 Feb 1958 (Sec'y 51–58, 115/R&D-1: Support of basic research); GAC 65, 20–22 Jul 1959, EHC.
91. Alvin Weinberg, "A Nuclear Journey through Europe," *BAS,* 10:6 (1954), 217; "News Roundup," *BAS,* 12:8 (1956), 318; Hewlett and Duncan, 435; Holl, 147, 190.
92. GAC 29, 15–17 Feb 1952, EHC; AEC 603.
93. AEC 603/9; GAC 36, 17–19 Aug 1953, EHC; Rabi to Strauss, 24 Aug 1953, GAC 36.
94. Rabi to Strauss, 23 Nov 1954, GAC 42, and 22 Dec 1954, GAC 43 (JCAE III, 33/GAC).
95. T. H. Johnson, proposed policy, 13 Dec 1954; AEC 603/23, 3 Feb 55; AEC meeting excerpt, 23 Feb 1955; AEC 603/28, 23 Mar 1956 and AEC meeting excerpt, 13 Mar 1958 (Sec'y 51–58, 139/R&D-6: Particle accels.).
96. GAC 47, 13–15 Dec 1955, EHC.
97. AEC 815, 7 Apr 1955 (Sec'y 51–58, 139/R&D-6: Particle accels., vol. 2).
98. AEC 925, 19 Jul 1956, McCool to Davis, 3 Aug 1956, AEC meeting minutes, 13 Sep 1956 and 22 Jan 1958, and Tammaro to Davis, 29 Jan 1958 (Sec'y 51–58, 100/PLB&L-48: Laboratories).
99. J. L. Morrill et al. in AEC 311/2, Jul 1957 (Sec'y 51–58, 55/O&M-2: Chicago Operations Office).
100. AEC, summary notes of meeting with ARMU, Inc., 14 May 1962 (Sec'y 58–66, 1404/5).
101. Seidel, "Home," 169n119; Wang, "PSAC and SLAC"; Greenberg, *Politics,* 226–245.
102. Seidel, "Home," 170–171; Rebecca S. Lowen, *Creating the Cold War University: The Transformation of Stanford* (Berkeley, 1997), 177–186.
103. McCool to Davis, 12 May 1958, and Davis to Libby, 12 Jun 1958 (Sec'y 51–58, 100/PLB&L-48: Laboratories).
104. Hewlett and Duncan, 491–493.
105. Percival Brundage to Charles Wilson, 19 Oct 1956; AEC 892/10, 14 Nov 56; and AEC minutes, 14 Nov 1956 (Sec'y 51–58, 176/BAF-2: 1958).
106. AEC 445, 18 Jun 1951 (Sec'y 51–58, 178/BAF-8: Joint Participation).

107. AUI, Board of Trustees, 18 Apr 1952.
108. Berkner to E. R. Piore, 3 Jan 1952, Dunbar to Berkner, 16 Jan 1952, Van Horn to Berkner, 24 Jan 1952, Berkner to Van Horn, 31 Jan 1952 (BNL-DO, II-4); Van Horn to files, 4 Apr 1952 (DBM, 18/MH&S-21: BNL).
109. Warren to Berkner, 26 Jun 1952 (BNL-DO, II-4); AUI, Ex Comm, 16 May 1952.
110. AUI, Board of Trustees, 18 Jul 1952.
111. R. R. Entwhistle to Weinberg, 3 Aug 1959, and Weinberg to Entwhistle, 19 Aug 1959 (ORNL-DO, Neutron Physics).
112. T. A. Welton to Charles Townes, 16 Feb 1961, and J. P. Ruina to Seaborg, 29 Aug 1961 (ORNL-DO, Physics).
113. GAC 60, 30–31 Oct and 1 Nov 1958, EHC.
114. Merrill Eisenbud to Haworth, 5 Jan 1959; Clinton Anderson and Melvin Price to McCone, 3 Feb 1959; and A. R. Luedecke to Anderson, 18 Feb 1959 (BNL-DO, II-22).
115. *The Future Role,* 105, 113, 122, 130, 139, 141; Froman on Los Alamos in AEC, "Summary Notes," 19 Nov 1959 (GM, 5672/13).
116. GAC 62, 9–11 Mar 1959, EHC.
117. GAC 66, 28–30 Oct 1959, EHC; McMillan in AEC, "Summary Notes."
118. GAC 62, 9–11 Mar 1959, EHC. One factor motivating the British: a housing shortage at Harwell.
119. Stanwood to Dunham, 30 Nov 1959 (DBM-DOE, 3374/16).
120. W. Johnson to McCone, 4 Aug 1959, GAC 65; John S. Graham to Luedecke, 17 Jul 1959 (DOE-DBM, 3374/16).
121. GAC 66, 28–30 Oct 1959, EHC; *The Future Role,* 12.
122. GAC 66.
123. In *Oxford English Dictionary,* 2d ed. s.v. "interdisciplinary," the first citation is to an article on sociology in 1937, followed by references in sociology and linguistics in 1956 and 1957.
124. Hilberry, Tape, and Weinberg in AEC, "Summary Notes," 19 Nov 1959 (GM, 5672/13); Bradbury in *The Future Role,* 141.
125. Alvin Weinberg, "Criteria for Scientific Choice," *Minerva,* 1:2 (1963), 159–171, on 166.
126. AEC Division of Biology and Medicine, "Role of AEC Laboratories," 17 Sep 1959 (DBM-DOE, 3375/1).
127. *The Future Role,* 11.
128. Haworth to G. W. Beadle, 17 Jul 1961 (ANL-PAB, 7/Review Comm.-Chem.). See also Seaborg and Wilkes, *Education and the Atom,* 36.
129. *The Future Role,* 13, 122, 161–198, 230–240.
130. Ibid., 33, 155; W. Johnson to J. T. Ramey, 14 Jun 1960, GAC 69, EHC, reprinted ibid., 243–246.

131. R. R. Coffin to files, draft 1 Jul 1960 (BOB, 12/Laboratories).
132. PSAC, "Scientific Progress, the Universities, and the Federal Government," 15 Nov 1960 (Sec'y 58–66, 1423/7).
133. GAC 81, 4–6 Oct 1962, EHC.
134. Weinberg statement in GAC 81, 4–6 Oct 1962, EHC.
135. Notes of meeting with ARMU.
136. GAC 81, 4–6 Oct 62, and GAC 82, 7–9 Jan 1963, both in EHC; Holl, 214–216.
137. R. C. Bynum, "Davis," in Verne Stadtman, ed., *The Centennial Record of the University of California* (Berkeley, 1968), 153–154.
138. Teller to McCone, 8 Mar 1960 (Sec'y 51–58, 60/O&M-2: San Francisco); "Use of the Livermore Laboratory," n.d. [ca. 1963], and Dwight Ink to John Pastore, 6 Jun 1963 (Sec'y 58–66, 1404/7); AEC 1023/22, 17 May 1963, Luedecke to Pastore, 17 Oct 1963, and A. W. Betts to AEC, 4 Dec 1963 (Sec'y 58–66, 1404/6).
139. AUI Ex Comm, 21 Sep 1962.
140. GAC 81, 4–6 Oct 1962, EHC.
141. Seidel, "Home," 171; Catherine Westfall, "The First 'Truly National Laboratory': The Birth of Fermilab" (Ph.D. diss., Michigan State University, 1988), 102–123.
142. GAC 74, 27–29 Apr 1961.
143. AUI Ex Comm, 20 Feb 1959, 16–17 Jul 1959, and 20–21 Jul 1961.
144. Weinberg quoted in Albert Teich and W. Henry Lambright, "The Redirection of a Large National Laboratory," *Minerva,* 14 (1976–77), 447–474, on 452.
145. William D. Carey, notes on meeting with PSAC, 31 May 1961, and Seaborg speech excerpts, 26 Apr and 9 Jun 1961 (BOB, 12/Laboratories).
146. GAC 73, 22–24 Mar 1961, EHC.
147. GAC 74, 27–29 Apr 1961, EHC.

6. ADAPTIVE STRATEGIES

1. Interview with Herbert York, 7 Nov 1997; Joan Lisa Bromberg, *Fusion: Science, Politics, and the Invention of a New Energy Source* (Cambridge, Mass, 1982), 27–28.
2. Sybil Francis, "Warhead Politics: Livermore and the Competitive System of Nuclear Weapon Design" (Ph.D. diss., MIT, 1996), 69.
3. GAC 41, 12–15 Jul 1954, EHC.
4. Thomas B. Cochran et al., *Nuclear Weapons Databook,* vol. 2 (Cambridge, Mass., 1987), app. B.
5. GAC 41, 12–15 Jul 1954, EHC.
6. Francis, "Warhead Politics," 130–131.
7. Ibid., 67–68.

8. Ibid., 116–117, 134–135.
9. Ibid., 26–27, 93, 96, 103, 135–137; GAC 49, 28–30 Mar 1956, EHC.
10. Ibid., 72–73, 98; Rabi in GAC 50, 16–18 Jul 1956, EHC.
11. JCAE, 83d Cong., 1st sess., *Atomic Power Development and Private Enterprise,* 247; Weinberg, 100–102, 109–131.
12. GAC, special meeting, 29–30 Dec 1947, EHC.
13. GAC 25, 15–17 Mar 1951, EHC.
14. ORNL, Research Committee, minutes, 3 Dec 1958 (ORNL-DO, Committees-Research Comm.).
15. AEC 532/10, 24 Sep 1953 (Sec'y 51–58, 141/R&D-7: CTR).
16. AEC minutes, 3 and 21 Sep 1953, AEC 532/15, 15 Jan 1954, and Strauss to Sterling Cole, 24 Feb 1954 (Sec'y 51–58, 141/R&D-7: CTR); Bromberg, *Fusion,* 40.
17. Bromberg, *Fusion,* 13–29, 79–81.
18. AEC 532/39, 15 Mar 1957 (Sec'y 51–58, 114/Reactor Dev.-1: CTR).
19. Bromberg, *Fusion,* 26, 28–29, 79; GAC 37, 4–6 Nov 1953, EHC, on Oak Ridge ion work.
20. W. Johnson to McCone, 6 Aug 58, GAC 1959; "Notes for Meeting with Mr. McCone," 5 Sep 1958 (BOB, 17/AEC FY60 General Budget).
21. Bromberg, *Fusion,* 117–118, 140–146, 154–167. The tokamak had a toroidal configuration with an axial magnetic field.
22. Donald J. Keirn to K. Fields, 14 Oct 1954; V. G. Huston (for Fields) to Bradbury and to York, 22 Oct 1954; Paul Fine to Keirn, 5 Nov 1954; York to Fields, 10 Nov 1954; AEC 855, 19 Aug 1955 (Sec'y 51–58, 52/MR&A-5: Rocket Propulsion).
23. Darol Froman to Huston, 16 May 1955, and R. E. Schreiber to Bradbury, 12 May 1955, in AEC 855.
24. York to Huston, n.d. [ca. mid-1955], in AEC 855.
25. AEC minutes, 23 and 27 Sep 1955 (Sec'y 51–58, 176/BAF-2: 1957).
26. GAC 47, 13–15 Dec 1955, and GAC 49, 28–30 Mar 1956, EHC.
27. GAC 50, 16–18 Jul 1956, EHC, and Rabi to Strauss, 30 Jul 1956, in GAC 50.
28. GAC 50, 16–18 Jul 1956, and Rabi to Strauss, 30 Jul 1956.
29. York and Bradbury to Keirn, 14 Aug 1956, in GAC 52, 17–19 Jan 1957.
30. Froman to Huston, 16 May 1955, and York to Huston, n.d. [ca. mid-1955], in AEC 855. The slow-bomb concept was rekindled in 1956 in the context of Project Plowshare, and again after Sputnik in Project Orion. York to A. D. Starbird, 15 Oct 1956 (Sec'y 51–58, MR&A-9–1: Non-military uses, EHC); Freeman Dyson, *Disturbing the Universe* (New York, 1979), 109–115, 127–129.
31. York and Bradbury to Keirn.
32. AEC minutes, 27 Sep 1955 (Sec'y 51–58, 176/BAF-2: 1957); GAC 47, 13–15 Dec 1955, EHC.

33. Froman to Huston, 16 May 1955, and York to Huston, n.d. [ca. mid-1955], in AEC 855.
34. AEC 855/2, 31 Oct 1955 (Sec'y 51–58, 52/MR&A-5: Rocket propulsion); AEC 939/25, 7 Feb 1958 (Sec'y 51–58, 177/BAF-2: 1959); Strauss to C. Anderson, 9 Nov 1956 (Sec'y 51–58, 51/MR&A-4: ANP).
35. AEC 939/25; AEC 17/127, 10 Feb 1957 (Sec'y 51–58, 51/MR&A-4: ANP); GAC 50, 16–18 Jul 1956, EHC.
36. AEC 17/127, 10 Feb 1957, and AEC minutes excerpt, 20 Dec 1956 (Sec'y 51–58, 51/MR&A-4: ANP); AEC minutes excerpt, 4 Jan and 16 Jan 1957, Fields to Carl Durham, 5 Feb 1957, and Strauss to Warren Johnson, 27 Apr 1957 (Sec'y 51–58, 52/MR&A-5: Rocket propulsion).
37. Froman to Huston, 16 May 1955, AEC 855; Davis to Fields, 21 Jun 1957, and AEC 17/134, 26 Jun 1957 (Sec'y 51–58, 51/MR&A-4: ANP); AEC 939/25.
38. Strauss to P. Brundage, 21 Dec 1956 (Sec'y 51–58, 176/BAF-2: 1958); Davis to Fields, 21 Jun 1957, and AEC 17/134, 26 Jun 1957 (Sec'y 51–58, 51/MR&A-4: ANP); AEC 939/25; GAC 54, 9–11 Jul 1957, EHC.
39. Van Horn to Haworth, 24 Apr 1956, and G. H. Vineyard to Van Horn, 15 May 1956 (BNL-DO, V-58); Leonard Link, "History of High Flux Reactor Development at Argonne," 4 Jan 1963 (ANL-DO, 5/AARR); GAC 49, 28–30 Mar 1956, EHC.
40. Haworth to Van Horn, 8 Aug 1952 (BNL-DO, I-10); AUI Ex Comm, 16 Mar 1956; Crease, 195–199; Laurence Passell, "High Flux at Brookhaven," Oct 1985, and BNL, "High Flux Research Reactor," Jul 1958, copies courtesy of Julius Hastings.
41. GAC 49, 28–30 Mar 1956, EHC; ORNL Policy Committee, minutes, 19 Sep and 7 Nov 1956 (ORNL-DO, Committees-Policy); ORNL Research Council, 22 Aug 1958 (ORNL-DO, Committee-Research Council); Weinberg, 90–91.
42. Link et al., "The Mighty Mouse Research Reactor," Mar 1957, ANL-5688; Link and J. T. Weills to Hilberry, 1 May 1957 (ANL-DO, 3/CP-5'); Link, "History."
43. "Discussion with Walt Hughes on CP-5" [probably by Link], ca. Jun 1958 (ANL-DO, 5/AHFR).
44. AEC 998/2, 6 May 1958 (Sec'y 51–58, 178/BAF: 1960).
45. R. M. Adams to J. R. Gilbreath, 7 Jul 1958, M. B. Powers to L. A. Turner, 11 Jul 1958, and Powers to Hilberry, 30 Jul 1958 (ANL-DO, 1/AARR); "Saga of Mighty Mouse," 26 Jun 1958 (ANL-DO, 5/AHFR).
46. Link, "History"; Link et al., "Argonne Reactor for Advanced Research (ARFAR)," 18 Nov 1958 (ANL-DO, 1/AARR).
47. L. M. Bollinger et al. to Hilberry, 27 Oct 1958 (ANL-DO, 1/AARR).
48. J. H. Williams to Hilberry, 12 Dec 1958 (ANL-DO, 1/AARR).
49. Link, "Impressions of Argonne's Status," 12 Dec 1958 (ANL-DO, 1/AARR).
50. E. C. Weber to J. R. Gilbreath, 30 Dec 1958, and W. M. Manning to Gilbreath, 5 Jan 1959 (ANL-DO, 1/AARR).

51. Link, "Impressions."
52. Manning to Gilbreath, 5 Jan 1959 (ANL-DO, 1/AARR).
53. Ibid.
54. L. J. Koch to Link, 15 Dec 1958 (ANL-DO, 1/AARR).
55. Weinberg, 92.
56. K. A. Dunbar and S. R. Sapirie to F. K. Pittman, 21 May 1959 (ANL-DO, 1/AARR); Williams to Tammaro and Luedecke, 4 Jun 1959 (Sec'y 58–66, 1318/9).
57. G. R. Ringo to Hilberry, 17 Dec 1958 (ANL-DO, 1/AARR); Ringo to Gilbreath, 11 Nov 1960 (ANL-DO, 5/AARR).
58. Crewe to G. W. Beadle, 7 Jan 1963, Crewe to Seaborg, 1 Mar and 6 Mar 1963, and Crewe to J. T. Conway [JCAE staff], 1 Mar 1963 (ANL-DO, 5/AARR).
59. JCAE, *AEC Authorizing Legislation, 1964,* 145–151; Holl, 235–241, 257–259.
60. Tammaro to Hilberry, 26 Nov 1957 (ANL-DO, 1/AARR).
61. Link, "History."
62. Link, "Mighty Mouse," 12 Mar 1957 (ANL-DO, 1/AARR); Link et al., "Mighty Mouse."
63. Dean memo, 10 Jun 1952, in Roger M. Anders, ed., *Forging the Atomic Shield: Excerpts from the Office Diary of Gordon Dean* (Chapel Hill, 1987), 265–270, on 269.
64. AEC minutes, 3 Sep 1953 (Sec'y 51–58, 141/R&D-7: CTR).
65. Hewlett and Holl, 274–276.
66. Bradbury to T. H. Johnson, 21 Nov 1955 (EOL, 32/12).
67. AEC 855.
68. Hewlett and Holl, 362, 477.
69. GAC 60, 30–31 Oct and 1 Nov 1958, EHC.
70. AEC minutes excerpt, 6 Nov 1957, and Melvin Price to Eisenhower, 24 Oct 1957 (Sec'y 51–58, 51/MR&A-4: ANP).
71. Norris E. Bradbury, interview by Arthur L. Norberg, 1976, TBL, 110.
72. GAC 24, 4–6 Jan 1951; "News Roundup," *BAS,* 13:2 (1957), 68.
73. AEC 934, 5 Sep 1956 (Sec'y 51–58, 105/PLB&L-50: Los Alamos).
74. Los Alamos press release, 9 Jan 1956 (LANL A-83–0005, 1/1); "The First 20 Years at Los Alamos," *LASL News,* 1 Jan 1963, 47–49; Bradbury interview, 79.
75. Edward Teller, "The Program of the Lawrence Radiation Laboratory at Livermore for the Period 1959–70," 26 Aug 1959 (JCAE III, 37/LRL), reprinted in *The Future Role,* 136.
76. GAC 75, 13–15 Jul 1961, EHC. The figures for 1961 come from estimates provided to the GAC; the figure for Livermore was 55 percent.
77. GAC 71, 24–26 Oct 1960; Pitzer to Seaborg, 2 May 1961, in GAC 74, and 19 Jul 1961, in GAC 75; GAC 75, 13–15 Jul 1961, EHC.
78. GAC 80, 9–11 Jul 1962, and GAC 84, 25–27 Apr 1963, EHC; Manson Benedict to Seaborg, 21 Feb 1963 (Sec'y 58–66, O&M-7: GAC corres., EHC).

79. Richard G. Hewlett and Francis Duncan. *Nuclear Navy, 1946–1952* (Chicago, 1974), 225–257; Francis Duncan, *Rickover and the Nuclear Navy: The Discipline of Technology* (Annapolis, 1990), chap. 8; William Beaver, *Nuclear Power Goes On-Line: A History of Shippingport* (New York, 1990).
80. Weinberg, "Some Problems in the Development of the National Laboratories," Dec 1955 (EOL, 32/12).
81. *The Future Role*, 37.
82. ANL Policy Advisory Board, minutes, 7–8 May 1962 and 16 Jan 1963, box 4; 16 Jul 1964, box 5; and 20 Jan 1966, box 6, ANL-PAB; C. Cohn et al., "Basic Material Resulting from ANL Rocket Study," May 1963, ANL-6656; M. Benedict to Seaborg, 24 Jul 1963, GAC 85; material for Lab Directors' meeting, May 1966 (GM, 5625/15).
83. AEC 827/7, 21 Nov 1955, and AEC 827/8, 15 Nov 1955 (Sec'y 51–58, 102/PLB&L-49: Midwestern Univs.).
84. Weinberg to A. H. Snell et al., 23 Oct 1953 (ORNL-DO, Physics Div.); ORNL, Policy Committee, minutes, 21 Apr 1954 (ORNL-DO, Meetings-Policy Comm.).
85. M. Stanley Livingston, ed., *The Development of High-Energy Accelerators* (New York, 1966), 303–314; Heilbron, Seidel, and Wheaton, 70–71.
86. E. Creutz et al. to C. E. Larson, 8 Aug 1955 (ORNL-DO, Advisory Committees-Electronuclear); Weinberg to T. H. Johnson, 26 Mar 1956 (ORNL-DO, Physics).
87. Report of NSF Advisory Panel on High Energy Accelerators, 4 Oct 1956 (Sec'y 51–58, 139/R&D-6: Particle accels.); Holl, 196–197; Peter Galison, *Image and Logic* (Chicago, 1997), chap. 6.
88. Weinberg to T. H. Johnson, 26 Mar 1956, and to Snell and R. S. Livingston, 29 Aug 1956 (ORNL-DO, Physics); GAC 49, 28–30 Mar 1956, EHC; McCool to Johnson, 31 May 1956, and Paul McDaniel to Libby, 11 Jun 1956 (Sec'y 51–58, 139/Particle accels.).
89. Joint ORINS-ORNL Accelerator Committee, report, 8 Oct and 12 Nov 1956 (ORNL-DO, Southern Regional Accelerator).
90. Weinberg to Libby, 23 May 1957 (ORNL-DO, Southern Regional Accelerator); AEC 603/25, 31 May 1957 (Sec'y 51–58, 139/R&D-6: Particle accels.).
91. Weinberg to Johnson, 26 Mar 1956 (ORNL-DO, Physics); Southern Regional Accelerator Committee, report, 14 Feb 1958 (ORNL-DO, Southern Regional Accelerator).
92. Luis Alvarez to McMillan, 17 Sep 1957, attached to Hugh Bradner to Weinberg, 24 Sep 1957, and Snell to Alvarez, 8 Apr 1957 (ORNL-DO, Southern Regional Accelerator); Galison, *Image and Logic*, 479–482.
93. Southern Regional Accelerator Committee, minutes, 14 Oct 1957, and 14 Feb and 21 Apr 1958 (ORNL-DO, Southern Regional Accelerator); C. E. Center to S. R. Sapirie, 7 Jul 1958 (BOB, 9/HEP: FY59–61); W. Johnson to Strauss, 28

Aug 1957 (GAC 54), 14 Oct 1957 (GAC 55), and [with quote] 17 Mar 1958 (GAC 57).

94. W. Johnson to Strauss, 17 Mar 1958, GAC 57; AEC 605/53, 28 Nov 1958, and McCool to John H. Williams, 18 Dec 1958 (Sec'y 58–66, 1424/1); Williams to Weinberg, 23 Dec 1958 (ORNL-DO, Southern Regional Accelerator).
95. Heilbron, Seidel, and Wheaton, 74–75; Ruth R. Harris and Richard G. Hewlett, "The Lawrence Livermore National Laboratory: The Evolution of Its Mission, 1952–1988," report prepared for Livermore, 21 Mar 1990, 6–7; AEC 29/113, 30 Apr 1957 (O&M-7: GAC corres., EHC).
96. H. V. Argo and F. L. Ribe, "Preliminary Study of 2-Bev Synchro-Cyclotron," 27 Oct 1955, LASL-HVA-FLR-1; Darragh E. Nagle, "A Proposed Multifrequency Accelerator," 13 Feb 1956, Los Alamos report LASL-DEN-2.
97. "Comments Arising from the Meeting of 31 October," 1 Nov 1956 (LANL A-91–011, 55/1).
98. Bradbury to K. F. Hertford, 23 Apr 1957 (LASL-DO, 410: Accelerators); L. Rosen, "Outline of the Proposal for a Meson Facility at Los Alamos," 16 Jul 1963, Los Alamos report LAMS-2935; M. Stanley Livingston, "Origins and History of the Los Alamos Meson Physics Facility," June 1972, Los Alamos report LA-5000.
99. Bradbury to Strauss, 7 Feb 1956 (LASL-DO, 410: Accelerators).
100. Nagle to John Bolton, 23 Jul and 15 Aug 1956 (LANL A-91–011, 55/1).
101. "Comments from Meeting"; Kellogg to Bradbury, 1 Nov 1956 (LASL-DO, 410: Accelerators).
102. David Hill, "Should Los Alamos Move Vigorously toward the Construction of a High Energy Accelerator?" 15 Feb 1957 (LANL A-91-011, 55/1).
103. John E. Brolley, Jr., to Froman, 28 Aug 1958 (LASL-DO, 310.1: P-Div. History).
104. H. H. Barschall to Kellogg, 27 Aug 1958 (LASL-DO, 310.1: P-Div. History).
105. Brolley to Froman, 28 Aug 1958, and Nelson Jarmie to Bradbury and Froman, 1 Dec 1958 (LASL-DO, 310–1: P-Div. History).
106. Handwritten notes, n.d. [ca. 1958] (LASL-DO, 310.1: P-Div. History).
107. GAC 60, 30–31 Oct and 1 Nov 1958, EHC.
108. Southern Regional Accelerator Committee, 27 Jul 1959, and Snell to V. E. Parker [SRAC member], 7 Dec 1959 (ORNL-DO, Southern Regional Accelerator).
109. Rosen, "Outline," and Livingston, "Origins."
110. Weinberg to McDaniel, 23 Aug 1962 (ORNL-DO, Electronuclear); Rosen, "Outline."
111. In 1965 the AEC's Division of Research divided the energy spectrum into three regions: low energy, below 50 MeV; medium energy, between 50 MeV and 1 BeV; and high energy, over 1 BeV. McDaniel to Weinberg, 25 Feb 1965 (ORNL-DO, Electronuclear).

112. *The Future Role,* 61.
113. AUI Ex Comm, 21 Sep 1956, and AUI Trustees, 17 Oct 1958.
114. AUI, Ex Comm, 18 Jan and 18 May 1951; Van Horn to Tape, 7 Jul 1952, Clarke Williams to C. R. Russell, 29 Sep 1952, and Tape to Van Horn, 4 Aug 1955 (BNL-DO, V-83).
115. AUI Ex Comm, 25 Feb 1955, and Board of Trustees, 21 Oct 1955; "Justification for an Accelerated LMFR Program," draft, ca. 25 Jul 1955, and Tape to Van Horn, 4 Aug 1955 (BNL-DO, V-83); *The Future Role,* 43; material for Lab Directors' meeting, May 1966 (GM, 5625/15).
116. Haworth, "The Future Role of Brookhaven National Laboratory," 5 Dec 1955 (BNL-DO, III-10).
117. Bradbury interview, 68–69.
118. Robert W. Seidel, "From Mars to Minerva: The Origins of Scientific Computing in the AEC Labs," *Physics Today* (Oct 1996), 33–39; N. Metropolis and E. C. Nelson, "Early Computing at Los Alamos," *Annals of the History of Computing,* 4 (Oct 1982), 348–357; Herman H. Goldstine, *The Computer from Pascal to von Neumann* (Princeton, 1972), 214–215, 225–226; AEC, *In the Matter of J. Robert Oppenheimer* (Cambridge, Mass., 1971), 654–656; Donald Mackenzie, "Nuclear Weapons Laboratories and the Development of Supercomputing," in *Knowing Machines: Essays on Technical Change* (Cambridge, Mass., 1996), 99–129.
119. Seidel, "Mars," 35–36; Rabi to Strauss, 9 Apr 1956, GAC 49; Goldstine, *Computer,* 332.
120. Tad Kishi, "The 701 at the Lawrence Livermore Laboratory," *Annals of the History of Computing,* 5 (Apr 1983), 206–210.
121. Goldstine, *Computer,* 329–330; Seidel, "Mars," 35.
122. AEC, *Seventh Semiannual Report* (Jan 1950), 86; Seidel, "Mars," 35; Holl, 122–125; Goldstine, *Computer,* 307n4; Creutz et al. to Larson, 8 Aug 1955 (ORNL-DO, Advisory Committees-Electronuclear); Johnson and Schaffer, 70–72. AVIDAC stood for Argonne's Version of the Institute [for Advanced Study]'s Digital Automatic Computer, ORACLE for Oak Ridge Automatic Computer Logical Engine.
123. Morse to Arnold N. Lowan, 3 Sep 1947 (BNL-DO, I-4); Haworth to Lowan, 22 Apr 1949 (BNL-DO, I-7); Haworth to L. Ridenour, 30 Nov 1951 (BNL-DO, V-16).
124. L. V. Berkner to files, 6 Jun 1952; Berkner to AUI Trustees, 9 Jun 1952; AUI computer conference, 11 Jul 1952; G. A. Kolstad to Haworth, 1 Aug 1952; C. F. Dunbar to Haworth, 14 Aug 1952; Haworth to T. H. Johnson, 30 Sep 1952; Johnson to Irving Kaplan, 12 Nov 1952; and Kolstad to Haworth, 21 Jan 1953 (BNL-DO, III-25); Tape to Van Horn, 26 Jun 1952 (BNL Central Records, DO-87); AUI, Ex Comm, 20 Jun, 17 Jul, and 16 Sep 1952.
125. Haworth to Kolstad, 2 Dec 1953, and Van Horn to Tape, 16 Mar 1954 (BNL-

DO, III-25); Haworth to Johnson, 31 Aug 1954 (BNL-DO, V-17); AUI Ex Comm, 16 Mar 1956; Haworth to Van Horn, 2 May 1956 (BNL-DO, I-14).

126. S. K. Allison, "Visit to BNL," 16 Sep 1957; H. B. Huntington to E. T. Booth (report of visiting committee), 1 Oct 1957; and report of the visiting committee of the Physics Dept., 1959 (BNL-DO, II-23).
127. ORNL Policy Committee, 16 Feb 1955 (ORNL-DO, Meetings-Policy Comm.).
128. *The Future Role;* GAC 68, 17–19 Mar 1960, EHC; Lawrence Radiation Laboratory, "Status Fiscal Year 1962," LBL Archives and Records; "Space Science at L.A.S.L.," Feb 1964 (LANL, A-83–0006, 8/3); Larry L. Deaven and Robert K. Moyzis, "The Los Alamos Center for Human Genome Studies," *Los Alamos Science,* 20 (1992), preface; Seidel, "Mars," 36.
129. Barton C. Hacker, *Elements of Controversy: The Atomic Energy Commission and Radiation Safety in Nuclear Weapons Testing, 1947–1974* (Berkeley, 1994), 161, 177–178.
130. Holl, 101; Albert H. Holland to C. N. Rucker, 22 Jul 1949 (ORNL-DO, Health Physics); BNL, Scientific Steering Committee, minutes, 4 Aug 1947 (BNL-DO, IV-19); Morse to L. R. Thiesmeyer, 18 Dec 1947 (BNL-DO, I-4).
131. AUI Ex Comm, 21 Nov 1952 and 20 Feb 1953; W. Kelley to John Bugher, 1 Dec 1952; Bugher to Kelley, 2 Feb 1953; and Maynard Smith to Walter Claus, 19 Feb 1953 (DBM, 21/BNL-1953).
132. ANL long-range program (quotes), submitted Mar 1959, in *The Future Role,* 111–112; ANL Policy Advisory Board, minutes, 13–14 Jul 1959 (ANL-PAB, box 2); AEC 956/9, 10 Nov 1959 (GM, 5672/13).
133. U.S. House, 83d Cong., 1st sess., *Second Independent Offices Appropriations for 1954,* 460; AEC, *Fourteenth Semiannual Report to Congress* (Jul 1953), 53–54.
134. Teller in *The Future Role;* Ralph Sanders, *Project Plowshare: The Development of the Peaceful Uses of Nuclear Explosions* (Washington, D.C., 1962), 109.
135. M. B. Rodin and D. C. Hess, "Weather Modification," Dec 1961, ANL-6444.
136. H. E. Landsberg, "Climate Made to Order," *BAS,* 17:9 (1961), 370–374; "The Weather Weapon: New Race with the Reds," *Newsweek,* 13 Jan 1958, 54–56; see also Spencer R. Weart, "Global Warming, Cold War, and the Evolution of Research Plans," *HSPS,* 27:2 (1997), 319–356, on 335; W. Henry Lambright and Stanley A. Changnon, Jr., "Arresting Technology: Government, Scientists, and Weather Modification," *Science, Technology, and Human Values,* 14:4 (1989), 340–359.
137. Teller, "We're Going to Work Miracles," *Popular Mechanics,* 113 (Mar 1960), 97–101, 278–282.
138. York to A. D. Starbird, 15 Oct 1956, and AEC 811/4, 26 Nov 1956 (Sec'y 51–58, MR&A-9–1: Non-military uses, EHC); Hewlett and Holl, 528–530.
139. Rabi to Strauss, 9 Apr 1956, GAC 49; GAC 57, 27–28 Feb and 1 Mar 1958,

EHC; W. Johnson to Strauss, 17 Mar 1958, GAC 57; GAC 58, 5–7 May 1958; W. Johnson to Strauss, 6 Jun 1958, GAC 58. MICE stood for Megaton Ice-Contained Explosion; the meaning of BATS is obscure. See also John McPhee, *The Curve of Binding Energy* (New York, 1974), 112–113; Ronald E. Doel, *Solar System Astronomy in America: Communities, Patronage, and Interdisciplinary Science, 1920–1960* (Cambridge, 1996), 177.

140. GAC 57, 27–28 Feb and 1 Mar 1958, EHC; W. Johnson to Strauss, 17 Mar 1958, GAC 57; GAC 58, 5–7 May 1958.
141. Hewlett and Holl, 544–545. The test finally proceeded, under Plowshare, as Project Gnome on 10 Dec 1961 in a salt dome in Carlsbad, New Mexico. It was unsuccessful but spectacular: radioactive steam broke containment and shot a 300-foot geyser from the access shaft. Hacker, *Elements*, 214–215.
142. GAC 60, 30–31 Oct and 1 Nov 1958, EHC.
143. Daniel Yergin, *The Prize: The Epic Quest for Oil, Money, and Power* (New York, 1991), 428–429, 694, 716–717; Report of Ad Hoc Plowshare Review Committee, 22 May 1972, GAC 120.
144. AEC 811/12, 6 Jun 1958 (Sec'y 51–58, MR&A-9–1: Non-military uses, EHC).
145. GAC 56, 21–23 Nov 1957, EHC.
146. AEC 811/12; Teller to A. D. Starbird, 15 Aug 1958, quoted in Dan O'Neill, "H-Bombs and Eskimos: The Story of Project Chariot," *Pacific Northwest Quarterly*, 85:1 (1994), 25–34, on 27. Teller and Livermore would revive the plan in 1969, this time for a harbor on Prudhoe to tap into North Slope oil. Scott Kirsch and Don Mitchell, "Earth-Moving as the 'Measure of Man': Edward Teller, Geographical Engineering, and the Matter of Progress," *Social Text*, 16:1 (1998), 101–134.
147. Frank Pace to Lilienthal, 23 Apr 1948; Lilienthal to Pace, 7 May 1948; Rowland Hughes to Strauss, 2 May 1955; Strauss to Hughes, 15 Jun 1955 (Sec'y 51–58, 88/PLB&L: Design and construction).
148. Wilson K. Talley and Carl R. Gerber, "Nuclear Explosives as an Engineering Tool," in Hans Mark and Lowell Wood, eds., *Energy in Physics, War, and Peace* (Boston, 1988), 221–236; Trevor Findlay, *Nuclear Dynamite: The Peaceful Nuclear Explosions Fiasco* (Sydney, 1990), 5; Hacker, *Elements*, 236–241; Hafstad to Seaborg, 4 Feb 1966, GAC 95; David R. Inglis and Carl L. Sandler, "Prospects and Problems: The Nonmilitary Uses of Nuclear Explosives," *BAS*, 23:10 (1967), 46–52. The Limited Test Ban Treaty of 1963 constrained Plowshare by requiring contained, underground tests; the AEC considered cratering tests as underground shots.
149. GAC 60, 30–31 Oct and 1 Nov 1958, EHC; Teller, "Program of Livermore."
150. Barton C. Hacker, "A Short History of the Laboratory at Livermore," Livermore, *Science and Technology Review* (Sep 1998), 12–20, 18; Harris and Hewlett, "Livermore," 11.

151. Frederick Reines, "Are There Peaceful Engineering Uses of Atomic Explosives?" *BAS,* 6:6 (1950), 171–172.
152. Frederick Reines, "The Peaceful Nuclear Explosion," *BAS,* 15:3 (1959), 118–122.
153. GAC 71, 24–26 Oct 1960.
154. D. J. Hughes to Samuel Goudsmit, 8 Feb 1960, G. N. Glasoe to Haworth, 25 Apr 1960, and Haworth to Van Horn, 9 May 1960 (BNL-HA, 4/Project Gnome); Gerald W. Johnson, "Nuclear Explosions in Science and Technology," *BAS,* 16:5 (1960), 155–161, on 160–161.
155. AEC meeting 1336, minutes excerpt, 21 Feb 1958 (Sec'y 51–58, 60/Space Task Force); W. Johnson to Strauss, 17 Mar 1958, GAC 57; Brian Balogh, *Chain Reaction: Expert Debate and Public Participation in American Commercial Nuclear Power, 1945–1975* (Cambridge, 1991), 174–176. The GAC in March 1958 had no members who had participated in the decision in 1947 to centralize reactor research, and only one member, Eger Murphree, had taken part in the criticism of a second weapons lab in 1951.
156. GAC 60, 30–31 Oct and 1 Nov 1958, EHC.
157. Herbert F. York, *Making Weapons, Talking Peace: A Physicist's Odyssey from Hiroshima to Geneva* (New York, 1987), 128–132, 148–150; Cochran et al., *Databook,* app. B.
158. Teller, "Program of Livermore."
159. "Subjects for Investigation in Connection with High Altitude Shots," n.d. (LASL-DO, 310.1: P-Div.).
160. Handwritten notes [ca. 1958] (LASL-DO, 310.1: P-Div.).
161. John E. Brolley, Jr., to Bradbury and Froman, 10 Mar 1958, and R. F. Taschek and W. T. Leland, 10 Apr 1958 (LASL-DO, 310.1: P-Div.).
162. H. H. Barschall to Kellogg, 27 Aug 1958, and Manley to Kellogg, 20 Nov 1958 (LASL-DO, 310.1: P-Div.).
163. "P-4 Activities since January 1958," 16 Mar 1959; Starbird to Teller, Bradbury, and Molmar, 13 Feb 1959, extract in Bradbury to Graves et al., 18 Feb 1959; R. B. Leachman to Kellogg, 7 May 1959; handwritten notes, 26 Aug 1959; and typed table of responsibility for various detectors (LASL-DO, 310.1: P-Div.).
164. Also known as Systems for Nuclear Auxiliary Power or Satellite Nuclear Auxiliary Power. Johnson and Schaffer, 118–121; Harold P. Green and Alan Rosenthal, *Government of the Atom: The Integration of Powers* (New York, 1963), 267n1.
165. O. E. Dwyer to Clarke Williams, 15 Oct 1959, 7 Apr 1960, and 12 May 1960 (BNL-DO, V-59).
166. H. J. Curtis, "Limitations on Space Flight Due to Cosmic Radiations," *Science,* 133 (1961), 312–316.
167. AEC General Manager to J. T. Ramey, 11 Jul 1962 (Sec'y 58–66, 1424/3); AEC 1088, 7 Sep 1961 (Sec'y 58–66, 1424/2).

168. R. S. Livingston to Weinberg, 24 Mar 1961 (ORNL-DO, Physics-General).
169. Snell, "The Oak Ridge Mc^2 Cyclotron Program," 12 Jan 1963 (ORNL-DO, Electronuclear); ORNL, "A Proposal for the Mc^2 Isochronous Cyclotron," 1 Nov 1963, ORNL-3540.
170. Pnina Abir-Am, "Themes, Genres and Orders of Legitimation in the Consolidation of New Scientific Disciplines: Deconstructing the Historiography of Molecular Biology," *History of Science,* 23 (1985), 73–117; Evelyn Fox Keller, "Physics and the Emergence of Molecular Biology: A History of Cognitive and Political Synergy," *Journal of the History of Biology,* 23:3 (1990), 389–409; Nicolas Rasmussen, "The Mid-Century Biophysics Bubble: Hiroshima and the Biological Revolution in America, Revisited," *History of Science,* 35 (1997), 245–293; Donald Fleming, "Emigré Physicists and the Biological Revolution," *Perspectives in American History,* 2 (1968), 152–189; Peter J. Westwick, "Abraded from Several Corners: Medical Physics and Biophysics at Berkeley," *HSPS,* 27:1 (1996), 131–162, on 150–154, 159–160.
171. Westwick, "Abraded," 150–154. "Biophysics" as used by lab scientists could refer to molecular biology, but the AEC defined it as "health physics" or radiological health protection.
172. Manley to Kellogg, 20 Nov 1958 (LASL-DO, 310.1: P-Div. History).
173. Bradbury to G. J. Keto, 12 Nov 1959 (Sec'y 58–66, 3374/16).
174. BNL Biology Department, "Long Range Plans," 27 Nov 1959 (DBM-DOE, 3374/16); "Laboratory Program Forecast" for Brookhaven and Oak Ridge [ca. 1959] (DBM-DOE, 3375/6).
175. AUI Ex Comm, 16–17 Jul 1959; D. E. Koshland, Jr., "A Program in Physics and Biology for Brookhaven National Lab," 1 Mar 1961 (BNL-HA, Haworth professional files); AUI Ex Comm, 13 Dec 1963.
176. AEC 855; Rabi to Strauss, 9 Apr 1956, GAC 49.
177. Alvin Weinberg, "Impact of Large-Scale Science on the United States," *Science,* 134 (1961), 161–164. Weinberg also criticized high-energy physics, "with its absence of practical applications and its very slight bearing on the rest of science" (164).
178. Johnson and Schaffer, 118–119.
179. GAC 69, 16–18 May 1960, EHC.

7. EXEMPLARY ADDITIONS

1. Peter J. Westwick, "Abraded from Several Corners: Medical Physics and Biophysics at Berkeley," *HSPS,* 27:1 (1996), 131–162, on 132–135.
2. J. Newell Stannard, *Radioactivity and Health: A History,* ed. Raymond W. Baalman (Springfield, Va., 1988); R. S. Stone, "General Introduction to Reports on Medicine, Health Physics, and Biology," in R. S. Stone, ed., *Industrial Medicine on the Plutonium Project* (New York, 1951), 1–16; Barton C. Hacker,

The Dragon's Tail: Radiation Safety in the Manhattan Project, 1942–1946 (Berkeley, 1987), 29–44; Hewlett and Anderson, 206–207; Holl, 10–12.

3. Hacker, *Dragon's Tail,* 52–54; Johnson and Schaffer, 21.
4. L. H. Hempelmann, "History of the Health Group (A-6)," 6 Apr 1946 (LANL A-84-019, 8/5); T. L. Shipman, "H Division Activities," 6 May 1969, LANL VFA-1741; Hacker, *Dragon's Tail,* 59–64.
5. S. Cantril and K. S. Cole, "Health Division Report for Month Ending September 15, 1942," quoted in Hacker, *Dragon's Tail,* 43–44; Stone to Hamilton, Cole, H. J. Curtis, K. Z. Morgan, et al., 14 Aug 1945 (EOL, 5/10).
6. Hamilton to J. W. Howland, 23 Apr 1946 (John H. Lawrence papers, TBL, 4/31).
7. Oppenheimer to Compton, 11 Feb 1944 ("not equipped"), HSPT-REL-94–879; Hempelmann to J. Kennedy and K. Bainbridge, 12 Apr 1944, HSPT-REL-94–175; Hempelmann to Oppenheimer, 16 Aug 1944, HSPT-REL-94–174; Oppenheimer to Hempelmann, 16 Aug 1944 ("urgent problems"), HSPT-REL-94–176; J. Kennedy et al. to Oppenheimer, 15 Mar 1945, HSPT-REL-94–187; Hempelmann to Oppenheimer, 26 Mar 1945, HSPT-REL-94–188; Oppenheimer to Col. S. L. Warren, 29 Mar 1945, HSPT-REL-94–189; Hacker, *Dragon's Tail,* 67–68.
8. Thomas L. Shipman, "H Division Activities," 6 May 1969, LANL VFA-1741; Stafford Warren, "Purpose and Limitations of the Biological and Health Physics Research Program," 30 Jul 1945 (EOL, 28/41).
9. Alice Kimball Smith, *A Peril and a Hope: The Scientists' Movement in America, 1945–47* (Chicago, 1965), app. A, 544–546.
10. Stone to Cole et al., 30 Jan 1945 (EOL, 28/41).
11. Stone, memo to files, 17 Aug 1945, in Hacker, *Dragon's Tail,* 51; Stone to Warren, 19 Jan 1946, ORO-38, and Stone to Warren, 23 Jan 1946 (EOL, 28/41).
12. H. J. Curtis and R. E. Zirkle to Stone, 10 May 1945, and P. S. Henshaw to Stone, 8 Jun 1945 (EOL, 29/36).
13. Stone to California Area Engineer, 12 Sep 1946 (EOL, 22/4).
14. Report of the Research Program Committee, 11 Jun 1945 (EOL, 29/36).
15. Hamilton to Stone, 31 Aug 1945, and to S. Allison, 11 Sep 1945 (EOL, 5/10).
16. Henshaw to Curtis, 17 Aug 1945, ES-208.
17. Paul McDaniel to T. S. Chapman, 17 Jul 1946 (EOL, 28/25).
18. E. O. Meals to District Engineer, 3 Sep 1946 (EOL, 28/25); Hamilton to Col. E. B. Kelly, 28 Aug 1946 (EOL, 28/25).
19. Groves to District Engineer, Oak Ridge, 21 Nov 1946, ORO-125; MED Medical Committee, minutes, 6 Dec 1946 (Sec'y 47–51, 26/Medical Committee).
20. Warren to Col. K. D. Fleming, "Recommendations of Medical Advisory Committee," 9 Sep 1946, HSPT-REL-94–557.
21. W. H. Zinn to Col. A. H. Frye, Jr., 31 Dec 1946, ORO-895; Hempelmann,

"Medical Research Program of Los Alamos Scientific Laboratory," n.d. [ca. late 1946] (LANL A-84-019, 9/2); Henshaw, "Proposed Training Program in Radiobiology," 8 Oct 1946, ORO-1482.

22. E. E. Kirkpatrick to Berkeley Area Engineer, 15 Oct 1946 (EOL, 28/25); MED Medical Advisory Committee, minutes, 6 Sep 1946 (Sec'y 47–51, 26/Medical Committee).
23. Interim Medical Advisory Committee, 23–24 Jan 1947 (Sec'y 47–51, 32/Interim Medical Advisory Committee).
24. Stafford Warren, report of 23–24 Jan 1947 meeting of Interim Medical Committee (Sec'y 47–51, 32/Interim Medical Advisory Committee).
25. Stafford Warren to Carroll Wilson, 7 Apr 1947, and Wilson to Warren, 30 Apr 1947, HSPT-REL-94–559 and 560; Report of the Medical Board of Review, 20 Jun 1947 (Sec'y 47–51, 64/Biol. and Med. Sciences).
26. Notes of conference at Oak Ridge, 1 Oct 1947 (UCh-VPSP, 32/5).
27. Atomic Energy Act, sec. 3(a)(1)(c) and (e), in Hewlett and Anderson, app. 1.
28. AEC, *Second Semiannual Report* (Jul 1947), 12.
29. Notes of Oak Ridge conference.
30. Report of Medical Board of Review.
31. ACBM 1, 12 Sep 1947, and ACBM 2, 11 Oct 1947 (GM, 5664/2); Wilson to AEC, 27 Jan 1948 (RRC-UT, MS-1067, folder 2); ACBM 12, 8–9 Oct 1948.
32. AEC, *Third Semiannual Report* (Feb 1948), 16.
33. AEC staff study of AEC advisory boards, 15 Mar 1955 (Sec'y 51–58, 66/O&M-7: General).
34. Alan Gregg to Wilson, 21 Dec 1948 (Sec'y 47–51, 31/ACBM sec. 2); E. W. Goodpasture to Lilienthal, 12 Oct 1949 (Sec'y 47–51, 31/ACBM sec. 3).
35. ACBM 2, 11 Oct 1947 (GM, 5664/2).
36. Zinn to Frye, 31 Dec 1946, ORO-895.
37. "Research Program, Contract W-7405-eng-48, 1947–1948," ORO-302; Stafford Warren to Wilson, 7 Apr 1947, HSPT-REL-94–559; UCRL 1947–48 budget summary, 19 Mar 1947, ORO-1397; M. E. Day to R. A. Nelson, 3 Jun 1948 (DC, 5/Personnel: Professional).
38. Morgan, "X-10 Health Physics Program for Fiscal Year 1947," 19 Sep 1946, ORO-1478; Forrest Western to C. N. Rucker, 3 Feb 1948 (ORNL-DO, Health Physics).
39. A. Hollaender to Barnett Cohen, 8 Nov 1946 (RRC-UT, MS-652, 10/53).
40. Weinberg, 67; Waldo Cohn, interview by Leland Johnson, 11 Jul 1991 (ORNL History Project, box 1).
41. Hollaender, "Outline of Research of Biology Division," 1 Dec 1946, ORO-295; MED Medical Committee, minutes, 6 Dec 1946 (Sec'y 47–51, 26/Medical Committee); Hollaender to J. H. Lum, 18 Feb 1947 (RRC-UT, MS-652, 9/4).
42. Hollaender to Shields Warren, 4 Feb 1948, ORNL/CF-48-2-346.

43. Hollaender to C. E. Larson, 16 Mar 1950 (ORNL-DO, Budget-Divisional).
44. Committee report quoted in Shipman, "H Division Activities," 20–21.
45. Hempelmann, "Medical Research Program," n.d. [ca. late 1946–early 1947] (LANL A-84-019, 9/2).
46. Shipman, "Health, Safety, and Biomedical Research at Los Alamos," 1948, HSPT-REL-94–1560.
47. "H-Division Annual Report of Research Activities," group H-4, 1 Dec 1947–1 Dec 1948, HSPT-REL-94–255.
48. AEC 228, minutes, 4 Jan 1949; Gregg to Wilson, 21 Dec 1948; and Lilienthal to Gregg, 11 Jan 1949 (Sec'y 47–51, 31/ACBM sec. 2); ACBM 13, 10–11 Dec 1948.
49. Lee DuBridge to K. T. Compton, 13 May 1946 (BNL-HA, Planning Committee folder); E. C. Shoup to F. D. Fackenthal, 14 Jul 1949 (BNL-DO, II-5).
50. BNL, "Initial Program Report," ca. Jan 1947 (BNL-HA).
51. "Report to the Program and Policy Committee," 23 Jul 1947 (DBM, 20/MH&S: BNL).
52. ACBM 7, 13 Mar 1948; ACBM 8, 23–24 Apr 1948; Shields Warren (quote) to E. Reynolds, 24 Mar 1948 (DBM, 20/BNL).
53. "Report of Conference of the Medical Advisory Board, BNL," 16 Jul 1947 (BNL-DO, II-12).
54. Morse to files, 29 Nov 1946 (BNL-DO, I-1).
55. AUI, Ex Comm, 16 Aug 1947; BNL, Policy and Program Committee, minutes, 22 Aug 1947 (BNL-DO, IV-16).
56. C. P. Rhoads to R. D. Conrad, 11 Oct 1948, and Conrad, "Radiobiological Institute," discussion notes, 20 Sep 1948 (BNL-DO, V-49).
57. Stuart M. Feffer, "Atoms, Cancer, and Politics: Supporting Atomic Science at the University of Chicago, 1944–1950," *HSPS,* 22:2 (1992), 233–261; Holl, 75.
58. Reynolds to Wilson, in AUI, Board of Trustees, 20 Sep 1948; AUI, Board of Trustees, 25 Oct 1947.
59. Crease, 62–66.
60. Reynolds to Shields Warren, 29 Mar 1948, and Warren to Morse, 24 Mar 1948 (DBM, 20/MH&S: BNL).
61. D. D. Van Slyke to Haworth, 26 Aug 1948 (BNL-DO, V-49).
62. Haworth, in AUI, Ex Comm, 17 Sep 1948; Lee Farr, "Proposals for Implementation," Oct 1948 (BNL-DO, V-49).
63. AUI, Board of Trustees, 14 Oct 1948; Van Slyke to Shields Warren, 10 Dec 1948 (BNL-DO, V-52).
64. Shields Warren, comments on Brookhaven, 29 Aug 1952 (DBM, 19/MH&S: BNL); Paul Pearson to files, 5 Jul 1951 (DBM, 18/MH&S-21: BNL).
65. Austin Brues to Pearson, 9 Dec 1949 (DBM, 18/MH&S-21: ANL 1950).

66. AEC, *Sixth Semiannual Report to Congress* (July 1949), 23–29, 40–46, 77, 101–109.
67. Paul Boyer, *By the Bomb's Early Light: American Thought and Culture at the Dawn of the Atomic Age* (New York, 1985), 107–121; Feffer, "Atoms"; Nicolas Rasmussen, "The Mid-Century Biophysics Bubble: Hiroshima and the Biological Revolution in America, Revisited," *History of Science,* 35 (1997), 245–293.
68. JCAE, 81st Cong., 1st sess., *Investigation into the United States Atomic Energy Project* (1949), 876–887.
69. Lilienthal, in JCAE, 81st Cong., 1st sess., *Atomic Energy Report to Congress* (Feb 1949), 13; Strauss, in U.S. House, 83d Cong., 2d sess., *Independent Offices Appropriations for 1955,* 2528–29.
70. U.S. House, 81st Cong., 1st sess., *Independent Offices Appropriation Bill for 1950,* 1198.
71. ACBM 5, draft minutes, 9 Jan 1948 (GM, 5664/2).
72. U.S. House, 83d Cong., 1st sess., *Second Independent Offices Appropriations for 1954,* 378–379; U.S. Senate, 83d Cong., 1st sess., *Second Independent Offices Appropriations for 1954,* 10–11.
73. Warren testimony in U.S. House, 82d Cong., 1st sess., *Independent Offices Appropriation for 1952,* 829.
74. "Atomic Medicine," *Time,* 7 Apr 1952, 64–66; John Lear, "Atomic Miracle," *Collier's,* 127 (21 Apr 1951), 15–17.
75. James T. Patterson, *The Dread Disease: Cancer and Modern American Culture* (Cambridge, Mass., 1987), 181.
76. U.S. House, 80th Cong., 1st sess., *Independent Offices Appropriation Bill for 1948,* 1538–49; Rep. Dirksen, 1539, 1540; Rep. Coudert, 1544; Rep. Andrews, 1542.
77. U.S. Senate, 80th Cong., 1st sess., *Independent Offices Appropriation Bill for 1948,* 52–53, 56–58.
78. DBM, report to GAC, 22 Nov 1947; Lilienthal to Hickenlooper, 15 Dec 1947 and 15 Jan 1948; AEC 26, 21 Jan 1948; John Z. Bowers to GAC, 29 Jan 1948; and Bowers to Strauss, 14 Jun 1948 (DBM, 16/MH&S-18: Cancer).
79. T. K. Glennan to M. Boyer, 21 Aug 1951, and T. Murray to Boyer, 23 Aug 1951 (DBM, 16/MH&S-18: Cancer).
80. ACBM 29, 27–29 Oct 1951, and Henry Smyth to Goodpasture, 28 Dec 1951, in ACBM 29; Rep. Charles Jonas in U.S. House, 83d Cong., 1st sess., *Second Independent Offices Appropriations for 1954,* 477.
81. Westwick, "Abraded," 140–141.
82. Report of the Medical Department Visiting Committee, 20–21 Sep 1951 (BNL-DO, II-21).
83. Weinberg to C. N. Rucker, 16 Jun 1948, ORNL/CF-48-6-196.

84. ANL, financial and operating statement, month of Jan 1950 (DBM, box 17).
85. UCRL, budget report FY1950 (DC, 3/Rad Lab: Budget); A. Tammaro to Pitzer, 4 Jan 51 (DC, 11/Finance-Admin.: Budget Biol. and Med.); W. B. Reynolds to L. W. Tuttle, 5 Jun 1950 (DBM, 22/Berkeley).
86. P. Sandidge, "Program for Fiscal Years 1948, 1949, 1950," 20 Sep 1947, ORNL/CF-47-10-28.
87. Form AEC-189s in "ORNL Research and Development Program from Fiscal Year 1954," ORNL/CF-52-4-50.
88. Form AEC-189s in "Program 6000, Biology & Medicine, U.C. Radiation Laboratory" (DBM, 22/Berkeley 1953).
89. Form AEC-189s in BNL Central Records, box MED-39, folders 1951, 1952, 1953.
90. John Gofman to Reynolds, 7 Jan 1950 (DC, 8/Gofman's program); Gofman to Warren, 21 Jan 1950, and John Derry to A. Saxe, 13 Feb 1950 (DBM, 22/Berkeley, Jan–Mar 1950); J. S. Robertson to Jones, 7 Aug 1950; Jones to Robertson, 14 Aug 1950; and Gofman to Robertson, 1 and 5 Sep 1950 (BNL Central Records, MED-12/BNL corres.).
91. J. Lawrence to Warren, 5 Apr 1950; C. L. Dunham to Lawrence, 11 Apr 1950; transcript of phone conversation, Gofman with C. J. Van Slyke, 4 May 1950; and Lawrence to Van Slyke, 4 May 1950 (DBM, 22/Berkeley).
92. C. E. Andressen to Reynolds, 6 Nov 1950 (DC, 7/Finance-Admin.: Budget, Internal Correspondence).
93. E. O. Lawrence to William Borden, 9 Aug 1950 (EOL, 32/26).
94. U.S. House, 82d Cong., 1st sess., *Independent Offices Appropriation Bill for 1952,* 833.
95. Gofman, interview by Sally Smith Hughes, 1985, TBL, 98.
96. Spencer R. Weart, "The Solid Community," in Lillian Hoddeson et al., eds., *Out of the Crystal Maze: Chapters from the History of Solid-State Physics* (New York, 1992), 617–669; Bernadette Bensaude-Vincent, "The Construction of a Discipline: Materials Science in the United States," *HSPS,* 31:2 (2001), 223–248.
97. GAC 24, 4–6 Jan 1951; Oppenheimer to Dean, 11 May 1951, in GAC 26.
98. GAC 44, 2–4 Mar 1955, EHC; Rabi to Strauss, 2 Apr 1955 (Sec'y 51–58, R&D-7); GAC 49, 28–30 Mar 1956, EHC; GAC 50, 16–18 Jul 1956, EHC; GAC 51, 29–31 Oct 1956, EHC; W. Johnson to Strauss, 28 Aug 1957, GAC 55; AEC 17/145, 18 Dec 1957 (Sec'y 51–58, 51/MR&A-4: ANP); Johnson to Strauss, 20 Feb 1958 (Sec'y 51–58, 139/R&D-6: Particle accels); AEC 152/92, 29 May 1958 (Sec'y 51–58, 113/Reactor Development-1: Policy).
99. Donald K. Stevens, "Fundamental Materials Research in the AEC," briefing for AEC, 9 Apr 1959 (Sec'y 58–66, 1423/9); GAC 62, 9–11 Mar 1959, EHC.
100. L. W. Nordheim to E. J. Murphy, 3 May 1946, ORNL/CF-46-5-83; Sandidge,

"Program"; Sidney Siegel, "Clinton Laboratories Physics of Solids Research," 13 Nov 1947, ORNL/CF-47-11-361.

101. Bill Thompson, "ORNL—1951," 17 Mar 1950 (ORNL-DO, Program); Frye to Weinberg, 15 Aug 1950; Weinberg to ORNL staff, 1 Nov 1950; and Weinberg to staff, 11 Jan 1952 (AMW, Metallurgy); Mike Wilkinson, "The Solid State Division, Oak Ridge National Laboratory: A Brief History, 1952–1995," n.d. [ca. 1996], ORNL Archives.
102. R. W. Dodson et al. to Weinberg, 2 Nov 1956 (ORNL-DO, Adv. Comm.-Chemistry); R. F. Christy et al. to Weinberg, 25 Aug 1956 (ORNL-DO, Adv. Comm.-Physics); Weinberg, "State of the Laboratory—1957," 30 Dec 1957, ORNL/CF-57-12-127.
103. Solid State Division, research personnel, 1 Jan 1960–13 Jan 1964 (ORNL-DO, Solid State Div.); G. M. Slaughter, H. W. Hayden, and W. D. Manly, "A History of the Metals and Ceramics Division at Oak Ridge National Laboratory," *Advanced Materials & Processes* (Jan–Feb 1995), reprints in ORNL Archives.
104. BNL, "Initial Program Report," n.d. [ca. early 1947] (BNL-HA); T. H. Johnson to Morse, 6 Oct 1947 (BNL-DO, V-63); Dodson to Morse, 27 Jun 1947 (BNL-DO, V-14).
105. Crease, 152–164, 177–181; "Researches of the Physics Department," Oct 1950 (BNL-DO, V-63).
106. AUI, Ex Comm, 15 Feb 1952; J. C. Slater, interview by C. Weiner, 7 Aug 1970, AIP.
107. "Proposal for Annealing," 4 Sep 1953 (BNL-DO, V-71); Crease, 177–181.
108. AUI, Ex Comm, 15 Feb 1952; list of programs, Physics Department, 28 Jan 1955, and "Research of Interest to AEC Research Division—Physics," n.d. [ca. 1957] (BNL-DO, V-64); AUI, Trustees, 27 Oct 1961.
109. G. J. Dienes to S. A. Goudsmit, 8 Sep 1958 (BNL-DO, V-64).
110. G. Vineyard, "Solid State at BNL," Jan 1962 (BNL-HA, Physics Dept. 1966–71); "Report of the Visiting Committee for the Physics Dept. for 1962–63" (BNL-HA, Physics Dept. Visiting Committee).
111. P. W. Levy to L. Parsegian, 17 Feb 1954, and H. B. Huntington to Goudsmit, 1 Apr 1957 (BNL-DO, V-64); AUI, Trustees, 27 Oct 1961.
112. Goudsmit to Haworth, 18 Apr 1957 (BNL-DO, V-64); Haworth to R. F. Bacher, 26 Nov 1958 (BNL-DO, I-16); AUI, Ex Comm, 15 Apr 1960; author interview with Gen Shirane, BNL, 1 Nov 1996.
113. ANL, Chemistry Division organization, 1 Jan 1947, Apr 1947, Jul 1947, and 8 Dec 1948, Chemistry Division files, NARA-GL; Oliver Simpson, interview by Krzysztof Szymborski, 13 Aug 1982, AIP.
114. Jürgen Teichmann and Krzysztof Szymborski, "Point Defects and Ionic Crystals: Color Centers as the Key to Imperfections," in Hoddeson et al., *Crystal Maze*, 236–316, on 279–280.

115. ANL, Chemistry Division organization, 1 May and 1 Aug 1949, 1 Feb 1950, Chemistry Division files, NARA-GL; Simpson, interview by Szymborski, 13 Aug 1982, AIP; Charles Delbecq and Philip Yuster, joint interview by Szymborski, 24 Mar 1982, AIP.
116. See ANL Metallurgy Program files, NARA-GL.
117. ANL, Metallurgy and Solid State Review Committee, 6 Jan and 6 Jul 1959 (ANL-PAB, 10/Solid State-Metallurgy).
118. ANL, Policy Advisory Board, minutes, 21–22 Jan 1959 (ANL-PAB, box 2).
119. ANL Policy Advisory Board, minutes, 15 Apr 1959 and 13–14 Jul 1959 (ANL-PAB, box 2); Solid State and Metallurgy Review Committee, 7 Jul 1960 (ANL-PAB, 10/Solid State-Metallurgy).
120. ANL, Physics and Applied Math Review Committee to Kimpton, 2 Oct 1959 (ANL-PAB, 9/Physics).
121. Paul Leurgans, "Formation of a Solid State Group," n.d. [ca. 1955] (LASL-DO, 310.1: P-Div.).
122. Handwritten notes [ca. 1958?] (LASL-DO, 310.1: P-Div.).
123. David L. Hill, "Should Los Alamos Move Vigorously toward the Construction of a High Energy Accelerator?" 15 Feb 1957 (LANL A-91-011, 55/1); H. H. Barschall to J. M. B. Kellogg, 27 Aug 1958, and J. H. Manley to Kellogg, 20 Nov 1958 (LASL-DO, 310.1: P-Div. History). Hill put solid-state physics, in addition to accelerators, at the forefront.
124. Manley, "Physical Research at Los Alamos," June 1960 (LASL-DO, 310.1-P-Div. History).
125. Reynolds to T. H. Johnson, 21 Jan 1955, and R. L. Thornton, "Long Range objectives of the Radiation Laboratory," 30 Nov 1955 (EOL, 22/8); "Statement of Long Range Objectives, Berkeley Chemistry Division," 3 Nov 1955 (EOL, 24/13); UCRL, physical research program book, 1 Jun 1956 (EOL, 22/9).
126. Robert W. Seidel, "A Home for Big Science: The AEC's Laboratory System," *HSPS*, 16:1 (1986), 135–175, on 159; GAC 44, 2–4 Mar 1955, EHC.
127. GAC 80, 9–11 Jul 1962, EHC. High-energy physics had 96 Ph.D.s and 74 graduate students on the experimental side, 23 Ph.D.s and 25 graduate students on theory.
128. Haworth to Van Horn, 28 Feb 1950 and 29 Jun 1950 (BNL-DO, I-8).
129. Haworth to Bradbury, 29 May 1958 (BNL-DO, I-16); John Yarnell to Kellogg, "Solid State Physics in P-2," 12 Nov 1958 (LASL-DO, 130.1: P-Div. History).
130. Simpson to D. Nagle, 27 Oct 1961 (LANL A-91-011, 55/1); ANL Review Committee for Solid State and Metallurgy, 10–11 Apr 1962 (ANL-PAB, 10/Solid State-Metallurgy).
131. Seaborg, "The Value of the Interdisciplinary Laboratory," remarks at the dedication of the Inorganic Materials Research Laboratory, 16 Jul 1965 (author files).

132. T. Lauritsen to G. W. Beadle, 28 Jun 1963 (ANL-PAB, 9/Physics).

133. ANL Policy Advisory Board, minutes, 15 Apr 1959 (ANL-PAB, box 2).

134. ANL Policy Advisory Board, transcript minutes, 21–22 Jan 1959 (ANL-PAB, box 2).

135. John Willard to Winston Manning, 12 Feb 1959, and Fred Wall to Manning, 27 Feb 1959 (ANL-PAB, 7/Rev. Comm.-Chem.); H. R. Crane to L. R. Kimpton, 2 Oct 1959, and W. C. Parkinson to Beadle, 6 Jul 1961 (ANL-PAB, 9/Physics).

136. GAC 44, 2–4 Mar 1955, EHC; AEC 872, 17 Oct 1955, and AEC meeting 1158, minutes excerpt, 20 Dec 1955 (Sec'y 51–58, 66/O&M-7: Committees and Boards); GAC 49, 28–30 Mar 1956, and GAC 50, 16–18 Jul 1956, EHC; Rabi to Strauss, 9 Apr 1956, GAC 49; Rabi to Strauss, 30 Jul 1956, GAC 50.

137. Stevens to J. H. Williams, 11 Feb 1959, in AEC 102/33, 27 Feb 1959, and W. B. McCool to Williams, 12 Feb 1959 (Sec'y 58–66, 1423/7); AEC 851/8, 2 Apr 1959 (Sec'y 58–66, 1385/O&M-7: Committees and Boards, vol. 1); Federal Council for Science and Technology (FCST), Coordinating Committee on Materials Research and Development, 28 Apr 1959, in AEC 1023, and R. N. Kriedler to Williams, 11 May 1959 (Sec'y 58–66, 1404/3); K. A. Dunbar to Williams and F. K. Pittman, 22 Jun 1959 (ANL-DO, 1/AARR).

138. AEC minutes, 12 May 1959, AEC 1023, 18 Jun 1959, McCool to Williams, 23 Jun 1959, AEC 1023/2, 11 Aug 1959 (Sec'y 58–66, 1404/3); Ames Laboratory, "Special Report on the Ames Laboratory Materials Research Program," Nov 1960 (Sec'y 58–66, 1404/8).

139. BOB Military Division to Elmer Staats, 31 Mar 1960, and Hugh Loweth to Staats, 29 Jul 1959, 6 Aug 1959, and 4 Apr 1960 (BOB, 8/Research programs-FY60).

140. R. R. Coffin to files, 11 Apr 1960, J. W. Clark to Staats, 25 Apr 1960, and FCST, minutes, 26 Apr 1960 (BOB, 8/Research programs-FY60); AEC 1023/18, 24 Apr 1962 (Sec'y 58–66, 1404/5).

141. ORNL, physical research program for FY1958, ORNL/CF-56-3-4.

142. Author interview with Gen Shirane, 1 Nov 1996, BNL.

8. EPILOGUE

1. Shields Warren, quoting ACBM 12, to Carroll Wilson, 14 Oct 1948 (Sec'y 47–51, 31/ACBM); Fred Schlemmer to R. C. Muir, 17 Sep 1948, in ACBM 11.

2. Stu Hight, Feb 1954, in Necah Furman, *Sandia National Laboratories: The Postwar Decade* (Albuquerque, 1989), 638, 667–678.

3. GAC 60, 30–31 Oct and 1 Nov 1958, EHC.

4. GAC 69, 16–18 May 1960, EHC.

5. GAC 82, 7–9 Jan 1963, and GAC 84, 25–27 Apr 1963, both in EHC.

6. Thomas B. Cochran et al., *Nuclear Weapons Databook,* vol. 2, *U.S. Nuclear Warhead Production* (Cambridge, Mass., 1987), 19.
7. Sybil Francis, "Warhead Politics: Livermore and the Competitive System of Nuclear Weapon Design" (Ph.D. diss., MIT, 1996), 155–156.
8. Richard T. Sylves, *The Nuclear Oracles: A Political History of the General Advisory Committee of the Atomic Energy Commission, 1947–1977* (Ames, 1987), 266–268; Harold Orlans, *Contracting for Atoms* (Washington, D.C., 1967), 23–26; Michele Stenehjem Gerber, *On the Home Front: The Cold War Legacy of the Hanford Nuclear Site* (Lincoln, Neb., 1992), 140.
9. AEC, *Atomic Energy Commission Research and Development Laboratories: A National Resource,* TID-26400 (Sept 1973).
10. S. L. Fawcett to Weinberg, 4 Jan 1967 (AMW, National laboratories).
11. AEC, *Laboratories.*
12. GAC 85, 18–20 July 1963, EHC.
13. "A-Installation Demanded for Midwest," *Washington Post,* 21 Sep 1965, A4.
14. Joan Lisa Bromberg, *Fusion: Science, Politics, and the Invention of a New Energy Source* (Cambridge, Mass, 1982), 3, 140–142, 173–174.
15. Charles G. Manly, "Notes of Meeting with ANL Officials," 12 Feb 1968 (GM, 5624/12).
16. J. J. Liverman, remarks for Program Directors' meeting, 13 Feb 1973 (BNL-HA, 33/Laboratory Directors' meetings).
17. Robert W. Miller, *Schedule, Cost, and Profit Control with PERT* (New York, 1963), 3–4, 28; Harry F. Evarts, *Introduction to PERT* (Boston, 1964), 1.
18. AEC, *Annual Report for 1962* (Washington, D.C., Jan 1963), 87–88.
19. Miller, *Control with PERT,* 7–10, 12.
20. John M. Jordan, *Machine-Age Ideology: Social Engineering and American Liberalism, 1911–1939* (Chapel Hill, 1994); David F. Noble, *America by Design: Science, Technology, and the Rise of Corporate Capitalism* (Oxford, 1977), 257–320. See also Stephen P. Waring, *Taylorism Transformed: Scientific Management Theory since 1945* (Chapel Hill, 1991); and Robert Lilienfeld, *The Rise of Systems Theory* (New York, 1978).
21. Herbert A. Simon, *The New Science of Management Decision* (New York, 1960); Thomas P. Hughes, *Rescuing Prometheus* (New York, 2000), 141–166; Walter A. McDougall, *The Heavens and the Earth: A Political History of the Space Age* (New York, 1985), 307.
22. See also Peter Galison, *Image and Logic* (Chicago, 1997), 606–609.
23. L. C. Teng, "Report of Progress on the Zero Gradient Synchrotron," 16 Oct 1962 (ANL-PAB, 9/PAB and HEP); Donald Hagerman, "Interim Report on the Construction of the LAMPF" (LANL A-91–011, 69/13); M. Stanley Livingston, "Origins and History of the Los Alamos Meson Physics Facility," LA-5000 (June 1972), 37–41.

24. Bradbury to lab directors, 27 May 1963 (AMW, Lab Directors' Meetings); Bradbury to lab directors, 4 Mar 1970, and Bradbury to Goldhaber, 20 Jan 1967 (BNL-HA, 33/Lab Directors' Meetings).
25. George Vineyard to R. S. Hansen, 9 May 1973, and Weinberg to R. B. Duffield, 5 Apr 1972 (AMW, Lab Directors' Meetings).
26. Sylves, *Oracles,* 48–49, 90.
27. U.S. Census Bureau, *Statistical Abstracts of the United States* (1999), 882.
28. Johnson and Schaffer, 126.
29. Johnson and Schaffer, 129; Albert H. Teich and Mark E. Rushefsky, "Diversification at Argonne National Laboratory," unpublished manuscript (May 1976), 9; AEC *Laboratories,* 5. I thank Albert Teich for a copy of his manuscript and permission to cite it.
30. AEC 102/39, 4 Mar 1964 (Sec'y 58–66, 1423/8).
31. Daniel J. Kevles, *The Physicists: The History of a Scientific Community in Modern America* (New York, 1978), 393–409; Spencer R. Weart, *Nuclear Fear: A History of Images* (Cambridge, Mass., 1988), 309–374; Brian Balogh, *Chain Reaction: Expert Debate and Public Participation in American Commercial Nuclear Power, 1945–1975* (Cambridge, 1991), 221–301; Thomas Raymond Wellock, *Critical Masses: Opposition to Nuclear Power in California, 1958–1978* (Madison, 1998).
32. Johnson and Schaffer, 126.
33. Barton C. Hacker, "A Short History of the Laboratory at Livermore," Livermore, *Science and Technology Review* (Sep 1998), 12–20, on 18.
34. Crease, 358.
35. Holl, 271, 309.
36. Vineyard to National Science Board, 1976, quoted in Teich, "Bureaucracy and Politics in Big Science: Relations between Headquarters and the National Laboratories in AEC and ERDA," in U.S. House, 95th Cong., 2d sess., *The Role of the National Energy Laboratories in ERDA and Department of Energy Operations,* 380.
37. Darol Froman to Alfred D. Starbird, 4 May 1959, in Francis, "Warhead Politics," 153.
38. John Lewis Gaddis, *Strategies of Containment: A Critical Appraisal of Postwar American National Security Policy* (Oxford, 1982), 198–236.
39. Thomas Cochran et al., *Nuclear Weapons Databook,* vol. 1., *U.S. Nuclear Forces and Capabilities* (Cambridge, Mass., 1984), 8–13; Francis, "Warhead Politics," fig. 2. For spending, see fig. 1.1 in Cochran et al., *Databook* 2:3. Budgets would not approach the levels of around 1960, in constant dollars, until the late Reagan years.
40. GAC 125, minutes, 9–11 June 1973.
41. L. Hafstad to Seaborg, 13 Jan 1965, GAC 91.

42. Joan Lisa Bromberg, *The Laser in America, 1950–1970* (Cambridge, Mass., 1991), 214–218, 238–243; Ruth R. Harris and Richard G. Hewlett, "The Lawrence Livermore National Laboratory: The Evolution of Its Mission, 1952–1988," report prepared for Livermore, 21 Mar 1990, 17–18. For laser isotope separation Livermore used atomic vapor while Los Alamos used molecular uranium hexafluoride.
43. AEC, *Annual Report to Congress for 1964*, 74–77.
44. AEC, *Laboratories*, 17, 39.
45. AEC, *Annual Report for 1961*, 25–29, and AEC, *Annual Report for 1964*, 90.
46. Holl, 265–280.
47. Holl, 270.
48. AEC 99/39, 25 Jun 1965 (Sec'y 58–66, 1402/10); Division of Reactor Development and Technology, evaluation of Los Alamos, 1967 (GM, 5627/20).
49. Holl, 234, 244.
50. Johnson and Schaffer, 135–138; Weinberg, 125–131; Holl, 333, 340, 354–355.
51. AEC, Division of Research, discussion paper on Ramsey panel report, draft 14 May 1963 (HEP 57–64, 3/GAC-PSAC:1–5/63).
52. Catherine Westfall, "The First 'Truly National Laboratory': The Birth of Fermilab" (Ph.D. diss., Michigan State University, 1988). Berkeley's conservative design of the 200-BeV device perhaps reflected its position as the senior accelerator lab, similar to the role of Los Alamos for weapons and Argonne for reactors.
53. Crease, 365–367.
54. Alexander Zucker and Arthur H. Snell, "Meson Factories?" n.d [ca. late 1962] (HEP 57–64, 3/GAC-PSAC: 1–5/63); L. Rosen, "Nuclear Physics Applications of a High-Flux Meson Facility," 2 Dec 1963 (Los Alamos report LAMS-3030).
55. ORNL, "A Proposal for the Mc^2 Isochronous Cyclotron," 1 Nov 1963 (ORNL-3540).
56. Rosen to J. M. B. Kellogg, 16 May 1962 (LANL, A-91-011, 204/4); Rosen, "Outline of the Proposal for a Meson Facility at Los Alamos," 16 Jul 1963 (Los Alamos report LAMS-2935).
57. Snell to Edward Reynolds, 9 Apr 1964 and 5 May 1964, Leon Lederman to Reynolds, 14 Apr 1964, and Reynolds to Snell, 27 and 28 Apr 1964 (ORNL-DO, Physics-General); High Energy Study Group, report to AUI Trustees, 2 Jan 1964 (LANL, A-91-011, 57/4); AUI Trustees, 18 Oct 1963, 17 Jan and 17 Apr 1964; AUI Ex Comm, 20 Mar 1964.
58. Midwestern Universities Research Association, "The MURA 10-BeV FFAG Accelerator as a Pion Factory," 22 Mar 1963 (HEP 57–64, 5/MURA-1963).
59. Weinberg to P. W. McDaniel, 1 June 1964; Bradbury to McDaniel, 24 June 1964; Weinberg to McDaniel, 27 Aug 1964 (ORNL-DO, Physics-General).

60. D. Nagle, G. Wheeler, J. P. Blewett, and L. Smith to G. Kolstad, 3 Dec 1963; Kolstad to Nagle et al., 6 Jan 1964; Nagle to Kolstad, 10 Feb 1964; Blewett to Kolstad, 6 Apr 1964; McDaniel to Blewett, 22 Apr 1964; minutes of Linac Coordinating Committee, 7–8 May 1964 and 17 Jun 1964 (LANL, A-91-011, 57/4).
61. "Meson Factories," report of Bethe panel, Mar 1964 (LANL, A-91-011, 57/5); Livingston, "Origins of LAMPF."
62. Wright H. Langham and David E. Groce, "A Proposal for a Biomedical Addition to the Los Alamos Scientific Laboratory's High-Flux Meson Physics Facility," July 1970 (Los Alamos report LA-4490P).
63. "The Los Alamos Meson Facility Proposal," n.d. [ca. 1963] (LANL, A-91-011, 204/4); Rosen, presentation for Rep. Craig Hosmer, 20 Jan 1967 (ibid., 141/4); Lawrence Cranberg, "The Utilization of Neutrons from a Proton Linac Beam Dump," 24 May 1968, and D. R. F. Cochran to Cranberg, 26 Jun 1968 (ibid., 55/4); L. Agnew, summary of telephone conversation with F. Tesche, 25 Aug 1969 (ibid., 133/5); Ralph R. Fullwood, "Uses for the Weapons Neutron Research Facility," app. 2 to Rosen, statement to JCAE, 3 Mar 1970 (ibid., 141/5).
64. "Meson Facility Proposal."
65. Rosen et al., "Proposed Organization for the Management of LAMPF as a National Facility," 10 Aug 1966; G. A. Kolstad to Rosen, 10 Aug 1966; V. Telegdi, "Comments on Kolstad's Memorandum," 14 Sep 1966 (LANL, A-91–011, 57/4). SLAC, another accelerator supposedly for outside users, gave almost 60 percent of its beam time to Stanford researchers (GAC 125, 9–11 Jul 1973).
66. Weinberg to McDaniel, 14 Jan 1969 (ORNL-DO, APACHE-1969); ANL, "Midwest Tandem Cyclotron: A Proposal for a Regional Accelerator Facility," June 1969 (ANL-7582).
67. "Summary of Chapter 3 of the Apache Proposal (Plus 2 Remarks about ORNL and ANL)," n.d. (ORNL-DO, APACHE); ANL, FY1971 budget (ANL-DO, Program Budgets).
68. Snell to J. L. Fowler et al., "Design Optimization for a Medical Heavy-Ion Accelerator," 7 July 1971 (ORNL-DO, Neutron Physics).
69. R. F. Taschek to H. M. Agnew, 3 Apr 1972; Snell, handwritten memo to Weinberg [quote], 30 Jun 1972; Snell to Weinberg, 17 Jul 1972; Weinberg to Daniel Miller, 20 Jul 1972; Taschek to Miller, 7 Aug 1972; Snell to Weinberg, 22 Jul 1972; Snell to Miller, 5 Sep 1972 (ORNL-DO, National Heavy Ion Lab).
70. E. G. Pewitt (Argonne) and McDaniel in Notes of Coordination Meeting, 28 Apr 1965 (BNL-HA, 31/14′ bubble chamber).
71. R. R. Wilson to Goldhaber, 19 Dec 1968; E. Goldwasser to Goldhaber, 31 July 1970; Goldwasser to Shutt, 26 Aug 1979; and Wilson to Shutt, 28 Aug 1970 (BNL-HA, 6/Director's Office-NAL).

72. AUI Trustees, Committee on BNL, 1 Apr 1969.
73. Richard Wilson to Goldhaber, 13 Jul 1970; Norman Ramsey to V. Weisskopf, 6 Aug 1970 (BNL-HA, 33/High energy research policy).
74. AUI Ex Comm, 16 Apr 1970; "BNL Policy for High Energy Research at Other Laboratories," 1970 (BNL-HA, 33/High energy research policy).
75. AEC 102/39, 4 Mar 1964; S. G. English to AEC staff and field offices [quotes], 28 Apr 1964; English, memo for Commissioners, 13 Jun 1964; AEC 102/41, 26 Oct 1964 (Sec'y 58–66, 1423/8).
76. Weart, *Nuclear Fear,* 325; Rachel Carson, *Silent Spring* (Boston, 1962), 37.
77. Weinberg, 150.
78. Albert H. Teich and W. Henry Lambright, "The Redirection of a Large National Laboratory," *Minerva,* 14 (1976–77), 447–474, on 463.
79. Ibid., 464–465.
80. Stanley I. Auerbach, "A History of the Environmental Sciences Division of Oak Ridge National Laboratory," n.d. (ORNL/M-2732); G. D. Kerr et al., "A Brief History of the Health and Safety Research Division at Oak Ridge National Laboratory," July 1992 (ORNL/M-2108).
81. Teich and Rushefsky, "Diversification," 11.
82. Ibid.; Holl, 246–257.
83. AUI Trustees, 20–21 Oct 1966 and 20 Apr 1967.
84. GAC 125, 9–11 Jul 1973; Heilbron, Seidel, and Wheaton, 104.
85. AEC, *Laboratories,* 20; Harris and Hewlett, "Livermore," 15.
86. David J. Rose, "New Laboratories for Old," *Daedalus,* 103:3 (1974), 143–155, on 145.
87. Senate bill S. 3410, in Teich and Lambright, "Redirection," 464.
88. JCAE, *Authorizing Appropriations for the Atomic Energy Commission for Fiscal Year 1971,* quoted in Rose, "New Laboratories," 145–156.
89. Rose, "New Laboratories," 146. The bill for environmental labs passed the Senate but not the House; President Nixon had already threatened to veto it.
90. Holl, 282–284; Daniel Yergin, *The Prize: The Epic Quest for Oil, Money, and Power* (New York, 1991), 588–632.
91. AEC, *Annual Report to Congress* (1972), 3.
92. Teich and Rushefsky, "Diversification," 35–37; Holl, 287–288.
93. Ralph G. Scurlock, ed., *History and Origins of Cryogenics* (Oxford, 1992), esp. Scurlock, "Introduction," 1–47, on 39–44, and F. G. Brickwedde, E. F. Hammel, and W. E. Keller, "The History of Cryogenics in the U.S.A., Part I—Cryoengineering," 357–468, on 400–410; G. Seaborg to John Conway, 9 Feb 66 (Sec'y 58–66, 1424/5).
94. E. B. Forsyth, ed., "Report on Superconducting Electrical Power Transmission Studies," Dec 1971 (BNL 16339), BNL and AUI, "Superconducting Electrical

Power Transmission System," proposal to NSF, 1 May and 1 Oct 1972 (BNL-HA, Haworth files, box 8); AEC, *Laboratories,* 13, 40, 43.

95. Holl, 476–480.
96. Teich and Rushefsky, "Diversification," 38–39; Holl, 312–314.
97. AEC, *Laboratories,* 26.
98. Teich and Rushefsky, "Diversification," 33–35; Holl, 288–289.
99. J. T. Ramey to John Pastore, 23 Sep 1964, and Roger Revelle to Donald Hornig, 23 Mar 1964, in JCAE, 88th Cong., 2d sess., *Use of Nuclear Power for the Production of Fresh Water from Salt Water,* on 10–11, 39; AEC, *Annual Report to Congress for 1964,* 104–106; Teich and Lambright, "Redirection," 454–457.
100. G. W. DePuy and L. E. Kukacka, eds., "Concrete Polymer Materials," Dec 1973 (BNL 50390); AUI and BNL, "Work for Other Federal Agencies," 1971 (BNL-HA, Haworth files, 10/"Financial support").
101. Weinberg, 144–147; "Water to Cool the Middle East," *Life,* 18 Aug 1967, 4.
102. R. F. Hibbs to Manson Benedict, 27 May 1974 (ORNL-DO, A0.6.3-Advisory Council).
103. English, memo for Commissioners.
104. H. G. Vesper to J. Schlesinger, 24 Aug 1972, GAC 121.
105. Kevles, *Physicists,* 411.
106. AUI Ex Comm, 18 Sep 1969.
107. Crease, 350.
108. Teich and Rushefsky, "Diversification," 2, 42–44.
109. AUI, "Summary of Financial Support from Other Organizations," 30 Jun 1970, and "Work for Other Federal Agencies," 1971 (BNL-HA, Haworth files, 10/Financial support).
110. Elmer Staats to Chairmen, House and Senate Committees on Appropriations and Armed Services, 29 Sep 1976 (JPL 150, 1/10, Jet Propulsion Laboratory Archives, Pasadena).

CONCLUSION

1. Alfred D. Chandler, *Strategy and Structure: Chapters in the History of the Industrial Enterprise* (Cambridge, Mass., 1962), 13–17.
2. Paul Forman, "Behind Quantum Electronics: National Security as Basis for Physical Research," *HSPS,* 18:1 (1987), 149–229, on 229, 153–154. See also Stuart W. Leslie, *The Cold War and American Science: The Military-Industrial Complex at MIT and Stanford* (New York, 1993).
3. Daniel J. Kevles, "Cold War and Hot Physics: Science, Security, and the American State, 1945–1956," *HSPS,* 20:2 (1990), 239–264, on 262.

4. Forman, "Behind Quantum Electronics," 219.
5. Raemer Schreiber, "What Happened to LASL?" draft, Nov 1991, LANL VFA-1240; Louis Rosen to Darol Froman, 15 May 1958 (LASL-DO, 310.1: P-Div.).
6. GAC, 3–4 Jan 1947, quoted in Robert W. Seidel, "A Home for Big Science: The AEC's Laboratory System," *HSPS,* 16:1 (1986), 135–175, on 142.
7. Richard T. Sylves, *The Nuclear Oracles: A Political History of the General Advisory Committee of the Atomic Energy Commission, 1947–1977* (Ames, 1987), 226.
8. Ibid., 228 and 290n33. Warren Johnson of Chicago, a partisan of Argonne, participated in the same accelerator discussion after admitting his lack of objectivity.
9. Warren Johnson to John McCone, 15 May 1959 (KP, 11/GAC May 1959), and McCone to Johnson, 10 Jul 1959 (KP, 11/GAC July 1959).
10. Alvin Weinberg, "Oak Ridge National Laboratory," *Science,* 109 (11 Mar 1949), 245, 248.
11. Franz Simon in 1932, quoted in J. L. Heilbron and Robert W. Seidel, *Lawrence and His Laboratory: A History of the Lawrence Berkeley Laboratory,* vol. 1 (Berkeley, 1989), 36; David E. Nye, *American Technological Sublime* (Cambridge, Mass., 1994).
12. Du Pont, for example, in 1951 spent about $72 million on R&D and employed 3,376 professional R&D staff at several sites; more than half of the staff, however, focused on production and sales support as opposed to research. David A. Hounshell and John Kenly Smith, Jr., *Science and Corporate Strategy: Du Pont R&D, 1902–1980* (Cambridge, 1988), 328, 335.
13. Michael Crow and Barry Bozeman, *Limited by Design: R&D Laboratories in the National Innovation System* (New York, 1998), 80.
14. Report of the Visiting Committee for Physics, 1952 (BNL-DO, II-23); AUI, Ex Comm, 18 Jun 1954.
15. See also John Krige, "The Ppbar Project, II: The Organization of Experimental Work," in John Krige, ed., *History of CERN,* vol. 3 (Amsterdam, 1996), 251–274, on 255.
16. John L. Heilbron, "Creativity and Big Science," *Physics Today* (Nov 1992), 42–47, on 45; Owen Hannaway, "Laboratory Design and the Aim of Science: Andreas Libavius versus Tycho Brahe," *Isis,* 77 (1986), 585–610; cf. Jole Shackelford, "Tycho Brahe, Laboratory Design, and the Aim of Science: Reading Plans in Context," *Isis,* 84 (1993), 211–230.
17. Charles Coulston Gillispie, "Science and Secret Weapons Development in Revolutionary France, 1792–1804: A Documentary History," *HSPS,* 23:1 (1992), 35–152, on 37.
18. A. Hunter Dupree, *Science in the Federal Government* (Baltimore, 1986), 275; Rexmond C. Cochrane, *Measures for Progress: A History of the National Bu-*

reau of Standards (Washington, D.C., 1966), 68; David Cahan, *An Institute for an Empire: The Physikalisch-Technische Reichsanstalt, 1871–1918* (Cambridge, 1989), 223; Edward Pyatt, *The National Physical Laboratory: A History* (Bristol, 1983), 220; Daniel J. Kevles, *The Physicists: The History of a Scientific Community in Modern America* (New York, 1978), 66–67, 81, 190; Robert W. Seidel, "Editor's Foreword," *HSPS*, 18:1 (1987).

19. Alex Roland, *Model Research: The National Advisory Committee for Aeronautics, 1915–1958*, 2 vols. (Washington, D.C., 1985); Harold Orlans, *Contracting for Atoms* (Washington, D.C., 1967).
20. Nathan Reingold, "The Case of the Disappearing Laboratory," in *Science, American Style* (New Brunswick, N.J., 1991), 224–246; Ronald C. Tobey, *The American Ideology of National Science, 1919–1930* (Pittsburgh, 1971), 53–58.
21. Alfred D. Chandler, *Scale and Scope: The Dynamics of Industrial Capitalism* (Cambridge, Mass., 1990), 15.
22. V. Telegdi, "Comments on Kolstad's Memorandum of August 10, 1966," 14 Sep 1966 (LANL A-91-011, 57/4); AUI, Ex Comm, 15 Mar 1963 and 15 Oct 1964; H. G. Vesper to J. Schlesinger, 24 Aug 1972, GAC 121; J. Fowler to McDaniel, 6 May 1971 (ORNL-DO, Physics); A. Zucker to Weinberg, 17 Nov 1972 (ORNL-DO, National Heavy Ion Lab).
23. J. B. Ball et al. to A. H. Snell, 5 Feb 1971 (ORNL-DO, Physics); Weinberg to A. K. Kerman, 6 Oct 1971 (ORNL-DO, PES 18.5, 1974–75); ANL Chemistry and Physics Divisions, "Midwest Tandem Cyclotron: A Proposal for a Regional Accelerator Facility," June 1969, ANL-7582. Oak Ridge later renamed it the Holifield Heavy Ion Research Facility, after Chet Holifield, the chairman of the congressional Joint Committee.
24. Weinberg to Sen. Howard Baker, 4 Apr 1972 (ORNL-DO, National Heavy Ion Lab).
25. Arnold Kanter, *Defense Politics: A Budgetary Perspective* (Chicago, 1979); Sybil Francis, "Warhead Politics: Livermore and the Competitive System of Nuclear Weapon Design" (Ph.D. diss., MIT, 1996), 25–27; William M. Evan, *Organization Theory: Structures, Systems, and Environments* (New York, 1976), 126.
26. Margaret Gowing, *Britain and Atomic Energy, 1939–1945* (London, 1964), 321–338, and *Independence and Deterrence: Britain and Atomic Energy, 1945–1952* (London, 1974), 2:203–261. Britain's decision to pursue its own nuclear weapons spun off programmatic labs for production reactors at Risley and weapons design at Aldermaston (intended as a British Los Alamos), but these labs remained centralized and focused on a single program. Gowing, *Independence*, 2:442–454.
27. Lloyd Berkner, in AUI Trustees, 17 Oct 1952.
28. Spencer R. Weart, *Scientists in Power* (Cambridge, Mass., 1979), 219–239.
29. Cf. Sharon Traweek, *Beamtimes and Lifetimes: The World of High Energy Phys-*

icists (Cambridge, Mass., 1988), 126–156, comparing the competitive, laissez-faire approach of American high-energy physicists to the *sensei* model of mutual responsibility and generational interdependence of their colleagues in Japan.

30. C. E. Falk, in notes of Coordination Meeting, 28 Apr 1965 (BNL-HA, 31/14′ bubble chamber)
31. Robert Galvin et al., *Alternative Futures for the Department of Energy National Laboratories* (Washington, D.C., 1995). Crow and Bozeman, *Limited by Design,* app. 2, lists nineteen panel reports on national labs produced since 1975.

INDEX